Pilot's Guide
to Weather Reports, Forecasts,
and Flight Planning

TAB
PRACTICAL
FLYING SERIES

Pilot's Guide to Weather Reports, Forecasts, and Flight Planning

2nd Edition

Terry T. Lankford

TAB Books
Imprint of McGraw-Hill

New York San Francisco Washington, D.C. Auckland Bogotá
Caracas Lisbon London Madrid Mexico City Milan
Montreal New Delhi San Juan Singapore
Sydney Tokyo Toronto

pbk   4 5 6 7 8 9 10 FGR/FGR 9 9 8
hc   2 3 4 5 6 7 8 9 FGR/FGR 9 9 8 7 6

Library of Congress Cataloging-in-Publication Data

Lankford, Terry T.
    Pilot's guide to weather reports, forecasts, and flight planning/
  by Terry T. Lankford. -- 2nd ed.
      p.   cm.
  Includes index.
  ISBN 0-07-036427-3 (pbk.)   ISBN 0-07-036426-5
  1. Meteorology in aeronautics. I. Title.
TL556.L36 1995   95-1617
629.132'4--dc20   CIP

Acquisitions editor: Jeff Worsinger
Editorial team: Robert E. Ostrander, Executive Editor
               Norval G. Kennedy, Book Editor
Production team: Katherine G. Brown, Director
               Rose McFarland, Desktop Operator
               Nancy K. Mickley, Proofreading
               Jodi L. Tyler, Indexer
Design team: Jaclyn J. Boone, Designer   PFS
          Katherine Stefanski, Associate Designer   0364273

Contents

Appendices

Preface

I OBTAINED A PRIVATE PILOT CERTIFICATE IN 1967 THROUGH AN AIR FORCE aero club in England. (It's a certificate because the FAA cannot spell license.) Back in the states, with the G.I. Bill, I obtained a commercial certificate and flight instructor certificate along with an instrument rating. Subsequently, as a full-time flight instructor, I earned an FAA Gold Seal. I have owned two airplanes; both were Cessna 150s that took me across the country twice. I have also flown in Canada and Mexico. I don't fly as much as I'd like, but manage to keep my instrument rating and instructor certificate current.

The certificate and ratings qualified me for my present position with the FAA as an Air Traffic Control Specialist (Station), although I prefer the reference flight service station (FSS) specialist.

Only once as flight service specialist did I become an air traffic controller. While at the Lovelock, NV, FSS a call came in from a VFR pilot caught in clouds at 13,000 feet. In such cases the control facility (approach control or center) usually provides assistance. Coordinating with Oakland Center, the controller replied: "OK, you've got 14 and below. Keep me advised." Instant air traffic controller.

I've been an FSS specialist for 20 years, which has brought together two interests: aviation and weather.

With all the books and articles written about aviation weather, it might seem that everything has been said. As a flight instructor, however, I discovered that very little practical information was available for pilots on the subject of meteorology. Certain texts and periodicals contain excellent points and suggestions; most references merely paraphrase government manuals. Some actually include incorrect and misleading information; many say nothing.

I find it exasperating to read an article with meteorological terms that espouses the need to understand weather, then fails to include a definition of or explain application of a term. Virtually no references present an understanding of information available from the FSS weather briefer; requirements and responsibilities of the briefer are overlooked, as well. Nor do they present practical ways of translating, interpreting, updating, and applying information.

The 1992 National Aviation Weather Program Plan identified a number of unmet users needs. Among these was the necessity to improve the aviation weather education of pilots and weather briefers.

As an FSS specialist, I've become aware of requirements and procedures that the FAA seems to want to keep secret. Most requirements and procedures have to do with how FSS specialists and air traffic controllers do their jobs. Much of this book is dedicated to points glossed over, or ignored altogether, in the *Airman's Information Manual*.

Pilots within the general aviation community have an especially close relationship with the FSS specialist. The FSS specialist, whether during preflight or in-flight, has at his or her fingertips a wealth of information. The pilot, for whom the FSS exists, is an integral part of the aviation weather system. Pilots provide feedback in the form of pilot reports that are essential for the meteorologist to verify the forecast and optimum system operation. When the pilot or specialist forgets or does not understand the requirements and responsibilities of the other, the system can fail, sometimes with fatal results.

Accidents related to weather continue to take their toll; analysis often reveals an inadequate, misinterpreted, or misunderstood weather briefing. With FSS automation, automated weather observations, and increased availability of direct user access terminals and other commercial weather information systems, pilots will be required to read, interpret, and apply weather information on their own. A sound background in meteorology and the products available to aviation is essential to safety.

I am greatly indebted to many people, too numerous to be individually listed, for their generous help, guidance, and advice. Among them are the meteorologists of the National Weather Service (NWS) at the FAA's Mike Monroney Aeronautical Center in Oklahoma City, local and regional NWS offices, and flight service station specialists that I have been privileged to know, especially at the Oakland, CA, FSS. I am also indebted to the pilots who have allowed me to assist them and in turn provided me with the best education possible. This book is dedicated to these people.

Introduction

THE FIRST EDITION OF THIS BOOK WAS PUBLISHED IN 1990. SINCE THEN THERE have been several major changes in weather report and forecast format criteria. Changes have occurred with the implementation of automated weather observing systems. The format of AIRMETs, the area forecast, and terminal forecasts (FT) has significantly changed. The pilot report format, weather retrieval, and the notice to airmen system have experienced minor changes. With continuing flight service station consolidation, changes have occurred in FSS communications and procedures. In 1996, the United States will convert to the international METAR code for surface observations and terminal aerodrome forecasts will replace domestic FTs.

Weather affects a pilot's flying activity more than any other physical factor. Most pilots agree that weather is the most difficult and least understood subject in the training curriculum. Many pilots are intimidated by weather. In spite of these facts, or because of them, weather training for pilots consists of bare bones, only enough to pass the written test, while weather-related fatal accident statistics remain relatively unchanged.

Air Force pilot training contains a mere 15 hours of formal weather instruction, as compared to 50 or 60 hours in the past. Navy pilot weather training has been reduced by 25 percent to 30 hours, with no refresher. Army aviator weather training consists of about the same number of hours. My FAA-certificated primary ground school included a mere nine hours of weather. Ironically, Federal Aviation Regulations perpetuate this trend. Student, private, recreational, and commercial airplane pilots are only required to obtain and use weather reports and forecasts, recognize critical weather situations, and estimate ground and flight visibility. An additional requirement of forecasting weather trends on the basis of information received and personal observations is required for an instrument rating. Only the airline transport pilot certificate requires the applicant to have any in-depth meteorological knowledge.

A major revision to the practical test standards in 1984 required that applicants exhibit "knowledge of aviation weather information by obtaining, reading, and analyzing . . ." reports and forecasts, and make ". . . a competent go-no-go decision based on the available weather information." The practical test standards do reference AC 00-6 *Aviation Weather* (basic meteorological theory) and AC 00-45 *Aviation Weather Services* (a basic discussion of weather reports and forecasts), but neither reference relates weather to flight situations or application to a go-no-go decision. Most flight tests are given during good weather, and pilots requesting checkride briefings all too often have little idea of the information required or the presentation format of the FSS briefer; this requirement would seem to have little practical consequence.

This situation evolved because virtually all flying, except military and the airlines, was visual when the regulations were originally written. The aircraft of the day had neither the performance nor the instruments to take on weather. When weather was encountered, pilots simply landed in the nearest field. Aeronautical experience regulations that specify 40 hours to become a private pilot have been woefully inadequate for years due to the constant increase of additional requirements. Many of the people who write regulations and perform flight tests are former military pilots with minimal weather training because after graduation they were under direct control of older, more experienced pilots and gained experience in a controlled environment.

Accident reports and commentaries frequently refer to a pilot's poor judgment, namely the failure to reach a sound decision. Pilot judgment is based on training and experience. Training is knowledge imparted during certification, flight reviews, and seminars, which is supplemented by literature. Experience can be best defined when the test comes before the lesson; unfortunately, failure can be fatal. Pilot applicants have only their instructor to prepare them to make competent go-no-go decisions.

Odds are that little if any judgment training occurred unless situations were actually encountered. General aviation training runs the entire gamut from the flight school that prohibits its instrument students and instructors from flying in the clouds, to the instructor in a Cessna 182 who requests a vector into icing conditions to demonstrate the effects of ice to the student. (Neither instance seemingly exhibits sound judgment.) The fact remains that the least experienced pilots have minimal weather training. Following certification there is no requirement for additional or refresher weather instruction.

An essential part of flight preparation concerns the weather. No matter how short or simple the mission, Federal Aviation Regulations place the responsibility for flight preparation on the pilot, not the NWS forecaster and not the FSS briefer. To effectively use the available resources, a pilot must understand what weather information is available, how it is distributed, and how it can be applied to a flight situation.

Information in this book will help any applicant pass written examinations and practical flight tests, and help him or her operate safely and efficiently within an ever-increasingly complex environment. Student, recreational, private, and commercial pilots will be able to develop abilities to recognize and avoid critical weather situations. The student pilot will have access to the knowledge of the airline transport pilot.

Pilots will learn to obtain weather reports and forecasts, through the FAA and other sources, then interpret and apply information to flight situations. They will develop an understanding of the principles of forecasting, and apply forecasts and observations to the flight environment.

Pilots will also come to understand weather collection and distribution, symbols and contractions, charts and forecasts, and how terrain affects weather. They will be able to recognize cloud forms, conditions conducive to icing and other meteorological hazards, and the use of pilot weather reports.

Technical meteorological concepts and terms in this book are translated into language any pilot can easily understand. On the other hand, such subjects as vorticity, microbursts, and upper level weather systems will not be omitted because of complexity. Explanations of weather reports and forecast go beyond decoding and translating, to interpreting and applying information to actual flight situations. Discussions include applying weather information to VFR as well as IFR operations, and high-level as well as low-level flights.

A thorough understanding of the basics is essential: a sound foundation for the novice and a practical review for the experienced pilot. Judgment and application of judgment, plus a knowledge of the aviation weather system and its relation to air traffic control, are essential to a safe, efficient flight.

Weather reports, forecasts, and flight planning are interrelated; therefore, it might be necessary for the reader to complete the book to realize the most thorough understanding of the subject.

The book begins with a detailed description of surface weather reports including methods and criteria used by the weather observer. Failure to understand the observer's requirements often leads to pilot misunderstanding and unwarranted observer criticism. Limitations on observations are discussed with practical guidance on interpreting information.

Surface reports are related to hazards such as thunderstorms, turbulence, icing, low ceilings, fog, precipitation, haze, smoke, dust, winds, low-level wind shear, and high density altitude. The automated weather observing system is discussed, examining its inherent advantages and limitations.

The United States is scheduled to convert domestic weather reports to the international, or METAR, code in 1996. METAR format and codes are included in this book.

Most pilots are unaware of the contributions that they can make with pilot weather reports and the impact that their reports can have on the system. Inaccurate or overestimated reports of turbulence and icing cause misleading advisories to be issued. Pilots making these reports gain unrealistic confidence in their ability to handle severe conditions. Pilots who are receiving unwarranted advisories that are based upon overestimated pilot reports might conclude that all advisories are pessimistic. When severe conditions develop, these pilots can be led down the primrose path to disaster. Pilot report criteria are discussed in detail, along with distribution and application.

The methods, criteria, and limitations of forecasts are discussed. Numerous misunderstandings and misconceptions of forecast products and terminology, which are potentially catastrophic to the uninformed pilot, are explained. The book tells how the pilot can interpret and apply forecasts to specific flight situations; interpretation and application are extremely important when the pilot is using the direct user access terminal system (DUATS or DUAT) for a briefing.

Approximately one dozen different forecasts are written for aviation; each has criteria, purpose, and limitations; often they overlap and might appear to be inconsistent. Analysis, however, most often reveals the forecasts are consistent within the scope of each product. Each forecast is analyzed and put into perspective. The pilot can then apply the array of forecasts available with respect to regulations, aircraft performance, and his or her ability to efficiently operate within the system to make sound go-no-go decisions. Also in 1996, the National Weather Service will convert from the present domestic terminal forecast to the international or terminal aerodrome forecast (TAF). Like METAR, TAFs are discussed in detail.

Beyond written weather reports and forecasts, a multitude of charts are available for a briefing. Each chart has a specific purpose and use, but not without limitations. All charts are discussed and analyzed with respect to their individual significance and application.

The charts discussed are those available from the National Weather Service. (Many were obtained through Cyclogenesis Inc. Weather Systems, 327 Rivershore Dr., Elk Rapids, MI 49629.) Weather charts obtained through direct user access terminals and other private vendors may not be as detailed, but their interpretation and application are similar.

Three chapters are devoted to obtaining, updating, and using weather and aeronautical information. Discussions will help the pilot efficiently obtain weather and flight planning information that has become increasingly difficult with flight service station consolidation.

Navigation principles and techniques are not within the scope of this book. The final chapters are devoted to applying information to VFR and IFR flight planning situations, and the services available from the FAA's flight service stations. With direct user access terminals, pilots not only obtain weather and notices to airmen, but also file flight plans through a computer terminal. On the surface, this might seem a simple task, but it can be complicated and frustrating to the pilot; DUAT has the potential to cause trouble if a pilot does not obtain certain information or misinterprets obtained information.

Text references are made to various cloud forms that frequently appear in weather reports; photographs illustrate certain cloud formations. A detailed color cloud chart is available from the National Weather Association, 6704 Wolke Court, Montgomery, AL 36116-2134. Ask for "The Cloud Chart."

The following chapters, told with a little humor, hopefully, explain how to use the weather reporting system, then translate and apply the information to a flight situation. Incidents are not intended to disparage or malign any individual, group, or organization; the sole purpose of all information is illustration.

Examples of weather reports and forecasts are real and taken off the weather circuit; none have been created. Weather report phraseology used in the text is taken from FAA Handbook 7110.10 Flight Services. This is the same phraseology pilots can expect in FSS radio communications and broadcasts.

Great effort has been made to ensure that information is current and accurate, as of the time of writing; however, especially in aviation and aviation weather, the only thing that doesn't change is change itself.

1
Surface
observations (SA)

ASK MOST PILOTS ABOUT CODED WEATHER REPORTS AND YOU'LL HEAR: "Why do I have to read the weather? When I call flight service, they read and explain the reports." Pilots have been objecting to coded weather information since its inception in the early 1930s, but with the *direct user access terminal system* (DUATS) (pronounced dew-ahts) and commercially available weather systems, pilots will have to read and interpret weather on their own. Although the ability to read reports is essential, whether obtained from an FSS or other source, the skill and knowledge to interpret information is crucial. A pilot's knowledge of weather takes on greater significance with DUATS and flight service station (FSS) consolidation.

Whether you are a beginner—even Steve Canyon started as a student pilot—or an old hand, examine the following reports taken on July 17, 1988, at Newark, NJ (EWR) and Philadelphia, PA (PHL). Translate and interpret the observations; you might make some notes. When you've finished the chapter, see if your interpretation and understanding have changed.

EWR SP 2240 W0 X 0T+RW+ 3545Q60/989/R04VR45 T+ OVHD MOVG ESE FQT
   LTGICCCCG
EWR RS 2253 –X E15 OVC 3T+RW 102/75/70/0905/983/R5 T+ B10 MOVG ESE
   FQT LTGICCCCG VSBY HIR W RB07 PK WND 3560/
PHL RS 2250 W6 X 1TRW+F 102/77/74/3618G24/983/R27RVR30V60+ TB25
   ALQDS MOVG SE FQT LTGICCG RB38 PRESFR
PHL SP 2256 W5 X 3/4TRW+F 2921G28/989/R27RVR08V60+ WSHFT 55 T
   ALQDS MOVG SE FQT LTGICCG PRESRR

Most weather reports are history by the time they are transmitted and available. In the teletype era, reports were often an hour and a half old. The FAA introduced a computer system in 1978 known as Leased Service A (LSAS). With this and other computerized systems, most reports are available within 20 minutes of observation.

The accuracy and validity of observations, and therefore their usefulness, depends on many factors: who is taking the observation, what is the extent of the observer's training and experience, and what type of equipment, if any, is available.

The National Weather Service certifies weather observers; however, the most accurate, valid, and detailed observations come from NWS, military, FSS, tower, and other observers, in that order because NWS and military observers are professionals located at major airports with the latest equipment. At FSS and tower locations, weather observations are a secondary duty, and the equipment is generally not as sophisticated. As for the others, they are contract observers, retired air traffic controllers, local airport personnel, or even the fire department, often with little or no equipment, and minimal training and experience.

This is not to say that many of them don't provide quality observations, but in many cases they simply don't have the equipment, training, or experience. And the FAA and NWS have decided that automated weather observing systems will be the standard at major airport and automated flight service station (AFSS) locations. (The effects of automation on the quality of observations are examined later in this chapter.)

Reports begin with a three-letter location identifier (my abbreviation is "LCID"; the FAA doesn't specify one). Most reports use three letters, but some consist of letters and numbers; S06 is Mullan, ID. Official LCIDs are contained in *FAA Handbook 7350.5 Location Identifiers*, available for sale from the U.S. Superintendent of Documents, Government Printing Office. Pilots also have access to LCIDs from aeronautical charts, FSSs, DUAT, and the *Airport/Facility Directory*. Appendix A of this book contains LCIDs for most weather-reporting locations.

An abbreviated reference to the type of observation follows the station's identifier. Two basic types of reports are *record* and *special*. SAs (record hourly airway surface observations) are taken each hour. Whenever a significant change occurs, an SP (non-hourly special airway observation) is recorded. A complex criterion determines the requirement for specials. Generally, they're required when the weather improves to or deteriorates below specified minimums for visual flight rules (VFR) or approach and landing instrument flight rules (IFR). Specials are also required for the beginning, ending, or change in intensity of thunderstorm activity. For severe weather, such as tornadoes, a single-element urgent special (USP) can be issued.

XYZ USP 1527 TORNADO W MOVG E SPECIALIST HDG S

When special criteria are met at the normal time of observation, the report is identified as RS (hourly record special airway observation). USP, RS, and SP reports alert all concerned to potentially significant changes. Specials are not available for all locations; these reports will normally carry the remark NO SPL. In such cases, significant changes can occur without expeditious notification.

Some locations take supplemental, nonscheduled observations. These are encoded SW (supplemental surface observations) and are available only on an irregular basis.

Report type (SA, SP, SW, etc.) determines distribution priority and storage time. For example, USPs receive immediate distribution to all circuits, SAs are replaced hourly, and SWs are stored for three hours unless replaced or updated. Carefully note the SW observation time because the report might be hours old.

Following observation of the last element, usually atmospheric pressure, the official time of observation is recorded (OAK SA 1956). Times are *coordinated universal time* (UTC), commonly referred to as ZULU or Z.

SKY COVER

Sky cover refers to clouds or obscuring phenomena as seen by an observer on the ground from horizon to horizon. Sky cover reported on SAs is the summation of layers based on specific criteria. Conditions aloft, as seen by a pilot, can be substantially different. The summation principle with terms like "obscuring phenomena" and "obscuration" is complex and often misunderstood.

Refer to Table 1-1 for sky-cover definitions. Notice that even a report of clear (CLR) does not guarantee a cloud-free sky. In such cases, however, most observers will

Table 1-1. Sky cover.

Symbol		Definition
CLR	Clear	Less than $\frac{1}{10}$ sky cover.
SCT	Scattered	$\frac{1}{10}$ to $\frac{5}{10}$ sky cover.
BKN	Broken	$\frac{6}{10}$ to $\frac{9}{10}$ sky cover.
OVC	Overcast	More than $\frac{9}{10}$ sky cover.
X	Sky obscured	All of the sky hidden by surface-based obscuring phenomena.
Note: BKN, OVC, and X represent a ceiling.		
–SCT	Thin scattered	
–BKN	Thin broken	At least $\frac{1}{2}$ of the layer is transparent.
–OVC	Thin overcast	
Note: –BKN and –OVC do not constitute a ceiling.		
–X	Sky partially obscured	From $\frac{1}{10}$ to less than $\frac{10}{10}$ of the sky is hidden by surface-based obscuring phenomena.

add a remark, such as FEW CU (few cumulus) or ST NW (stratus northwest). Variable sky cover describes a situation that varies during the period of observation, normally the 15-minute period preceding time of observation: 15 SCT.../SCT V BKN (scattered layer variable to broken). Variability alerts pilots, briefers, and forecasters to rapidly changing conditions over the airport.

Sky-cover heights on surface observations are always reported above ground level (AGL). Here, variability indicates a rapid fluctuation in ceiling height: M10V BKN.../ CIG08V12 (measured ceiling 1,000 variable broken . . . /ceiling variable between 800 and 1,200). A variable ceiling below 3,000 feet must be reported, and a variable ceiling above 3,000 feet is reported only if considered operationally significant.

In Fig. 1-1, the observer sees $\%_0$ cloud cover at 1,000 feet AGL and reports 1,000 scattered (10 SCT). Another $\%_0$ cloud cover is measured at 1,800 feet AGL. According to the summation principle, a total of $\%_0$ ($\%_0 + \%_0 = \%_0$) sky cover exists; a measured ceiling 1,800 broken (M18 BKN) is reported. The observer sees the remaining sky covered by cloud at 3,000 feet AGL, and reports 3,000 overcast (30 OVC). The observer is unable to determine the extent of higher layers and reports them as continuous.

Summation of cloud cover

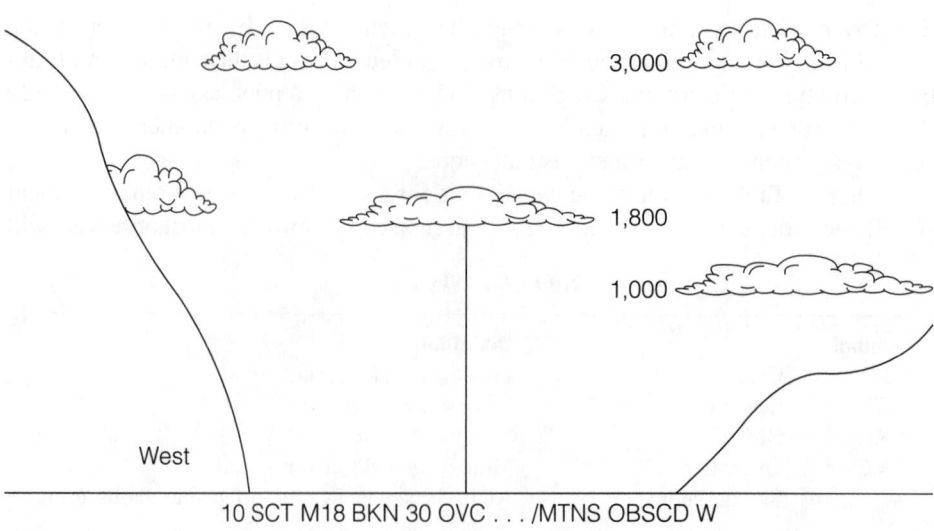

3,000

1,800

1,000

West

10 SCT M18 BKN 30 OVC . . . /MTNS OBSCD W

Fig. 1-1. *Cloud cover reported on SAs is the summarization of cloud layers as seen by an observer on the ground.*

This principle has led many a pilot to mistakenly question the accuracy of observations. Pilot reports (PIREPs) are usually the only means of determining the extent of upper layers and tops. A PIREP describing conditions in Fig. 1-1 might read SCT 012/

4

SCT 025/SCT 035 (tops of scattered layers at 1,200, 2,500, and 3,500 feet). Given the summation principle, the ground observation and pilot report are perfectly consistent.

Sky cover is always an estimate. As an FSS briefer for some 15 years, I have on occasion briefed doom and gloom only to find a bright, beautiful day. It would seem that certain observers might use logical criteria. "The ceiling of the room is opaque; therefore, the observation with one cloud is scattered, two is broken, and three is overcast with breaks!" (That was one attempt at humor.)

Notwithstanding the previous observation, care must be exercised climbing or descending VFR through a broken deck. Although it might be possible to safely negotiate the layer, several factors must be considered. Can appropriate distance from clouds as specified in Federal Aviation Regulation (FAR) Part 91's basic VFR weather minimums be maintained? Is the weather improving or deteriorating? We don't want to get caught on top or between layers. Is the area congested with other VFR traffic or aircraft operating IFR? The criteria in FAR 91 are minimums; they do not necessarily equate to "safe." What alternates are available, if needed? Positive answers are required before an attempt. Any negative or uncertain responses would indicate a no-go decision.

How many times have you heard or said to yourself, "Well, I think I'll go take a look." There is no inherent danger in this practice as long as the pilot knows when to quit. So the question becomes "When do I turn around?" When the thought, "I wonder if I should?" first occurs, that's the time to execute the 180. Ironically, most primary training sets the stage for an accident. Many instructors simulate the student flying into clouds, then require them to extricate themselves, when the better behavior would be to turn around before entering IFR weather. Flying into clouds or weather that is less than VFR does not satisfy the requirement to recognize critical weather situations. Remember the all-too-common accident scenario, "Continued VFR . . . ."

CEILING

The lowest broken, overcast, or obscured layer, not thin or partial, is classified as a *ceiling*. When an observer can see through one half or more of a layer, it is reported as thin (-BKN or -OVC). A thin layer does not constitute a ceiling because a pilot flying over the deck would see the ground. The pilot could penetrate the layer, assuming required horizontal visibility for that airspace could be maintained. Referring to FAR 91's basic VFR weather minimums, this requires 1 statute mile in Class G airspace, 3 statute miles within Class B, C, D, and E airspace, or 5 statute miles at or above 10,000 feet mean sea level (MSL).

Ceilings are designated estimated, measured, or indefinite. Although seemingly simple and self-explanatory, even a measured ceiling can be misleading, and an indefinite ceiling is complex and often misunderstood. Like many specialties, aviation weather has its own language. For communication and understanding to take place, each party must use the same definitions. Take the pilot who called unicom and asked, "What's the ceiling?" After a pregnant pause the operator replied, "I think it's oak."

An estimated ceiling (E) is just that, a more or less educated guess by the observer. But the fact remains it's just an estimate based on the observer's training and experience. Estimated ceilings must always be viewed with caution, especially at night or close to minimums. In addition to an educated guess, pilot reports, ceiling balloons, surrounding terrain, or a calculation from temperature/dew point might be employed to determine an estimated ceiling.

Ceiling balloons are relatively accurate for low clouds below about 2,000 feet; however, balloons are affected by strong winds and precipitation. A balloon that is filled with helium to lift a known weight is released. The observer times the balloon until cloud penetration. The elapsed time is converted into height. At tower and FSS locations where observations are secondary, this time-consuming procedure is rarely used.

A pilot report can be one of the most accurate means of determining the ceiling, assuming the pilot actually penetrates the clouds, or climbs or descends through a scattered or broken layer; otherwise, it's just a pilot's guess. As pilots, we should always pass cloud bases to the tower or FSS, especially when different from reported. But keep in mind that we reference cloud bases to a pressure altimeter set to read MSL and the observer reports clouds AGL. This explains some apparent inconsistencies between observer- and pilot-reported cloud bases.

Where available, surrounding terrain, buildings, or towers are usually quite accurate for determining cloud bases in the immediate vicinity, but they might not be representative of surrounding areas.

An estimate for the height of convective clouds can be determined from temperature/dew point. This procedure is based on surface heating of a parcel of air that rises and cools at the dry adiabatic lapse rate. When temperature reaches dew point, saturation occurs and clouds form at the *lifted condensation level* (LCL). When the layer is produced by surface heating, apply the following formula: $T - Td \times 2.2 \times 100 =$ cloud base AGL. In this formula $T =$ temperature in degrees Fahrenheit, $Td =$ dew point in degrees Fahrenheit, and 2.2 and 100 are constants. For example, an SA reports 70/60: temperature minus dew point (70°F – 60°F) equals 10°F; apply the formula $10 \times 2.2 \times 100 = 2,200$, and the approximate cloud bases are 2,200 AGL.

As another example, use this observation for Denver:

DEN SA 2350 90 SCT 150 SCT E250 BKN 45 075/91/44/1610/001

Applying the formula to the DEN SA, the cloud base works out to be approximately 10,000 AGL ($91 - 44 \times 2.2 \times 100 = 10,000$). This is consistent with the observer's report of 9,000 scattered. An article in a popular aviation magazine claimed this procedure could be used for any cloud layer. It's important to remember the procedure only applies to convective clouds produced by surface heating.

A measured ceiling (M) is the most accurate means of determining cloud height. A *rotating beam ceilometer* (RBC) or a *ceiling light* is normally used. The projector portion of the RBC shines a beam of light along the cloud base. When directly over the detector, a photoelectric cell activates. The observer reads cloud height directly off the indicator located in the tower, FSS, or weather office. A much less sophisticated system, the ceiling light, projects a vertical beam; using a *clinometer*—which is colloquially known as the Coke bottle—the observer measures the angle from the ground to the point where the light reflects off a cloud along a predetermined baseline. From a chart, the observer converts angle into cloud height. (From personal experience, I can say this is no fun in rain or snow.) Both methods measure cloud height at only one point above the airport. This is why ceilometers are usually located at the approach end of the primary instrument runway.

A measured ceiling might not be representative of surrounding conditions, especially at night or during low visibility when the observer cannot see the whole sky. Refer to Fig. 1-1; if the observer were unable to see the scattered layer at 1,000 feet, a ceiling of 1,800 feet would be reported. Or if the ceilometer went through a hole in the 1,800-foot layer, the observer might report a measured ceiling of 3,000 feet overcast. Such errors can and do occur because of limitations on equipment and the observer.

An indefinite ceiling (W), the least understood ceiling designator, is even defined incorrectly in some aviation texts. Technically, an indefinite ceiling is the vertical visibility upward into a surface-based obscuring phenomenon that completely conceals the sky: the distance at which a pilot can expect ground contact when looking straight down on descent, or the point at which the ground disappears on climbout.

So what's an obscuring phenomenon? Anything that prevents the observer from seeing all or part of the sky. It is most often caused by fog, but snow, smoke, or even heavy rain can cause this condition: W0 X...RW+ (indefinite ceiling zero, sky obscured by heavy rain showers).

The accuracy of an indefinite ceiling depends on the observer and available equipment. Whether the value is determined by a ceilometer, pilot report, balloon, or just a guess by the observer, it is reported as indefinite. In Fig. 1-2, the observer has either measured or estimated the vertical visibility as 200 feet. The sky is completely obscured. The observer, unable to determine cloud layers above, reports an indefinite ceiling 200, sky obscured (W2 X).

Indefinite ceilings are most often associated with IFR conditions. Presume that a destination is reporting an indefinite ceiling 200 and that 200 feet is the decision height (DH) for the instrument landing system (ILS) approach. Assuming the observation is accurate, at DH the pilot should be able to look straight down and see the approach lighting system (ALS); however, he or she would not necessarily be able to see the runway due to increased slant-range distance. This is illustrated in Fig. 1-2. In fact, slant-range visibility could be less than vertical visibility! That's why approach lights are considered part of the ILS, and minimums increase when they're out of service.

Would the conditions in the previous paragraph preclude a legal landing? Not necessarily, as long as the provisions of FAR 91 are met. A legal landing requires three conditions:

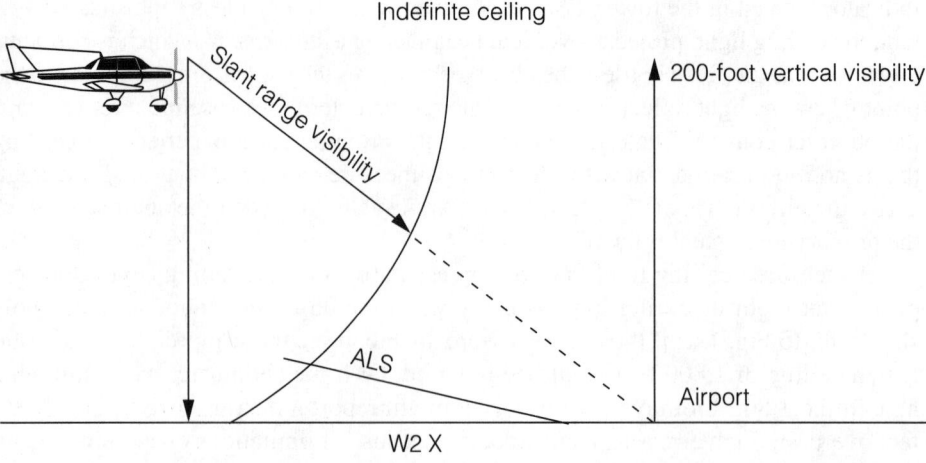

Fig. 1-2. *An indefinite ceiling represents the vertical visibility into a surface-based obscuring phenomenon.*

- The aircraft must remain in a position from which a normal descent to landing can be made.
- The flight visibility must not be less than that prescribed for the approach.
- The runway environment (approach lights, threshold, runway, etc.) remains distinctly visible.

Any requirement not met or lost after DH requires an immediate missed approach.

How about VFR with an indefinite ceiling of 1,000 feet? Even though ceiling and surface visibility might technically meet legal minimums, flight visibility might be substantially less. Now consider the possibility that the observer might not have ceiling height-finding equipment; the reported ceiling might only be a guesstimate. Good operating practice dictates extreme caution operating close to minimums. If IFR, a legal alternate (or two) or an escape route to VFR conditions are a must!

Sky partially obscured (-X) is also easily misunderstood. A partial obscuration indicates from $\frac{1}{10}$ to less than $\frac{10}{10}$ of the sky is hidden by a surface-based obscuring phenomenon. Precipitation, haze, smoke, and fog usually cause this condition. The observer, unable to evaluate the hidden portion of the sky, must consider it opaque when applying the summation principle.

In Fig. 1-3, half ($\frac{5}{10}$) of the sky is hidden. The remarks portion of the report describe the extent of the obscuration (/F5). In the example, fog obscures $\frac{5}{10}$ of the sky. The observer can see $\frac{2}{10}$ cloud cover at 3,000 feet. Applying the summation principle, a total of $\frac{7}{10}$ of the sky is covered or hidden; therefore, an estimated ceiling of 3,000 feet must be reported, even though only $\frac{2}{10}$ cloud cover exists, as shown in the illustration.

The summation principle is not always correctly applied. A report from a contract observer read ABC SA 1845 -X 10 SCT E25 OVC.../F5. Because half of the sky is ob-

Partial obscuration

3,000 feet

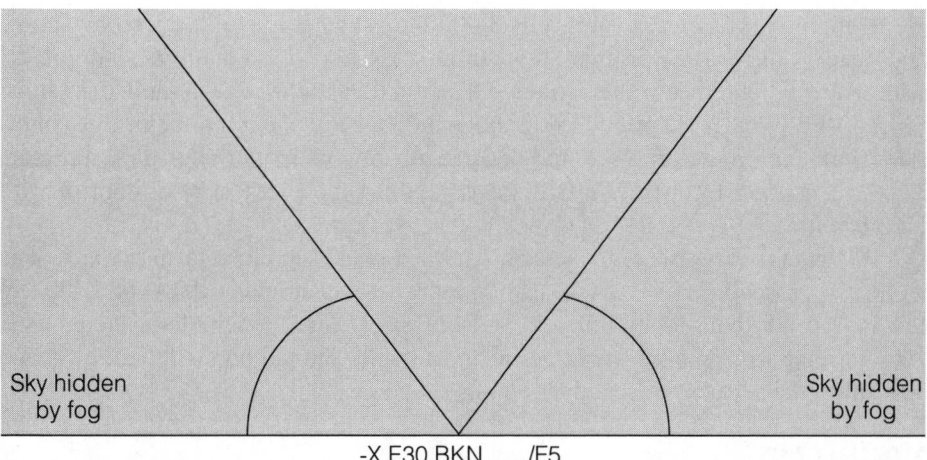

-X E30 BKN . . . /F5

Fig. 1-3. *A partial obscuration indicates that the observer cannot evaluate the entire sky. When a large amount of the sky is obscured, cloud cover amounts might not be representative.*

scured by fog and at least ¹⁄₁₀ cloud cover exists at 1,000 feet, the observer should have reported -X E10 BKN 25 OVC. The observer cannot determine the extent of the 1,000-foot layer above the obscured portion of the sky. It might be quite a shock finding an unexpected 1,000-foot broken layer!

A partial obscuration may be reported without cloud layers, for example -X.../HK3. Remarks indicate ³⁄₁₀ of the sky is hidden by haze and smoke. A partial obscuration in itself must not be confused with a ceiling. Although with a large amount of sky hidden, cloud coverage amounts might not be representative, and slant-range visibility can be less than the reported surface visibility.

Normally, a pilot can expect ground contact while flying in areas with a reported partial obscuration. This is why they are not considered ceilings; however, the same principles apply to a partial obscuration as thin cloud layers. Penetrating a partial obscuration VFR requires the appropriate horizontal visibility be maintained for that airspace. A partial obscuration with visibilities less than basic VFR can often be safely negotiated under the provisions of special VFR (discussed in chapter 14).

Ceiling search: After being briefed that his destination was reporting sky partially obscured, visibility 2 in fog and haze (-X 2FH), the Beechcraft Baron pilot emphatically demanded to know the ceiling. He stated he must have this information to determine IFR minimums. There was no ceiling. IFR minimums, in this case, would be based upon visibility alone. The pilot could expect to maintain ground contact throughout the approach, sighting the airport at about 2 miles.

At Redding, CA, the NWS observer made the following report:

RDD SP 1650 W30 X...KH

The reported vertical visibility is 3,000 feet, sky completely hidden by smoke and haze. A pilot flying in this area can only expect to maintain ground contact within about 3,000 feet of the surface. This situation could be extremely dangerous for the VFR pilot. If the pilot climbs above 3,000, he or she might be in IFR conditions without ground contact and most probably without the natural horizon. Pilots attempting to operate in similar conditions have lost aircraft control with fatal results.

IFR pilots must also exercise caution when operating close to minimums. The VOR-A approach to Ukiah, CA, has a minimum descent altitude (MDA) of 3,400 feet (2,784 AGL). Ukiah was reporting 20 SCT E50 BKN. Sure enough, the scattered layer was at the missed approach point. During the miss, the aircraft broke into the clear and was able to land. Luck: Never count on this happening.

VISIBILITY

Visibility is a measure of the transparency of the atmosphere. During the day, visibility represents the distance at which predominant objects can be seen; at night, visibility is the distance that unfocused lights of moderate intensity are visible. One NWS observer was quite perplexed when the tower always reported increased visibility after sunset. The reason was the change in criteria for the observation. Pilots should note that daytime values do not necessarily represent the distance that other aircraft can be seen. At night, especially under an overcast, unlighted objects might not be seen at all, and there might be no natural horizon.

SAs report prevailing visibility in statute miles (sm). That is, the greatest visibility equaled or exceeded throughout at least half the horizon circle, which need not be continuous. Prevailing visibility less than 3 miles that rapidly increases and decreases during the observation is reported as variable: 1V.../VSBY3/4V11/2 (visibility 1 statute mile variable . . . /visibility variable between three-quarters and one and one-half). Variable visibility has the same implications as variable sky condition and ceiling conditions rapidly changing at the airport.

Figure 1-4 illustrates a reported prevailing visibility of 4 sm. Because one sector has a visibility of only 2 sm, which is operationally significant, remarks contain VSBY N2 (visibility north two). Sectors might also exist with visibility greater than prevailing. These values might be reported in remarks. This accounts for some apparent inconsistencies. In Fig. 1-4, a pilot approaching the airport in the sector where the visibility is 6 sm might question the report. The observation remains consistent within the definition of prevailing visibility.

LOL SA 1555 CLR 35.../VSBY S 1/4

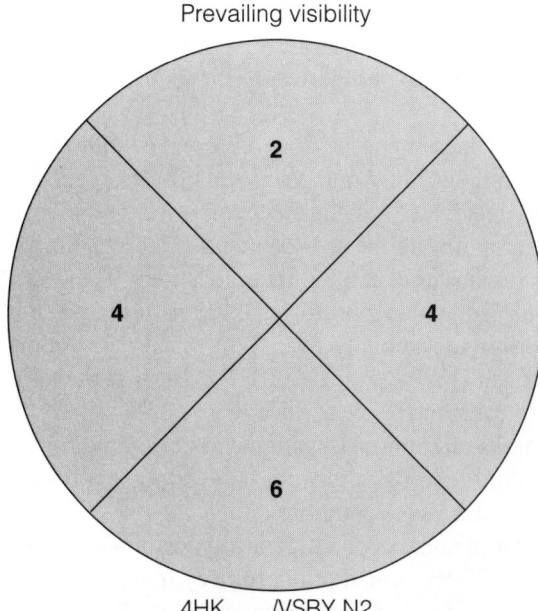

Prevailing visibility

Fig. 1-4. *Prevailing visibility represents the greatest distance that prominent objects can be seen throughout at least half the horizon circle, which need not be continuous.*

4HK . . . /VSBY N2

Could this be valid? Yes, indeed. A fog bank was south of the airport. A little while later, as the fog continued to move over the airport, CLR3/4F.../VSBY NE 35 was reported.

Conditions can be quite variable and change rapidly in areas affected by stratus and fog. The following SAs for Crescent City, CA, illustrate this point.

CEC SA 1755 -X 3 SCT 3F.../F1 VSBY E 35 S-N 3/4
CEC SP 1805 -X E3 BKN 3F.../F2 BKN V SCT VSBY N-E 35 S-NW 3/4
CEC SP 1810 -X 3 -BKN 25.../F1 VSBY S-NW 1

Conditions went from VFR to IFR and back in 15 minutes, which emphasizes the need for frequent weather updates. A VFR pilot should consider approaching or departing Crescent City from or toward the east, where remarks indicate good visibility. VFR flights to this airport must be planned for fuel to destination, a solid VFR alternate, and the required reserve. FARs don't require VFR alternates, but good judgment does.

At one Southern California airport, it appears local pilots estimate prevailing visibility in the following manner: "one-quarter mile to the north, one-quarter mile to the east, one-quarter mile to the south and one-quarter mile to the west. Great! Prevailing visibility is 1 mile, it's VFR." This is not only incorrect but extremely dangerous.

Pilots are required to be trained in estimating visibility while in flight because FARs specify visibility requirements in relation to flight visibility. To legally operate VFR where the visibility was reported as 2 sm in Fig. 1-4, a pilot must have flight visibility of three miles. If this cannot be done, the pilot would have to obtain an IFR clearance, or request a special VFR clearance, or depart the area. And for landing IFR, visibility must not be less than that prescribed for the approach.

Reported surface visibility comes into play when a pilot plans to take off or land, or enter the traffic pattern VFR within Class B, C, D, and E airspace. Surface visibility must be at least 3 sm. Or if surface visibility is not reported, flight visibility must be at least 3 sm.

SAs report surface visibility, which does not necessarily represent conditions at altitude. Visibilities aloft are most often reduced by rain, snow, dust, smoke, and haze. And with reduced visibilities, usually less than 5 sm, the apparent visibility looking toward the sun can be almost nil! Pilot reports are the only source of flight visibility.

NWS or FSS personnel take observations at certain tower locations. These facilities are not usually collocated. Tower personnel report tower-level visibility when less than 4 sm. Because visibility can differ substantially over short distances, a complicated formula determines whether tower (TWR) or surface (SFC) visibility prevails.

The remarks portion of the report contains the other value (CLR 11/2F.../SFC VSBY 2; -X IF.../TWR VSBY 0). These remarks alert pilots to variable visibility over the airport. It is not unheard of for the tower to report low visibility prevailing less than 3 or even 1 sm, which prevents VFR or special VFR operations.

Two other visibilities might appear in remarks of SAs: runway visibility (RVV) and runway visual range (RVR). RVV and RVR, located at the runway touchdown point, apply to instrument approach minimums found on IFR approach charts. They are surface measurements, not slant-range. A report indicating landing minimums does not necessarily mean that visibility exists at the DH or MDA. RVV and RVR reflect the fact that visibility can vary substantially from the normal point of observation (tower, FSS, or weather office) to the runway touchdown zone. RVV and RVR report a 10-minute average transmissometer value.

The transmissometer transmitter projects a beam of light toward the receiver. A photoelectric cell measures the amount of light reaching the receiver. This value is electronically converted into visibility and displayed at appropriate locations (tower, FSS, weather office, or a combination).

RVV refers to visibility measured in miles and fractions along a specific runway: RVV30R11/4 (Runway 30 Right, visibility 1¼).

RVR, also associated with a specific runway, reports the maximum distance that high-intensity runway lights can be seen, measured in hundreds of feet: RVR28L24V60+ (Runway 28 Left visual range variable between 2,400 and more than 6,000 feet).

RVV and RVR will only appear on the SA when values are continuously recorded and the following criteria are met:

- RVV must be less than 2 miles or prevailing visibility less than approach minimums
- RVR must be 6,000 feet or less, or prevailing visibility 1 mile or less

Because many locations have only instantaneous readouts, RVV or RVR values might not appear on the observation. This is because RVV and RVR values change relatively rapidly; also, instantaneous readouts are not considered significant for transmission and might even be misleading. (RVV to be deleted from SAs in 1995.)

ATMOSPHERIC PHENOMENA

Atmospheric phenomena—weather and obstructions to vision—follow prevailing visibility using standard aviation weather contractions contained in Table 1-2 and appendix A. This can be the most significant portion of the report.

Table 1-2. Weather and obstructions to vision.

Weather and precipitation

Funnel cloud	FUNNEL CLOUD	IPW	ICE PELLET
Waterspout	WATERSPOUT		SHOWERS
Tornado	TORNADO	S	SNOW
Volcanic ash	VOLCANIC ASH	SW	SNOW SHOWER
T+	SEVERE	SP	SNOW PELLETS
	THUNDERSTORM	SG	SNOW GRAINS
T	THUNDERSTORM	IC	ICE CRYSTALS
Liquid		**Obstructions to vision**	
R	RAIN	BD	BLOWING DUST
RW	RAIN SHOWER	BN	BLOWING SAND
L	DRIZZLE	BS	BLOWING SNOW
Freezing		D	DUST
ZR	FREEZING RAIN	F	FOG
ZL	FREEZING DRIZZLE	GF	GROUND FOG
Frozen		IF	ICE FOG
A	HAIL	H	HAZE
IP	ICE PELLETS	K	SMOKE

Weather

Because of its awesome potential, tornadic activity is always spelled out: TORNADO, WATERSPOUT, or FUNNEL CLOUD. Fair-weather waterspouts form over warm and shallow coastal waters. They are much smaller and less intense than the average tornado and tend to form in unstable air beneath developing cumulus.

VOLCANIC ASH is also spelled out. Volcanic ash became a significant weather phenomenon in the continental United States with the eruption of Mount St. Helens in 1980. Volcanic ash consists of fine particles of rock powder blown out from the volcano. The particles remain suspended in the atmosphere for long periods, extend well into the flight levels, and might drift thousands of miles. Volcanic ash can be extremely destructive to aircraft leading edges, windscreens, and engines.

Tornadoes are destructive, but thunderstorms have an even greater impact on aviation operations. Thunderstorms contain practically every aviation hazard; they form in lines or clusters and can even regenerate.

A thunderstorm (T) is reported when thunder is heard, or overhead lightning or hail (A) is observed. Thunderstorm and associated weather observed away from the station appear in remarks: /CB NE (cumulonimbus northeast). A report of CBs implies "thunderstorm," sometimes translated by briefers as "thunderstorm clouds . . . ." Convective activity reported at or near the station is a clue to possible low-level wind shear (LLWS), or microbursts. (Convective LLWS and microbursts, which produce the most severe wind shear threat, are discussed in chapter 9.)

At the Ontario, CA, airport, the FSS was responsible for weather observations; however, because the FSS observation point was poor, the tower controllers would report the visibility. One night while working the mid shift I thought I heard thunder. I asked the tower if they heard thunder or saw lightning. "Oh yeah," the controller said, "I've been watching it for the last couple of hours." Well, so much for the system. The FAA has taken steps to ensure that tower controllers immediately report this type of activity.

When surface winds gust to 50 knots or more, or hail ¾ inch or greater accompanies a thunderstorm, the storm is reported as severe (T+). Not a heavy thunderstorm: T+ has a specific definition with ominous implications.

Precipitation occurring at the station is assigned an intensity level: heavy (+), light (–), or with the absence of a symbol, moderate. Hail size appears in remarks: /HLSTO 1/2 (hailstones one-half inch in diameter) or <1/4 (less than ¼ inch).

The weather contraction TRW- is sometimes translated as a light thundershower or a light thunderstorm and rain showers; neither translation is correct. The correct translation of TRW- is thunderstorm accompanied by light rain showers. There is no such thing as a *light* thunderstorm!

Drizzle (L), rain (R), or snow (S) indicate a stable air mass. Falling from stratiform clouds (stratus, altostratus, nimbostratus, or stratocumulus) precipitation is usually steady, can be widespread, and is usually not heavy in intensity. Whereas showery precipitation such as rain showers (RW), snow showers (SW), and the like fall from unstable air. Showers that fall from cumuliform clouds (cumulus or cumulonimbus) are usually brief and sporadic and might be heavy in intensity.

Drizzle indicates a relatively shallow cloud layer. A cloud thickness of 4,000 feet is usually required to produce precipitation. Drizzle restricts visibility to a greater degree than rain because it falls in stable air often accompanied by fog, haze, and smoke. Snow grains (SG), the solid equivalent of drizzle, are small, white, and opaque grains of ice.

Snow can fall about 1,000 feet below the freezing level before melting. Snow can often begin with temperatures of 36°F; it's even possible to see snowflakes at temperatures around 50°F. This only occurs when the air is very dry. Snow begins to melt when it falls into air that is warmer than 32°F. The water evaporates and cools the air. Evaporation cools the snow, which retards melting. Water vapor is added to the air, which increases dew point. Finally, the air cools and becomes saturated at 32°F.

Dry snow does not lead to the formation of aircraft structural ice. However, wet snow, which contains a great deal of liquid water, produces structural icing. WET SNOW can appear in remarks.

Snow pellets (SP) are small, white, and opaque grains of ice that form when ice crystals fall through supercooled droplets and the surface temperature is at or slightly below freezing. Falling from cumuliform clouds, snow pellets are more prone to cause structural icing than snow grains.

Ice crystals (IC) might appear suspended and fall from a cloud or clear air. They frequently occur in polar regions in stable air and only at very low temperatures. Ice crystals are not assigned an intensity.

Ice pellets (IP), formerly sleet, are grains of ice consisting of frozen raindrops, or largely melted and refrozen snowflakes. They fall as continuous or intermittent precipitation. Ice pellet showers (IPW) are pellets of snow encased in a thin layer of ice formed from the freezing of droplets intercepted by the pellets, or water resulting from the partial melting of the pellets. Ice pellets do not bring about the formation of structural ice, except when mixed with supercooled water. Ice pellets or ice pellet showers frequently indicate areas of freezing rain above.

Freezing rain (ZR) and freezing drizzle (ZL) are caused by liquid precipitation falling from warmer air into air that is at or below freezing. Droplets freeze upon impact. Structural icing can be expected while flying through freezing precipitation. As stated in the U.S. Air Force manual, *Weather for Aircrews*: "Freezing precipitation is probably the most dangerous of all icing conditions. It can build hazardous amounts of ice in a few minutes and is extremely difficult to remove." Freezing rain can flow back along the aircraft, covering the static port with the resultant loss of accurate pitot-static instruments: airspeed, altimeter, and vertical velocity.

Aircraft without a heated pitot and alternate static source, especially in IFR conditions, would be in serious trouble. Another significant factor, especially for aircraft without ice protection equipment, is that accumulated ice could be carried all the way to the ground making landing extremely hazardous. It cannot be overemphasized that this hazard can affect VFR as well as IFR operations. If this phenomenon is encountered in aircraft without ice protection equipment, virtually the only option and certainly the safest is to fly into warmer air and land. Pilots who fly into ice, with aircraft not certified for flight in icing conditions, must have the right stuff. Because, in every sense of the word, they become test pilots.

Forward visibility is reduced when flying through precipitation, but sideward and downward visibility tend to remain relatively unaffected. The more intense the precipitation, the greater the reduction in visibility. It is possible to have VFR surface visibilities reported with flight visibility less than VFR. Figure 1-5 illustrates how precipitation, especially when heavy, can dramatically reduce visibility and obscure terrain. These showers should be avoided. Flying through steady precipitation is generally smooth; showers tend to be turbulent. Avoiding showers will not only result in improved forward visibility, but a smoother flight.

Remarks of NWS, FSS, and military observations contain the time that a weather phenomenon begins or ends. For example, EWR RS 2253.../T+ B10 reveals that a severe thunderstorm began 10 minutes after the hour (2210). More often, these remarks refer to precipitation: SFO RS 2350.../RB2252E02LB15LERB27RELB42 (rain began

Fig. 1-5. *Precipitation, especially when heavy, can dramatically reduce visibility and obscure terrain. These showers should be avoided.*

at 2252, ended at 2302, drizzle began at 15, drizzle ended, rain began, etc.) Well, you get the idea. The purpose of these remarks is climatological, but they do alert pilots, briefers, and forecasters to weather, which is often significant, occurring at the station.

Obstructions to vision

Obstructions to vision caused by fog, haze, dust, and smoke are reported with visibilities less than 7 miles. When these phenomena exist with visibilities 7 miles or greater, a remark might describe the condition: /H ALQDS (haze all quadrants). Briefers sometimes use the term "unrestricted" to describe visibilities of 7 miles or greater. This sometimes causes confusion. A pilot who is told "visibility unrestricted" might respond, "What about the haze?" Thus, the phrase "visibility unrestricted" does not imply that smoke, haze, dust, or even fog are not present, it just implies that visibility is 7 miles or greater.

Fog (F), ground fog (GF), and ice fog (IF) describe the same condition, a cloud on the ground. Ground fog, normally less than 20 feet deep, reduces visibility horizontally rather than obscuring the sky. Ground fog is usually localized, formed by radiational cooling, and tends to dissipate rapidly when clearing begins. Ice fog forms in cold weather at temperatures around −20°F from radiational cooling and exhibits the same characteristic as radiation fog.

Radiation fog forms when air cools from contact with the ground and becomes saturated. This occurs at night and tends to be most dense around sunrise. Clear skies, light winds (less than 5 knots), high relative humidity, and stable air are favorable conditions for the formation of radiation fog. Low water-vapor content in upper layers increases radiational cooling; dry air aloft enhances the formation of radiation fog. Overcast skies, strong winds, low relative humidity, and unstable air prevent or retard its formation.

Radiation fog tends to be patchy and shallow with lowest visibility around sunrise, usually burning off by midmorning. It tends to form in valleys after moisture has been added at the surface from passing storms. As high pressure with its clear and stable conditions builds into an area, circumstances are right for the formation of radiation fog. This condition can become persistent in California's Central Valley during winter and early spring. After frontal passage, a strong inversion locks moisture at lower levels, and radiation fog forms. Zero-zero conditions over widespread areas can persist for days or even weeks until the moisture evaporates or another storm system moves through the area.

IFR pilots normally will have no difficulty operating in radiation fog as long as they don't mind flying above zero-zero surface conditions. Landing minimums might not prevail until late morning or afternoon, if at all, especially in California's Central Valley. The VFR pilot will be delayed until the condition dissipates. Many Central Valley pilots routinely move their aircraft to mountain airports above the fog layer during winter months.

Advection fog forms when moist air moves over colder ground or water. The air cooled from below becomes saturated. Advection fog can form underneath an overcast. This is a persistent condition along the Pacific Coast during the summer months. The prevailing onshore flow moves the layer into coastal sections and valleys. It is usually deepest and farthest inland at sunrise and retreats toward the ocean during the day. Figure 1-6 illustrates the effectiveness of satellite imagery for determining the extent of the layer; the imagery is especially useful for areas without a weather-reporting service. Winds of 5 to 15 knots tend to cause low ceilings rather than fog.

VFR pilots planning flights into or out of coastal areas should plan arrivals and departures during afternoon hours. If this is not possible, moving the aircraft to an airport a few miles inland will often allow a morning departure.

Fig. 1-6. *Visual satellite imagery, available at many flight service stations, reveals the extent of status layers.*

Upslope fog forms as air is forced upward, expands, and cools at a relatively constant rate. Moist air must be forced upslope, which requires a wind of 5 to 15 knots. This condition occurs during winter and spring in the Midwest where terrain rises steadily from the Gulf of Mexico to the Rockies. It can be widespread and will persist as long as favorable conditions continue. During upslope-fog conditions, the VFR pilot is pretty much out of luck. The IFR pilot might not be much better off. He or she might encounter IFR landing minimums, but the condition can often exist over an area that is the size of several states. A legal IFR alternate airport might be beyond the range of the aircraft.

Rain-induced fog, also known as frontal fog, occurs when warm rain falls through cooler air, evaporates, and condenses to form fog, which can become dense and will persist as long as the rain continues. Winds must generally be light. This condition is usually associated with stationary, warm, or shallow cold fronts. A potential for clear icing exists in areas of rain-induced fog.

Steam fog develops as cold air moves over warm water. Evaporation from the water takes place, and saturation occurs. Low-level turbulence develops as the warm water heats lower levels, and a shallow layer of instability is created. Also known as evaporation fog, steam fog occurs in cold climates over lakes, such as the Great Lakes, in the autumn.

Haze (H) is caused by the suspension of extremely small, dry particles invisible to the eye, but sufficiently numerous to reduce visibility. Haze combined with smoke (K) often describes conditions in metropolitan areas. (The contractions are sometimes whimsically translated as "Hack and Kough.") Figure 1-7 was taken over Atlanta. Large anticyclones, which are high-pressure cells, can dominate the southeast United States, trapping haze and pollutants, especially in industrial areas. Above the haze layer, visibilities are unrestricted and temperatures cool, resulting in a much more comfortable flight.

Fig. 1-7. *Large anticyclones, which are high-pressure cells, can dominate the southeastern United States and trap haze and pollutants, especially in industrial areas as shown by this photograph of Atlanta. Above the haze layer, visibilities are unrestricted and temperatures decrease, which results in a much more comfortable flight.*

Strong inversions over cities or industrial areas can trap haze and smoke, reducing visibility to less than 1 mile. Landmarks will almost be invisible. Dense haze often appears solid in the distance, as if the haze were a cloud layer; this occurs with high relative humidity. When the sun strikes the layer, light waves scatter, causing the layer to appear white. This accounts for apparent inconsistencies in surface observations. SAs might report clear or partially obscured skies with visibilities 3 to 5 miles in haze and smoke. A pilot looking into the sun often has no forward visibility or natural horizon. Pilots caught in such situations, technically VFR, have become spatially disoriented and lost control of the aircraft.

Haze and smoke are usually restricted to an area below 5,000 feet, although they can extend to above 10,000 feet. During the devastating September 1987 forest fires in California, smoke tops were reported as high as 19,000 feet, with visibilities aloft zero. Figure 1-8 is a visual-satellite image that shows dense smoke caused by the fires; note the gray areas over the central and northern California interior and southern Oregon.

Inversion-induced wind shear turbulence develops along the boundary between cool air trapped near the surface and warm air aloft. The turbulence tends to be strongest in valleys during morning hours. Moderate or greater turbulence might be encountered when the airplane penetrates the layer. After an initial outside air temperature rise during the climb through the haze, the air on top will be clear and cool. It is possible to have several haze or smoke layers trapped within inversions aloft. In the Los Angeles Basin, a haze boundary often develops with a *Santa Ana*, which is a warm, dry foehn wind condition. Warm and dry desert air overruns haze that is trapped in cooler and moist marine air; moderate or greater wind shear turbulence can be expected when penetrating the transition zone. Takeoff and landing can be hazardous with the boundary in the vicinity of the runway; the boundary is often marked by a distinct transition between clear and hazy air. PIREPs are usually the only source for determining the tops of haze layers.

Dust (D) and blowing dust (BD) are combinations of fine dust or sand particles suspended in the air. The material can be raised to above 16,000 feet by the wind. Figure 1-9 shows blowing dust in California's Mojave Desert. Surface and airborne visibilities can be at or near zero. Because of its fine particles, dust can remain suspended for days after the wind subsides.

Blowing sand (BN) is made up of particles larger than dust and usually remains within a few hundred feet of the surface. It can also reduce visibility to near zero, but visibility improves rapidly when the wind subsides and the particles fall back to the surface. A pilot approaching Lovelock, NV, was skeptical of a reported visibility of 2 miles in blowing sand and reported his flight visibility was 20 miles. He concurred upon landing, however, stating the top of the blowing sand was at 200 feet AGL.

Blowing snow (BS) produces the same characteristics as blowing sand when strong winds blow over freshly fallen snow. Visibility can be near zero close to the surface, with rapid clearing after the wind subsides.

Fig. 1-8. *Gray areas of the central and northern California interior and southern Oregon reveal dense smoke caused by forest fires. Smoke has caused widespread IFR conditions from the surface to well above 10,000 feet.*

Every year, pilots become lost and even lose control of the aircraft when flying in reduced visibilities. Conditions often improve at a higher altitude. Above the layer, slant-range visibility is usually greater with a distinct horizon preventing disorientation. The seemingly obvious assumption that the closer you are to the ground, the better to see it, usually isn't true in this case.

Fig. 1-9. *Fine dust or sand particles can be raised by the wind to reduce visibilities to zero on the surface and aloft.*

ATMOSPHERIC DATA

Our working definition of atmospheric data will be sea-level pressure, temperature, dew point, and altimeter setting. Full reports contain these elements, in a numeric series, following atmospheric phenomena. Realize that not all aviation weather reporting locations observe all elements.

Sea-level pressure

NWS, FSS, military, and certain contract observations report sea-level pressure in millibars. This three-number code follows visibility and obstructions to vision (...1TRW+ "102"/77/74...). Used to prepare surface analysis charts, sea-level pressure must never be substituted for the altimeter setting in millibars. Sea-level pressure is not reported by limited aviation weather reporting stations, which are towers and most contract observers. The pressure appears only on hourly reports.

The code contains the last three digits of the sea-level pressure to the nearest tenth of a millibar, decimal point omitted. Because average pressure is 1013.2 millibars, prefix the code with a 9 or 10, whichever brings it closer to 1000.0. The sea-level pressure code group in the previous paragraph is 102. Prefix the group with a 10, which decodes a sea-level pressure as 1010.2 millibars.

Temperature and dew point

Temperature and dew point are reported in degrees Fahrenheit, except Canada, Mexico, and most of the rest of the world use Celsius (C). The reports are important beyond the personal comfort values of aircraft passengers because high and low temperatures affect aircraft operation and performance.

Temperature and dew point follow sea-level pressure or atmospheric phenomena (...1TRW+ 102/"77"/"74"/... or ...1TRW+ "77"/"74"/...). Certain locations only report temperature, while others contain neither. When reported, temperature always precedes dew point. Dew point can never be higher than temperature; a dew point that is greater than the temperature usually results from equipment malfunction or transposition of numbers upon transmission. Some observers have yet to learn this fact and will report dew point higher than temperature, rather than missing (/M/). Decoding the example, the temperature is 77 and dew point 74.

High temperatures at high-altitude airports produce high density altitude. Atmospheric pressure and humidity are also factors; however, temperature and elevation are paramount. Surface-temperature forecasts are not normally available, but maximum temperatures usually occur during the middle or latter part of the afternoon. Arrival and departure times must be planned based upon aircraft performance.

I have flown Cessna 150s out of Bryce Canyon, UT (Elev. 7,586), and Grand Canyon, AZ (Elev. 6,606), and a Cessna 210 out of Mammoth Lakes, CA (Elev. 7,128). There is no additional hazard in such operations as long as we calculate and do not attempt to exceed aircraft performance. Multiengine pilots must consider that airport density altitude might be above the inoperative engine ceiling.

After calculating that the aircraft has sufficient performance for conditions, which is required by FAR 91, it's a good idea to determine an abort point on the runway. If the aircraft is not airborne with a positive rate of climb, this point should allow the pilot to come to a safe stop on the remainder of the runway. Remember, aircraft performance data are based upon a brand-new airframe and engine, plus perfect pilot technique.

Low temperatures cannot be ignored. Snow, snow melt, freezing rain, and frost produce structural ice that can be difficult to remove from a parked aircraft. Even a thin layer of frost can severely affect performance and must be removed before takeoff, according to regulations.

Four of us flew a Cessna 172 into the South Lake Tahoe Airport in November. After our stay, about ¾" of ice had formed on the airplane. Being naïve at the time about such conditions, I assumed it would blow off during the takeoff roll. An experienced tower controller suggested that we remove the ice. It had to be scraped off with a plastic scraper! I calculated we had between 300 and 400 pounds of ice on the airplane. I had no experience with this condition at the time. I shudder to think what would have happened during a takeoff in a low-performance airplane, 300 pounds over gross, at a high-altitude airport, with thick ice on the airfoils. The test almost came before the lesson.

Solutions to this problem include hangaring the aircraft to prevent ice formation, arranging for a deicing service, removing the ice, or planning departure for later in the day when the sun has melted the ice.

During taxi, avoid areas of slush. Slush thrown into wheel wells, wheel pants, and control surfaces can freeze, resulting in locked controls and frozen landing gear. With a descent through an icing layer and surface temperatures close to or below freezing, a pilot must be prepared for an approach and landing with airframe icing, possibly to an ice- or snow-covered runway.

If snow, ice, or slush are on the runway, aircraft control might be difficult, especially in high winds. Braking action might be reduced, resulting in a longer than normal ground roll. Without external windscreen deicing—that's deicing, not defrosting—a pilot could be faced (pardon the pun) with zero forward visibility during the landing. FAR 91 requires the pilot to consider, ". . . runway lengths at airports of intended use . . . ." Pilots operating in this environment must consider these factors when selecting destination and alternate airports, or even the advisability of making the flight.

Daytime heating causes rising air currents that produce thermal turbulence. Thermal turbulence usually occurs within 7,000 feet of the surface in stable or conditionally unstable air. This means that vertical movement requires an initiating force, in this case surface heating. In stable or extremely dry air, skies remain clear. If air parcels reach the lifted condensation level, saturation occurs, and stratocumulus, or fair-weather cumulus, clouds will form. These clouds are most often scattered and rarely become overcast. If the air becomes *conditionally unstable*—a parcel of air that becomes unstable if it is lifted to the level of free convection (LFC)—cumuliform clouds form, which can develop into air-mass thunderstorms.

Although thermal turbulence rarely becomes severe, it can be extremely uncomfortable and annoying. Because thermal turbulence is caused by surface heating, it can usually be avoided by flying before midmorning or waiting until late afternoon; otherwise, the only remedy is to climb above the turbulent layer, which might be marked by clouds. A word of caution to the VFR pilot: If you elect to fly above the clouds, be careful not to get caught on top. If the air mass becomes conditionally unstable, clouds can build at an alarming rate and close up even faster.

I was between Winslow, AZ, and Needles, CA, one afternoon flying a Cessna 150 from Kansas to California. Skies were clear, and winds aloft were light and variable. I had to climb to 12,500 to get smooth, cool air. Descending into Needles, the turbulence was continuous light-to-moderate below about 11,000 feet; surface winds were calm.

On another occasion, I was flying one afternoon between Oklahoma City and Amarillo in air that was conditionally unstable. As is my habit, I prefer to fly above the clouds in clear, smooth, cool air. The cumulus appeared to top out at about 9,000 feet. I wanted to climb to a cruising altitude of 10,500, but something was strange. I was in what appeared to be level flight, but only indicating 60 knots in my Cessna 150. A scan of the instrument panel revealed the problem. I was not in level flight, but in a climb attitude above a sloping cloud deck! Topping the clouds was impossible, and I was forced to descend and bounce the rest of the way to Amarillo at 4,500 feet.

Fog can form when air is cooled to its dew point or when moisture is added to raise the dew point or when air is cooled and moisture is added. A temperature and dew point within 5°F indicates the possible development of fog. FSS briefers normally provide temperature/dew point when the spread is this close. Some FAA aviation weather broadcasts use this criterion. The formation of fog, as noted above, requires more than a close temperature/dew point spread.

Wind

Wind direction reported on SAs is always true north and is given as the direction from which the wind is blowing, which is crucial for determining crosswind component, especially in areas with large magnetic variation. Wind direction, to the nearest 10°, is the first two digits of the wind group; ...116/85/48/"2712"/... reports the wind from 270°. If the direction fluctuates by 60° or more, variability appears in remarks: /WND 24V30 (wind direction variable between 240° and 300°). Variable wind direction can make takeoff and landing difficult, even at relatively slow speeds. Wind speed, reported in knots, is the second pair of digits in the wind group. In the example, wind speed is 12 knots.

Gusts (G) refer to rapid fluctuations in speed that vary by 10 knots or more; therefore, the report 18G24 reflects an average speed of 18 knots with fluctuations between 14 and 24 knots. Gustiness is a measure of turbulence: the greater the difference between sustained speed and gusts, the greater the turbulence. When a sudden increase of at least 16 knots, sustained at 22 knots or more for at least one minute occurs, a squall (Q) is reported: 3545Q60. Usually associated with thunderstorm activity, squall implies severe low-level wind shear as well as severe turbulence.

Wind direction, speed, and character (gusts or squall) must be considered when determining the crosswind component or the advisability of landing at a particular airport. And as well as gustiness, surface winds in excess of 20 knots indicate moderate or greater mechanical turbulence, especially over rough terrain. Favorable conditions for turbulence exist just before, during, and after storm passage, especially when winds blow perpendicular to mountain ridges. For example, consider the following report for Mammoth Lakes, CA (MMH), that occurred just after storm system passage with winds perpendicular to the rugged California Sierra Nevada Mountains.

MMH SA 1545 50 SCT E120 BKN 50 46/-3/2345G90/002

The MMH runway is 09-27 and magnetic variation is 15° east; runway headings are magnetic and SA winds true; to convert wind direction from true to magnetic, subtract easterly variation (east is least); therefore, the MMH wind is blowing from 215° magnetic (230 – 015 = 215). At an angle of 55° to the runway (270 – 215 = 55), this results in a 35-knot crosswind component for the sustained speed, and 70 knots for the gusts.

Obviously, this airport would not be suitable for landing. For one thing, the highway patrol closed the roads and no one could pick you up after the, umm, arrival. Most aircraft manuals specify maximum demonstrated crosswind component. Pilots constantly attempt to test these values; some pilots even succeed. Each pilot should know his or her limitations and those of the aircraft. As a flight instructor, I always give my students specific crosswind limitations, always with an alternate if the limitations are exceeded.

Wind shift describes a change in direction of 45° or more that takes place in fewer than 15 minutes: WSHFT 55 (wind shift occurred at 55 minutes past the hour). A wind shift of relatively light winds might only indicate a local change; in coastal areas, the shift often signals the advance or retreat of stratus or fog. In the Midwest, the shift might precede the formation or dissipation of upslope fog. Wind shift is usually a good indicator of frontal passage. In Southern California, a shift often indicates the advance or retreat of a Santa Ana condition.

Peak wind appears in NWS, FSS, and military remarks when speed exceeds 25 to 30 knots. The direction, speed, and time of occurrence are reported: PK WND 3560/40 (peak wind from 350° at 60 knots occurred at 40 minutes past the hour). Peak wind might substantially exceed the value in the body of the observation.

If either the wind vane (direction) or anamometer (speed) are out of service or unreliable, the wind group is reported as estimated: E2515 (wind estimated 250° at 15 knots). We have no way of knowing whether a failure of the direction or speed mechanism, or both, caused the estimate to be taken.

Altimeter setting

The last three digits of the altimeter setting are after the wind group. The normal range of altimeter settings varies from 28.00" to 31.00" of mercury; most altimeters are calibrated within this range. To decode an altimeter setting: When the first digit is an 8 or 9, insert a 2 (.../899 decodes 28.99"); when the first digit is a 0 or 1, insert a 3 (.../045 decodes 30.45").

(An extremely cold high-pressure area developed over Alaska in 1989 causing the altimeter setting to rise well above 31.00 inches. Because these values were not in the range of most altimeters, special emergency rules were enacted. The FAA responded to this occurrence by updating the AIM and amending the FARs to include appropriate guidance for any subsequent occurrence.)

Remarks

Remarks can be the most important part of the report; they follow the altimeter setting, separated by a solidus (/). Standard aviation weather contractions are used for remarks. (A list of contractions can be found in appendix A.) Phenomena within 10 nm of the station are indicated by VCNTY STN: beyond 30 nm by DSNT. NWS, FSS, and military observations might contain numerical codes describing meteorological and climatological conditions (/1001 211 45). These code groups describe weather, cloud type, pressure tendency and change, and maximum/minimum temperatures.

Remarks amplify information already reported, describe conditions observed but not occurring at the station, or contain information considered operationally significant. Routine remarks that amplify information have already been discussed. And recall that NWS, FSS, and military observers tend to do a better job with the remarks because of their training and experience, although they can get carried away.

The following observation was taken at March AFB, CA: RIV SA 1955.../CREPUSCULAR RAYS SW. *The Glossary of Meteorology* defines crepuscular rays as, "Literally, 'twilight rays'; alternating lighter and darker bands (rays and shadows) that appear to diverge in fanlike array from the sun's position at about twilight." Towering cumulus produce this effect, especially with haze in the lower atmosphere. This would seem a rather complicated way of saying HAZY TCU SW (hazy with towering cumulus southwest).

This observation came from a National Weather Service observer at Denver: DEN SA 2359.../DSIPTG GUSTNADO N (dissipating gustnado north). The phenomenon was a glorified dust devil. Gustnado describes a funnel cloud that develops along the gust front of a thunderstorm; the funnel cloud is not a tornado, which would warrant a single-element special. It is believed that the gustnado receives its initial rotation from the shift in wind directions across the gust front. Cold, dense air behind the gust front lifting the warm air ahead imparts a rotating motion in the wind shear zone.

Remarks can be ambiguous, as this report taken by an FSS illustrates: DEF SA 1755 -X E40 OVC.../F9 TOPS 015. This observation would seem to indicate a 4,000-foot ceiling with tops at 1,500 feet, which is impossible. Could the observer have meant 400 overcast? Upon checking, ⁹⁄₁₀ of the sky was obscured by fog and the observer could see cloud cover at an estimated 4,000 feet. The remarks should have read: /F9 TOPS OBSCN 015 (tops of the obscuration 1,500 feet).

The only way to clarify a report is to check with the observer. For individuals using DUAT, this will be all but impossible. An FSS will usually be able to check through its telecommunications system, but don't even ask unless some very serious (i.e., emergency) operational requirement exists.

Figure 1-1 contains an excellent example of an operationally significant remark. The observer has reported /MTNS OBSCD W (mountains obscured west). If an overcast ceiling of 1,500 feet exists, VFR flight below the clouds could be conducted; however, due to mountain obscurement, VFR flight into or out of the valley might not be possible. Pilots accustomed to flying over flatlands need to exercise extra caution when evaluating conditions in mountainous areas. Studying terrain is as important as checking the weather.

A sky condition example for Ukiah, CA, was presented earlier and a scattered cloud layer existed at the missed approach point. If the observation had carried the remark /MTNS OBSCD E-W (mountains obscured east through west), a pilot reviewing the approach and terrain could deduce that the scattered layer existed at the missed approach point.

This example also illustrates how observers indicate phenomenon coverage. The observer starts at true north and works clockwise. If the remark read /MTNS OBSCD E S-N, it would be translated as mountains obscured east and south through north. It's important to understand this convention to correctly relate weather in relation to the station.

I was over Palmdale on a flight to Van Nuys, CA, in a Cessna 150 that was not equipped for IFR. Coastal stratus obscured the mountains with Van Nuys reporting E30 OVC. Because of my experience and familiarity with the area, I knew where there was often a hole. On this occasion, the hole was there, and I proceeded; I had plenty of fuel to return to the desert. Another pilot attempting to depart the L.A. Basin under similar conditions was not so fortunate; the route wound through the mountains into a blind canyon, and the Cessna 182 came to rest on a 45° slope. The only casualty was the totaled airplane.

The following paragraphs describe cloud types that are significant to aviation and are routinely carried in remarks or observations.

Altocumulus castellanus (ACC) is a midlevel cloud that indicates moisture and vertical movement. ACC might indicate thunderstorm development. Showers falling from these clouds can evaporate before reaching the surface as illustrated in Fig. 1-10. This phenomenon appears in remarks as VIRGA. Evaporative-cooling turbulence develops in the vicinity of virga; precipitation evaporates and cools the air, causing downdrafts. A pilot penetrating these areas will encounter wind shear turbulence, which can be severe. Turbulence can be avoided by circumnavigation of these areas. At times, ice crystals or snowflakes can fall from cirrus clouds. The crystals or flakes can change directly from a solid to a gas as they fall into dry air and *sublimate*. These dangling white streamers are known as *fall streaks*.

Altocumulus floccus is a cloud with a cumuliform or rounded appearance; the lower portion is ragged and often accompanied by virga. Altocumulus floccus can evolve from altocumulus castellanus. Like altocumulus castellanus, the floccus clouds indicate moisture and instability at midlevels in the atmosphere.

Standing lenticular altocumulus (ACSL), *standing lenticular stratocumulus* (SCSL), and *standing lenticular cirrocumulus* (CCSL) indicate mountain wave activity. Lenticular clouds appear smooth and remain stationary to the observer; they might develop as horizontal bands produced by long ridges as in Fig. 1-11, or circular and stacked from isolated peaks as in Fig. 1-12. Although these clouds imply turbulence, the roughness will not always be found.

I encountered a mountain wave in California's Owens Valley while flying a Cessna 150. With cruise power and attitude, the airplane rode the wave at the rate of 500 feet per minute, from 8,500 feet to 13,500 feet and back down. The ride was absolutely smooth!

Notwithstanding the previous example, the November 1977 *Approach* magazine reported: "A Navy T-39 trainer was flying a low-level, high-speed navigational training route in mountainous terrain when it encountered severe turbulence. Gust acceleration loads were so high that aircraft design limits were exceeded, resulting in separation of the tail . . . ." All aboard were killed. The article went on to say, "Mountain waves should never be taken lightly. In addition to the T-39 crash, mountain waves were identified in the crash of a C-118 and in extensive damage to a B-52 . . . . While this type of turbulence is obviously critical to traditional low fliers like helicopters, all aircraft are susceptible."

Rotor, or bell-shaped, clouds that often appear as tubular lines of cumulus or fractocumulus clouds parallel to the ridge line underneath the lenticulars always imply severe or greater turbulence. One such situation occurred at Reno, NV, where surface winds were reported gusting to 73 knots. The pilot of a corporate jet reportedly abandoned the approach when all the bottles in the cabin's liquor cabinet broke. (A new definition for severe turbulence?)

Mountain waves develop when strong winds, usually 40 knots or greater at crest level, blow perpendicular to a mountain range. Speed usually increases with altitude in stable air. Mountain waves can cause sustained updrafts and downdrafts occasionally reaching 3,000 feet per minute. Effects of the wave might reach from the ground to 35,000 feet and extend hundreds of miles downstream. Altimeter errors might exceed 1,000 feet. And waves occasionally occur in clear air.

Fig. 1-10. *Altocumulus castellanus, especially with virga (rain that evaporates before reaching the surface), indicates moisture and vertical movement at middle levels. Turbulence and possible microbursts should be expected in the vicinity of virga.*

Fig. 1-11. *Altocumulus standing lenticular clouds develop as horizontal bands produced by long ridges.*

Fig. 1-12. *Altocumulus standing lenticular clouds become circular or stacked from isolated peaks.*

Figure 1-13, a visual satellite image, clearly shows the presence of wave clouds in northern Nevada and Utah. It illustrates the extent of wave activity. Chances are also good that wave conditions exist in the clear air over the northern Sierra Nevada Mountains and central Nevada.

To avoid the worst conditions, most authorities recommend that the pilot remain at least 5,000 feet above mountain crests. In the West, most light aircraft simply don't have that performance. This leaves three options: select a course with lower terrain, wait it out, or take a chance on getting your fillings knocked loose and maybe losing the airplane. Passenger comfort and safety should be the priority considerations.

Cumulonimbus mamma (CBMAM), previously called *mammatus*, result from severe updrafts and downdrafts. They are characterized by lobes that protrude from the bottom of the cloud as illustrated in Fig. 1-14. (The term is from a Latin word meaning udder or breast.) They indicate probable severe or greater turbulence, and often appear at the beginning of a squall. Avoid these areas.

Associated with convective activity, three additional cloud types can appear in SA remarks. A *shelf cloud* (layered and resembling shelves) can appear under a thunderstorm. A *wall cloud*, usually on the southwest edge of the thunderstorm, has a lowered base and indicates that the storm might be severe. A *roll cloud* appears as a detached and dense horizontal cloud at the lower front part of the main cloud. All three indicate thunderstorms and a potential for severe weather.

Dust devils are whirlwinds that form on clear, hot days with light winds. They have diameters of 10 to 50 feet and extend from the surface to several thousand feet. Wind speeds within the rotation vary from 25 to more than 75 knots. Dust devils are capable of substantial damage, but the majority are small. I was doing pattern work with a student at Lancaster's Fox Field in the California Mojave Desert when we flew into a dust devil. The encounter was equivalent to light-to-moderate turbulence; it shook the Cessna 150, and the low pressure in the vortex caused both windows to pop open!

Foehn wall describes the steep leeward boundary of flat, cumuliform clouds formed on the peaks and upper windward sides of mountains during foehn conditions, such as the Santa Ana of Southern California and the Chinook on the eastern side of the Rocky Mountains. This remark should alert the pilot to strong winds, possible turbulence, and wind shear.

BINOVC (breaks in the overcast) or the synonym HIR CLDS VSB (higher clouds visible) means from 95 percent to less than 100 percent of the sky is covered. Such a report should never imply VFR flight through a layer is probable or even possible; however, it might be the first indication of a layer beginning to dissipate.

PRESFR (pressure falling rapidly) and PRESRR (pressure rising rapidly) signify the approach or passage of a frontal system. Pressure rising rapidly accompanied by a wind shift might be reported as FROPA (frontal passage) or APRNT FROPA (apparent frontal passage). And, PRJMP (pressure jump) can indicate the approach of a prefrontal squall line.

31MY87 38A-1   01682 23174 WA2

Fig. 1-13. *Visual satellite imagery clearly shows the presence of wave clouds in northern Nevada and Utah.*

Fig. 1-14. *Cumulonimbus mamma are produced by the strong updrafts and downdrafts associated with a thunderstorm.*

Freezing level data (RADAT) formerly appeared as remarks in 00Z and 12Z SAs associated with radiosonde (upper air) observations. The RADAT has been removed from those SAs in the contiguous United States. At the present, RADAT information is only available from an FSS. In the future, DUATS might make this data available. These locations can be found in the back of the appropriate *Airport/Facility Directory*. The data consists of the freezing level in hundreds of feet MSL and relative humidity at that level: .../RADAT 32078 (32 percent relative humidity at the freezing level of 7,800 MSL). High relative humidity indicates moisture and the possibility of icing at and above the freezing level.

Multiple freezing levels can occur. Figure 1-15 illustrates three crossings. Note the surface temperature in the example, 42°F; the surface is above freezing. The first crossing occurs at 8,100 MSL (081), the second at 10,500 MSL (105), and the third at 12,400 MSL (124). The highest relative humidity, 96 percent, was measured at the middle (M) crossing (L represents lowest and H highest). In this example, air temperature that is above freezing occurs from the surface to 8,100 feet, and again between 10,500 and 12,400 feet. This could be significant if a pilot needs to fly an airplane at a specific altitude to shed ice.

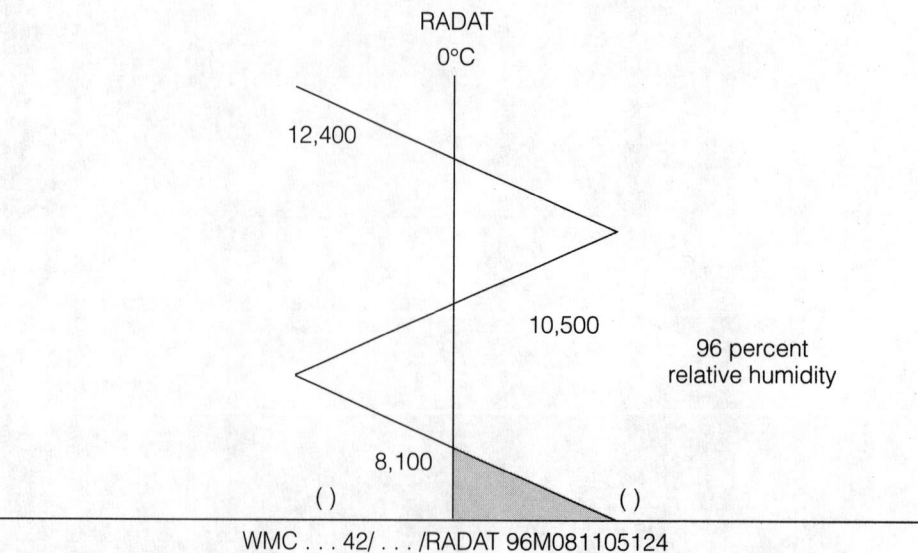

WMC . . . 42/ . . . /RADAT 96M081105124

Fig. 1-15. *Observed freezing level data used to appear in remarks of 00Z and 12Z SAs associated with radiosonde observations.*

Two crossings can occur with surface temperature below freezing: MFR...30/.../ RADAT 78H035097 (temperatures are below freezing from the surface to 3,500 feet MSL, with a layer above freezing from 3,500 to 9,700 feet; relative humidity of 78 percent occurred at 9,700 feet, the high crossing).

Other RADAT messages indicate the sounding was missed (MISG) due to equipment trouble or other factors, or it was delayed (DLAD). If the entire sounding is below 0°C, RADAR ZERO will be reported.

A word of caution: Lack of significant remarks does not infer the absence of hazardous weather. The omission of phenomena such as lightning, cumulonimbus clouds, and other significant phenomena is usually due to inadequate observer training.

AUTOMATED WEATHER OBSERVING SYSTEMS (AWOS & ASOS)

Approximately 1,600 FAA, NWS, Department of Defense (DOD), and locally operated *automated weather observing system* (AWOS) locations will be operational by the late 1990s. Most pilots will be dealing with AWOS to one degree or another on a continuous basis. AWOS and the National Weather Service equivalent *automated surface observing system* (ASOS) have the potential to provide an extra degree of safety or to lure the unsuspecting pilot to disaster.

Attempts at automated weather observations have existed for more than 30 years. Today's science strives to reach the accuracy and reliability of the human observers and reduce some of the limitations. Surface observations by humans are continuously decreasing. According to the FAA, AWOS has the requirement "To provide accurate, 24-hour airport weather information for pilot and air traffic control system personnel at over 1,000 airports with instrument approaches but without adequate weather observation capability." Private AWOS installations will also be purchased and maintained by local airport authorities.

The new technology, however, has its fault finders. FSS weather briefers and NWS forecasters are among the critics. Briefers are dubious about the quality and extent of these observations and any forecasts that are based upon these observations.

Norman Schuyler, pilot and former NWS meteorologist, stated in the November 1988 *Private Pilot*: "The Federal Aviation Administration is still determined to complete the (AWOS) program, a program that no one has a good word for."

Automated surface observations began with the automatic meteorological observing station (AMOS). Unfortunately, the system was only capable of reporting temperature, dew point, wind direction and speed, and barometric pressure. Human observers occasionally manually entered data to provide a complete report.

The automatic observing station (AUTOB), a refinement of AMOS, added sky condition, visibility, and precipitation reporting; however, AUTOB was limited to cloud amount and height measurements of 6,000 feet AGL and three cloud layers. Visibility values were reported in whole miles, to a maximum of 7.

AWOS is operationally classified into four levels. AWOS-A reports altimeter setting only; AWOS-1 (the equivalent of AMOS) reports temperature, dew point, wind, and altimeter setting; AWOS-2 adds visibility information; AWOS-3 adds sky conditions and ceiling.

AWOS equipment is normally installed near the touchdown zone of the primary instrument runway. These observations are updated minute-by-minute. Reports are broadcast by synthesized voice over selected frequencies. Eventually, approximately 110 hourly and special observations will automatically be transmitted over the FAA's telecommunications network.

Formatted into standard reports, AWOS and ASOS observations contain time of observation, sky condition, visibility, temperature, dew point, wind direction and speed, and altimeter setting. Some will report precipitation, thunderstorms (lightning), and density altitude. Although designed to be self-contained, human observers will be able to edit and supplement data. AWOS and ASOS reports use the same format and most of the same contractions as standard SA reports. Only the differences between the reporting systems appear in this section.

AWOS observations will be identified by the contraction AWOS (UKI SA 1235 AWOS). ASOS sites are identified by AO2 for unattended or AO2A for attended. (AO1—no precipitation discriminator.) AO2A does not necessarily mean that the observation contains augmented or supplemental data, only that an observer is present at the location, providing general oversight of the observation.

A laser ceilometer cloud height indicator (CHI) determines sky condition. Similar to the rotating-beam ceilometer, the cloud or clouds will reflect the laser. Heights will range from the surface to a maximum of 12,000 feet, depending on the sophistication of the system. Observed heights will be reported to the nearest hundred feet from the surface to 5,000, nearest 500 feet to 10,000, and nearest 1,000 above 10,000 feet. In general, as many as two cloud layers will be reported: lowest and highest. Higher layers may be added based upon satellite and other data. These systems have an advantage in that all heights are measured electronically rather than estimated by a human observer.

Obscurations will be estimated based upon visibility, temperature, and the computer cloud algorithm, which is a computer program for solving problems in a logical order. During evaluations, automated systems reported more obscurations than the human observer. The overreporting of obscurations has the potential of crying wolf. Automated systems will normally not report a partial obscuration. When reported, a partial obscuration might be indicated by the letter X with the amount of the obscuration omitted. Below is an example of the Warroad, MN, AWOS showing a partial obscuration.

RAD SA 1355 AWOS X M15 BKN 25 OVC 1 3/4 57/52/2212G17/960

A 30-minute history of cloud elements passing over the sensor determines sky cover (scattered, broken, or overcast) based on the computer algorithm. As noted in the September 1988 *ASOS Progress Report*, "Yes, this means that the algorithm assumes that the clouds are in motion above the CHI. That may not always be the case . . . . This automated approach to sky condition reporting has been tested by the NWS and the Air Force . . . with very favorable results." Automated systems tests indicated that at times, due to the algorithm, up to two-tenths coverage might not be reported. Automated systems commonly report scattered clouds when pilot reports indicate broken; however, this might be due to the human tendency to overestimate cloud coverage. Pilots should closely evaluate cloud coverage when making reports. Automated systems have an almost instantaneous ability to determine and report changing conditions.

Automated systems determine visibility from a backscatter device or transmissometer. Figure 1-16 shows the backscatter device in the center of the picture, with the CHI to the left. Values are reported from less than one-quarter (<1/4) mile to a maximum of 10 miles, with variability added as needed. This value might be considerably different from prevailing visibility. Proponents point out that automated systems will be more consistent, objective, standardized, continuous, and representative. This will

Fig. 1-16. *AWOS determines visibility from a backscatter device (center) and a laser cloud-height indication (left).*

certainly be true for IFR operations because the sensors are normally located at the approach end of the instrument runway.

Pilots and FAA controllers have criticized automated systems because they frequently report visibilities too high. Some pilots routinely cut the visibility report in half when light precipitation is in the area. Considerable errors have been noted in the presence of rain, snow, drizzle, and fog. Automated systems tend to be overly sensitive to ground fog. As visibility lowers, a complete obscuration is commonly reported.

Automated systems measure sky condition and visibility at only one point on the airport. Unfortunately that point might not be representative of surrounding conditions. If a single cloud remains stationary over the sensor, the system would report a ceiling when the layer might be scattered. Conversely, if the beam passes through a small hole in an overcast layer, the system would report clear. Meteorologists responsible for forecasts based upon automated systems are understandably concerned. Automated systems do not have the capability to observe the entire *celestial dome*, which is the sky viewed to the horizon in all directions. Proposals have been made to increase and disperse the number of sensors to help alleviate some drawbacks.

A weather report from an ASOS site in the mid-1990s will include rain, snow (snow, snow grains, snow pellets, ice crystals, and ice pellets), and drizzle. Light precipitation, which the sensor cannot identify, will be reported P- . Observations of freezing rain are normally not yet available. Some AWOS sites will report freezing rain, freezing drizzle, and hail as P (unknown type of precipitation). Sites with this capability will be indicated in the *Airport/Facility Directory* as "AWOS III P." Pilots are cautioned to recognize the inherent limitations of sensor technology. Obstructions to vision (fog and haze) from automated sites will also be reported based upon the temperature/dew point spread and the occurrence of precipitation.

ASOS is scheduled to accept lightning data and report the presence of thunderstorms in the vicinity of the airport. Some AWOS sites were being equipped with lightning detectors in the mid-1990s. A thunderstorm will be reported if lightning is detected within 5 nautical miles (nm) of the airport. Lightning detected between 5 nm and 10 nm will be reported as a thunderstorm or lightning in the vicinity; between 10 nm and 30 nm the report will be reported as lightning in the appropriate quadrant. AWOS sites with lightning detectors appear in the *Airport/Facility Directory* as "AWOS III T." (Sites with precipitation and lightning sensors are identified as "AWOS III P/T.")

ASOS will not provide the following data automatically. (The addition of enhanced precipitation and lightning sensors are planned for the future.)

- Tornado, funnel cloud, waterspout
- Thunderstorms
- Hail
- Ice crystals
- Snow pellets, snow grains, ice pellets
- Freezing rain
- Drizzle (might be reported as light rain)

- Volcanic ash
- Blowing snow, sand, dust, spray
- Smoke (might be reported as haze)
- Clouds above 12,000 feet
- Virga and distant precipitation
- Clouds obscuring distant mountains
- Visibility differences in sectors

Except from an augmented site, remarks will normally not appear on automated observations. Pilots can expect to see several automated remarks on ASOS reports. Most deal with climatological data, similar to standard SAs. The following, however, are significant to aviation:

- PWINO Precipitation identifier sensor not operational (ASOS).
- ZRNO Freezing rain sensor not operational—when available (ASOS). (FZRANO on METAR.)
- TNO Thunderstorm information not available (ASOS). (TSNO on METAR.)
- LTG DATA MISG Lightning data missing (AWOS).
- PCPNNO Precipitation data missing (AWOS).

The FAA views remarks as irrelevant and unnecessary to aviation operations. This would seem to be confirmed by the NWS's test at Topeka, KS. The study evaluated remarks added to ASOS reports by control tower personnel. The NWS concluded that this augmentation was unreliable, often not appropriate and inconsistent, and that remarks have value only if the observer is also a user. The conclusion also noted that augmentation was expensive. These conclusions, to say the least, are controversial.

Reports of cloud type, such as altocumulus castellanus, standing lenticular altocumulus, rotor clouds, and cumulonimbus, will not appear, nor will other weather phenomena, such as virga and weather not occurring at the station. A great number of pilots, briefers, and forecasters often view remarks as the most significant part of the report. With AWOS, pilot reports to flight watch or a local FSS will take on an even greater significance.

During the FAA's evaluation, commenters favored the AWOS by about two to one. "Pilots indicated foremost that 'availability' and 'currency' were the most favorable characteristic . . . . AWOS performance increased considerably during degraded weather conditions." The FAA concluded: "The demonstration program made evident that AWOS can consistently provide accurate real-time weather information directly to pilots." Planned enhancements include RVR and low-level wind shear alerting systems (LLWAS).

AWOS performs best in less than VFR conditions; a primary function of the equipment is to provide ceiling, visibility, and altimeter setting for IFR operations, but VFR flights will also benefit. For example, a pilot might cancel a flight rather than "taking a look" with IFR or marginal VFR reported. Accurate temperature, wind, and altimeter settings will increase flight safety. For example, consider the Mammoth Lakes report:

MMH SA 1545 50 SCT E120 BKN 50 46/-3/2345G90/002

The AWOS-1 broadcast from the Mammoth Airport would alert pilots that wind conditions were unsuitable. The pilot could then divert before encountering severe conditions rather than possibly arriving without enough fuel for a suitable alternate despite his or her best efforts to follow regulations.

Availability of the service, frequencies, and telephone numbers are advertised in the *Airport/Facility Directory*.

The advantages of AWOS are undeniable. Observations will be available for many more locations, on a continuous basis. This will allow commercial operations into airports previously inaccessible due to lack of weather data. Airports with part-time observations will be usable for IFR operations 24 hours a day. Blind reliance on AWOS, however, has the potential for disaster. Remarks indicating thunderstorms or mountain waves normally will not appear. A VFR pilot planning to land at an AWOS airport reporting clear would have quite a shock finding an overcast with breaks. With weather reported close to minimums (VFR or IFR), extra caution must be exercised, terrain considered, and suitable alternates available. The observation that "Weather reports might not be accurate, but they're official," should be considered sage advice.

U.S. METAR AVIATION WEATHER REPORTING CODES

The new METAR code is scheduled to replace the SA format in 1996. Ostensibly, METAR will standardize surface aviation reports with the rest of the world; however, individual countries may retain some of their old practices, as the United States has elected to do in several areas. For example, the United States will continue to use the current units of measurement rather than metric units, except for temperatures. The following discussion contains only changes from the current SA format and codes. Interpretation remains the same.

METAR reports will contain the following sequence of elements:

- Type of report
- Station designator
- Date and time of report
- Wind
- Visibility
- Weather and obstructions to visibility
- Sky condition
- Temperature and dew point
- Altimeter setting
- Remarks

There will be two types of reports. The *METAR* is a routine observation that is equivalent to SA and RS. The *SPECI* is a special METAR weather report that is equivalent to SP.

METAR codes use standard four-letter International Civil Aviation Organization (ICAO) identifiers. For the continental United States, four-letter code will be the domestic identifier prefixed with the letter K; San Francisco will be KSFO. The date and time of observation is transmitted as a six-digit date/time group appended with a Z to denote UTC. AUTO indicates an automated report.

Wind

Wind is reported as a 2-minute prevailing wind direction and speed using a five-digit group, six digits if the speed exceeds 99 knots. The first three digits indicate the direction that the wind is blowing from in relation to true north. The next two or three digits indicate speed. The units of measurement follow: KT is knots, KMH is kilometers per hour, and MPS is meters per second. In the United States, knots will be used: 34017KT (wind 340° at 17 knots). Gustiness is reported with a G (23020G30KT), like the SA code.

If wind direction varies by 60° or more with a speed greater than 10 knots, a variable group separated by a V will follow the prevailing group (34017KT 310V010). A calm wind is reported as 00000KT. The contraction VRB indicates a variable wind direction. VRB is included for winds less than 7 knots (VRB06KT). VRB might be used in special cases at higher speeds, such as wind shift when a thunderstorm passes over the station.

Visibility

In the United States, visibility will be reported as prevailing (as before) or its automated station equivalent, in statute miles. Directional variations in visibility will not be reported: 1/2SM (one-half statute mile), 7SM (7 statute miles), 15SM (15 statute miles), or m1/4 (less than ¼ statute mile). Visibility at a second site will appear in remarks: VIS2RY11 (visibility 2 statute miles at Runway 11).

Runway visual range reporting parameters will be the same as SA reports. When reported, RVR uses the following format. The letter R followed by the runway number, a solidus(/), and the RVR in feet: R29/2400FT (Runway 29 visual range 2,400 feet). The following will be added to the visual range as required:

- V Variability (R32R/1600V2400FT)
- D Down or decreasing (R12/D4000FT)
- U Up or increasing (R02L/U1600FT)
- N No change (R18/N1800FT)
- M Minus or less than (R22R/M1600FT)
- P Plus or more than (R36/P6000FT)

Weather and obstructions to vision. Weather and obstructions to vision in the METAR code represent a significant change from SA reports. In METAR, weather and obstructions to vision are reported in the following format: intensity, proximity, descriptor, precipitation, obstruction to vision, and other.

Refer to Table 1-3. The intensity of precipitation in METAR is indicated in the same way as on SAs, except it is located at the beginning of the group. It's important to remember that intensity refers to the precipitation, not the descriptor: +TSRAGR (thunderstorm with heavy rain and hail). Hailstone size will appear in remarks: GR 3/4. A severe thunderstorm is indicated by surface winds of 50 knots and/or hail ¾" in diameter.

Table 1-3.
METAR weather and obstructions to vision.

Intensity		IC	Ice crystals
—	Light	UP	Precipitation
No symbol	Moderate		(automated
+	Heavy		observation)
Proximity		**Obstructions to vision**	
VC	Vicinity	FG	Fog (visibility
Descriptor			less than ⅝)
TS	Thunderstorm	PY	Spray
DR	Low drifting	BR	Mist (visibility
SH	Showers		⅝ to 6)
MI	Shallow	SA	Sand
FZ	Freezing	FU	Smoke
BC	Patches	DU	Dust
BL	Blowing	HZ	Haze
Precipitation		VA	Volcanic ash
RA	Rain	**Other**	
DZ	Drizzle	SQ	Squall
GR	Hail (greater	SS	Sandstorm
	than ¼ in.)	DS	Duststorm
GS	Small hail/snow	PO	Dust/sand whirls
	pellets	FC	Funnel cloud
SN	Snow		Tornado
PE	Ice pellets		Waterspout
SG	Snow grains	+FC	Well-developed tornado

Proximity, or vicinity (VC), reports weather occurring in the vicinity of the airport. Vicinity is defined as between 5 and 10 statute miles from the center of the runway complex.

Descriptors apply to precipitation or obstructions to vision. Automated sites might use UP to report precipitation of unknown type. Fog (FG) is only reported when the visibility is less than ⅝ of a mile. With visibility between ⅝ and 6 miles, mist (BR) is used. Spray (PY) is only reported in combination with the descriptor "blowing" (BL). Blowing spray (BLPY) is water droplets torn by the wind from a body of water, generally from crests of waves, and carried into the air in such quantities that horizontal visibility is reduced to 6 miles or less.

As shown in Table 1-3, five categories of weather phenomena appear in the "other" group. A squall (SQ), reported in SAs as a Q in the wind group, is a strong wind characterized by a sudden onset, a duration of minutes, and then a sudden decrease in speed. Sandstorms (SS) and dust storms (DS) in METAR are differentiated from blowing dust and blowing sand by visibility. Sandstorms and dust storms will be reported only when the visibility is less than 5/8 of a mile. Dust or sand whirls (PO) are dust or sand raised by rapidly rotating columns of air; they are dust devils when well-developed. PO is the equivalent of DUST DEVIL in the remarks of an SA.

Table 1-4 compares METAR and SA significant weather terms.

Table 1-4. METAR-SA significant weather terms.

METAR	Domestic SA	METAR	Domestic SA
FC	Funnel cloud	BLDU	BD(5/16–1/2)
FC	Waterspout	DS	BD (less than 5/16)
FC	Tornado	BLSA	BN (5/16–1/2)
TS	T	SS	BN (less than 5/16)
RA	R	BLSN	BS
RASH	RW	DU	D
DZ	L	FG	F (less than 5/8)
FZRA	ZR	BR	F (5/8–6)
FZDZ	ZL	MIFG	GF
GR	A	HZ	H
PE	IP	FU	K
SHPE	IPW	SQ	Q in wind group
SN	S	PO	Dust devils in remarks
SHSN	SW		
GS	SP		
IC	IC		
SG	SG		

Sky condition

METAR sky-condition reports represent another significant change in criteria and format. Thin layers and partial obscurations will no longer be reported, except in remarks. The format will consist of the amount of cloud coverage, height of the layers above ground level, and cloud type when conditions warrant (CB and TCU).

The amount of sky cover will be reported in one-eighth increments. A clear sky is reported using the contraction SKC (sky clear); otherwise, sky condition contractions remain the same as SAs:

- SKC Clear (no clouds) or CLR for automated stations
- FEW (1/8 to 2/8 coverage)
- SCT Scattered (3/8 to 4/8 coverage)
- BKN Broken (5/8 to 7/8 coverage)
- OVC Overcast (8/8 coverage)

A ceiling is not designated. For aviation purposes, the ceiling will remain the lowest broken or overcast layer, or vertical visibility into an obscuration. Cloud heights are reported with three digits, in hundreds of feet. The ceiling at a second site will appear in remarks.

When towering cumulus (TCU) (cumulus congestus) or cumulonimbus (CB) clouds are present, they will be included in the sky condition section of the report: BKN035TCU (towering cumulus clouds with a ceiling 3,500 broken).

Vertical visibility into a total obscuration will be reported using the letters VV followed by the vertical visibility in hundreds of feet: VV002 (indefinite ceiling 200 feet).

Temperature, dew point, altimeter, and remarks

Temperature and dew point are reported using two digits in degrees Celsius. Temperature below zero will be prefixed with the letter M for minus: 20/15 (temperature 20°C and dew point 15°C) and 08/M03 (8°C and –3°C).

The altimeter, in inches of mercury, will be reported in a four-digit format prefixed with the letter A: A2992 (altimeter setting 29.92" of mercury). Sea-level pressure, when reported, appears in remarks: SLP132.

Remarks should be similar to their SA form, following the contraction RMK:URAN (rain of unknown intensity north). NWS observations will continue with various coded additive data.

The following illustrates how surface observations will most likely appear in METAR code.

```
SPECI KEWR 172240 03545G60KT 0SM R04/4500 +TSSHRASQ VV000CB
  A2989/RMK TS OVHD MOVG ESE FQT LTGICCCCG
METAR KEWR 172253 09005 3SM TSSHRA OVC015CB 23/21 A2983 RMK
  TSB10  SE MOVG ESE FQT LTGICCCG VSBY HIR W RAB07 PK WND 35060/40
METAR KPHL 172250 36018G24 1SM R27R/3000VP6000 +TSSHRA VV006CB
  22/20 A2983 RMK TSB25 ALQDS MOVG SE FQT LTGICCG RAB38 PRESFR
SPECI KPHL 172256 29021G28 3/4SM R27R/0800VP6000 +TSSHRABR
  VV005CB A2989 RMK WSHFT 55 TS ALQDS MOVG SE FQT LTGICCG PRESRR
```

Essentially the same information will be available in METAR as SA, except using some different codes and format. When it is implemented, the sooner that we get used to it, the better off we'll all be.

Weather observations are only as useful as a pilot's understanding. This requires a knowledge of the observer's methods plus limitations due to equipment, training, experience, and time of day. Pilots must not only be able to read and translate reports, but also interpret a report's meaning and significance, then apply that to a flight situation (VFR or IFR) and capability of the aircraft when taking into account the route's terrain and alternate airports that might be required.

Several surface observations were presented at the beginning of the chapter with a request to translate and interpret their meaning. A strong cold front was moving through the northeast and things were happening at Newark and Philadelphia:

```
EWR SP 2240 W0 X 0T+RW+ 3545Q60/989/R04VR45 T+ OVHD MOVG ESE
   FQT LTGICCCCG
EWR RS 2253 -X E15 OVC 3T+RW 102/75/70/0905/983/R5 T+ B10 SE MOVG
   ESE FQT LTGICCCCG VSBY HIR W RB07 PK WND 3560/40
```

The EWR SP 2240 reports an indefinite ceiling zero sky obscured, visibility zero in heavy rain showers and a severe thunderstorm. Rain has reduced both vertical and horizontal prevailing visibility to zero. The wind group reports peak gusts in squalls to 60 knots, which meets a criterion for a severe thunderstorm. Remarks indicate Runway 04 visual range is 4,500 feet, which is above landing minimums. A severe thunderstorm is overhead moving east-southeast with frequent lightning in cloud, cloud-to-cloud, and cloud-to-ground. There is a distinct possibility of severe low-level wind shear with microbursts. Nobody has any business flying in this. (The frequency, or occurence, of lightning is defined as follows: OCNL less than one flash per minute, FRQ about one to six flashes per minute, and CNS more than six flashes per minute.)

EWR has technically improved to VFR at 2253, although the severe thunderstorm that began 10 minutes before the last hour is still reported to the southeast. Moderate rain showers obscure ⁵/₁₀ of the sky. Quite probably the base of the clouds has not changed significantly since the previous special, but because the heavy rain showers have moved through, the observer can see the cloud base. The ceiling report should be considered with a little skepticism.

Be alert for the development of fog because temperature and dew point are within 5° and the wind is almost calm. Rain can suddenly cool the air and increase moisture near the surface, making fog likely. The report indicates that conditions are improving from the west: T+ MOVG ESE VSBY HIR W.

```
PHL RS 2250 W6 X 1TRW+F 102/77/74/3618G24/983/R27RVR30V60+ TB25
   ALQDS MOVG SE FQT LTGICCG RB38 PRESFR
PHL SP 2256 W5 X 3/4TRW+F 2921G28/989/R27RVR08V60+ WSHFT 55 T
   ALQDS MOVG SE FQT LTGICCG PRESRR
```

Philadelphia's 2250 record special reports an indefinite ceiling 600, sky obscured, visibility 1, with heavy rain showers, fog, and a thunderstorm. Apparently this is well above IFR landing minimums, but note the RVR. Runway 27R's visibility is variable between 3,000, which is close to minimums, and more than 6,000 feet. With a thunderstorm in progress in all quadrants, frequent lightning, and pressure falling rapidly, conditions will continue unsettled with a distinct possibility of surface wind shifts, LLWS, and microbursts. The temperature/dew point spread is 3°, rain is adding moisture and cooling the air, and despite the strong gusty winds, fog has formed. Pilots attempting an approach must be alert for severe wind shear and gusty crosswinds and be prepared to execute a go-around at any point on the approach if severe shear is encountered, as well as a possible missed approach.

A special was issued only 6 minutes after the record observation; conditions can change rapidly. Ceiling and visibility have deteriorated, along with the variability in the RVR; however, a wind shift occurred at 2255 and the pressure is rising rapidly. WSHFT and PRESRR are indicators of frontal passage and improving weather.

I flew a Cessna 182 from Springfield, IL, to Oklahoma City under conditions similar to those described in the previous SA examples; VFR conditions, with scattered thunderstorms, prevailed to St. Louis. (Rather than file IFR without thunderstorm detection equipment, I prefer VFR to see what I'm getting into.) I arrived in St. Louis just before the front. At the FSS, cell after cell was reported along the airway to Oklahoma City; this was no-go VFR or IFR. Several hours later, after the front passed, conditions improved rapidly, and the trip was completed.

A final caution and a theme that will be repeated: Never rely on a single piece of information. The preceding reports were taken and analyzed out of context. Observations have their place, but to be used effectively, surface reports, especially AWOS, must be viewed within the overall picture that includes weather advisories, synopses, PIREPs, and forecasts.

The weather condition WO X OF is sometimes translated in the vernacular: "WOKS-off." I imagine that even with METAR, we "old-timers" will still refer to zero-zero as WOKS-off.

2
Pilot reports
(UA, PIREPs)

Dᴵᴰ YOU EVER COMPLAIN ABOUT FORECASTS, OR THINK THAT THEY WERE prepared in a sterile environment, or conclude that pilots have no influence on forecast preparation? If we wish to participate, pilots can and do influence forecasts. According to the National Weather Service, PIREPs are the most important ingredient for AIRMETs, SIGMETs, center weather advisories (CWA), and winds and temperatures aloft forecast amendments. In addition to the influence on forecasts, certain phenomena such as cloud layers and tops, icing, and turbulence can only be observed by the pilot. Add the fact that the observational network has dwindled due to FSS/NWS consolidation and military base closures—AWOS remains untested on a large scale—and it becomes very apparent that the need for accurate pilot reports cannot be overstated.

Satellites, upper air reports, and radar supplement surface observations. Satellites only observe cloud tops; upper air observations are infrequent and widely spaced; radar only provides storm information. An urgent need exists for information on weather conditions at flight altitudes, along routes between weather reporting stations (especially in mountainous areas) and at airports without a weather reporting service. In many cases, the pilot is the best and only source of actual weather conditions.

The FAA recognizes the importance of PIREPs and has directed air traffic controllers and flight service specialists to actively solicit PIREPs, especially during marginal or IFR conditions and periods of hazardous weather. Pilots operating IFR must by regulation "report . . . any unforecast weather conditions encountered . . .." These reports are not only of value to other pilots, but to controllers, briefers, and forecasters alike.

PIREPs can be provided to any air traffic facility: center, tower, FSS; however, to ensure widest distribution it's best to report directly to flight service, preferably the *enroute flight advisory service* (EFAS), radio call "Flight Watch." (More information about Flight Watch is in chapter 12.)

PIREPs are transmitted under the location identifier for the surface report (SA) nearest the occurrence using the file type UA: SAC UA (Sacramento pilot report). Reports can be appended, however, to major hub locations to ensure greatest prominence and widest distribution. Unfortunately, this FAA handbook requirement is not always followed. The following PIREP illustrates a serious problem:

HHR UUA /OV MQO/.../TB SVR/ RM 2 N SBP

This report of severe turbulence occurred 2 miles north of the San Luis Obispo, CA, Airport (SBP); however, the person who transmitted the report—it could have been sent by NWS or FSS—appended it to the Hawthorne, CA (HHR) file. A pilot or briefer would only by accident see this report during a briefing to SBP. Some systems display PIREPs along with the associated SA; others don't. Pilots using DUAT or a commercial provider must consult the vendor to see how these messages are handled.

PIREP FORMAT

PIREPs are entered using the standard format illustrated in Fig. 2-1. There is no need to memorize the form because FSS/NWS specialists encode the report; however, an understanding of the format illustrates information needed and will assist in decoding and interpreting reports. Standard contractions contained in appendix A are used.

Report type: UUA or UA. An urgent pilot report (UUA) represents a hazard or potential hazard to flight operations. The message type UUA receives special handling and immediate distribution. Urgent PIREPs contain information on tornadoes, severe or extreme turbulence, severe icing, hail, low-level wind shear when the pilot reports an airspeed change of 10 knots or more, volcanic ash, or any other phenomena considered hazardous. Reports that are designated UA receive routine distribution.

Location: /OV. The location where the phenomenon was observed is reported in relation to a three-letter airport or radio navigational aid (NAVAID) identifier. Examples are:

- Airport /OV HAF (Half Moon Bay Airport)
- Fix /OV LAX (Los Angeles VOR); /OV LAX 060010 (Los Angeles 060° radial at 10 nautical miles)
- Between fixes SEA - BTG (Seattle VOR and the Battleground VOR); SLC 245080 - JNC 210040 (Salt Lake City 245° radial at 80 nm and the Grand Junction 210° radial at 40 nm)

PIREP FORM

Pilot Weather Report → *Space Symbol*

3-Letter SA Identifier

—— —— —— → 1. UA → —— UUA →

Routine Report   *Urgent Report*

2. /OV →	Location:
3. /TM →	Time:
4. /FL →	Altitude/Flight Level:
5. /TM →	Aircraft Type:

Items 1 through 5 are mandatory for all PIREPs

6. /SK →	Sky Cover:
7. /WX →	Flight Visibility and Weather:
8. /TA →	Temperature (Celsius):
9. /WV →	Wind:
10. /TB →	Turbulence:
11. /IC →	Icing:
12. /RM →	Remarks

FAA FORM 7110-2 (1-85) Supersedes Previous Edition

Fig. 2-1. *FSS and NWS specialists encode PIREPs using the PIREP form. Pilots need not memorize the form, but it does indicate information needed.*

Normally, if a PIREP contains conditions at an airport or a specific geographical location such as a mountain pass, the code for that airport or location will appear in remarks:

DEN UUA /OV DEN 301021 .../TP MAN/RM OG 1V5 WND 50-80G100+
BLOWING 3/4 GRAVEL

This report contains conditions observed on the Denver 301 radial at 21 nm. The remarks (/RM) indicate that the report refers to conditions on the ground (OG) at Boulder Municipal Airport (1V5). It seems the wind is gusting to more than 100 knots and blowing ¾-inch gravel around. This information would certainly indicate a no-go decision and illustrates the importance of such reports.

Decoding reports can be difficult without a copy of FAA Handbook 7350.5 *Location Identifiers*. Major identifiers can be found in appendix A, and DUATS has a decode command. Sectional and world aeronautical charts contain NAVAID identifiers. Sectional charts depict airport identifiers along with other airport data. If necessary, a pilot can always call an FSS for an explanation of a report's coding.

Time: /TM. Time of the pilot report, referenced to Coordinated Universal Time (UTC).

Altitude/flight level: /FL. The altitude, in hundreds of feet MSL, that the phenomenon was encountered. DURC (during climb) and DURD (during descent) might appear on the report.

Type aircraft: /TP. Standard contractions for aircraft types are used. Appendix A is not specifically intended to decode all aircraft types, but with a little practice the appendix can be used to decode UA /TPs. From time to time, this element will show /TP PUP (pickup truck), /TP CAR, /TP FBO (airport fixed-base operator), or as reported on the Denver PIREP, /TP MAN.

Sky cover: /SK. Sky cover describes the amount of clouds (CLR, SCT, BKN, or OVC), bases and tops, in hundreds of feet MSL. Remember that sky cover is always an estimate, based on the observer's training and experience; in this case, the pilot's training and experience as an observer. If more than one layer is reported, the layers will be separated by a solidus: 055 SCT 070/105 BKN 130 (bases of a scattered layer 5,500, tops 7,000 feet; bases of a higher broken layer 10,500, tops 13,000 feet).

It's important to remember that PIREP bases and tops are always MSL. This accounts for some perceived surface observation errors. The observation in chapter 1, Fig. 1-1 reported 10 SCT M18 BKN 30 OVC. If field elevation is 2,500 feet, a pilot flying in the area might report 035 SCT/043 SCT/055 SCT. The surface observation and PIREP are perfectly consistent because the surface observer used the summation principle and the field elevation is 2,500 feet.

Weather: /WX. Weather encountered and flight visibility are reported in this element. Flight visibility should be reported to the nearest whole statute mile: FV01, FV05. Unrestricted visibility will be encoded FV99, which has a potential for ambiguity. Does it mean "visibility 99 miles or greater" or "seven miles or greater"? Pilots should report specific values to 100 miles to eliminate any possible misunderstanding. Weather, when reported, uses standard contractions and appears following the visibility.

Air temperature in Celsius: /TA. Self-explanatory.

Wind: /WV. Wind direction and speed are encoded using three digits to indicate direction and three digits to indicate speed: /WV 360020 (observed wind 360° at 20 knots). In spite of the FAA's *Flight Service* handbook, which directs that wind be reported in relation to magnetic north, direction should be reported in relation to true north, which is consistent with other reports and forecasts. Speed is in knots.

Turbulence: /TB. The intensity and altitude (when different from /FL) of turbulence appears in this element. Clear air turbulence (CAT) and CHOP should be added when appropriate. (Both terms will be defined in a subsequent subsection of this chapter.) When turbulence has been forecast, but reports indicate smooth, /TB NEG is entered; therefore, the proper interpretation is smooth, rather than turbulence that bounces the aircraft down.

Icing: /IC. The intensity, type (clear, rime, or mixed) and altitude, when different from /FL, are entered in this element.

Remarks: /RM. This element reports low-level wind shear, convective activity, dust storms and sandstorms, surface conditions at airports, or other information to expand or clarify the report.

Remarks of some PIREPs have read: /RM SMOKE OVER NWS BUILDING DRIFTING EAST—SOMEONE THINKING; /RM VFR NOT RECOMMENDED—THREE AIRCRAFT COULD NOT MAINTAIN VFR DUE TO ICING IN CLOUDS. Something is rather odd; oh well, I guess it *is* difficult to maintain VFR in the clouds when you're icing up. Consider /RM HAD TO CLIMB TO FL200 TO REMAIN VFR—NOW LEAVING FREQ TO CONTACT ZOA. Because this pilot is already 2,000 feet into Class A airspace, leaving the frequency to contact ZOA (Oakland Center) seems like a "right-good" idea.

Cloud bases and tops, temperatures, and even winds can be measured. Flight visibility and weather are direct observations; however, intensities of turbulence and icing are some of the most misunderstood quantities in aviation. That's because they're subjective, usually based upon the pilot's training and experience. (The introduction to this book touched on training and experience.) Take for example the rather shaky voice that called flight service to report "moderate to severe turbulence." The specialist asked the novice pilot if he "actually lost control of the aircraft." The pilot replied, "Well, no." The specialist then asked, "Would it be okay if we called it light to moderate turbulence?" The pilot agreed.

TURBULENCE

The intensity of turbulence is to some degree affected by aircraft type and flight configuration. U.S. Air Force studies have shown the following to generally increase the effects of turbulence:

- Decreased weight
- Decreased air density

- Decreased wing sweep angle
- Increased wing area
- Increased airspeed

Classifications for the intensity of turbulence can be found in the *Airman's Information Manual* and *Aviation Weather Services*; however, I prefer the following.

Light. Light is a turbulent condition during which your coffee is sloshed around, but doesn't spill, unless the cup is too full. Unsecured objects remain at rest; passengers in the backseat are rocked to sleep.

Moderate. Moderate is a turbulent condition during which even half-filled cups of coffee spill. Unsecured objects move about; passengers in the backseat are awakened by a definite strain against their seat belts.

Severe. Severe is a turbulent condition during which the coffee cup you left on the instrument panel whizzes past the passengers in the backseat. The aircraft might be momentarily out of control, but you don't let on. Anyone not using a seat belt is making a futile attempt to find the seat belt and fasten it.

Extreme. Extreme is a rarely encountered turbulent condition during which aircraft control might be impossible. It is usually associated with rotor clouds in a strong mountain wave or a severe thunderstorm. Structural damage is possible. Passengers become concerned while you are frantically trying to maintain control of the airplane.

The following reports illustrate mountain wave activity.

RNO UUA /OV FMG 330025/TM 2345/FL105/TP BE35/TB SVR/RM TMPRY LOST CONTROL...PILOT CUT ARM IN TURBC...RTNG TO RNO.

A SIGMET was in effect for severe turbulence, Reno surface winds were out of the west gusting to 27 knots, and winds across the Sierra Nevada mountains were gusting to near 50 knots. The pilot had the clues, but elected to go. From the PIREP it would appear he or she regretted the decision.

Another report had a somewhat cryptic tone:

RNO UUA /OV FMG 270012.../TP C404 /TB EXTREME 130-110 MDT SVR 110-090 /RM EXPERIENCING STRUCTURAL DAMAGE.

This pilot was fortunate to have returned safely to make this report.
The airlines are not immune to mountain waves:

DEN UUA /OV DEN 313047/TM 0158/FL350/TP L101/TB SVR/RM SVR MTN WAVE PLUS AND MINUS 6000 FPM.

This Lockheed Tristar (L101) at 35,000 feet over the Rockies experienced severe turbulence and 6,000 feet per minute updrafts and downdrafts. This illustrates the severity of mountain waves and the fact that they can extend to the stratosphere.

Turbulence that is encountered in clear air, not associated with cumuliform clouds, usually above 15,000 feet, and associated with wind shear should be reported as *clear air turbulence* (CAT). Slight, rapid, and somewhat rhythmic bumpiness without appreciable changes in altitude or attitude defines *chop*. Because chop does not cause appreciable changes in altitude or attitude, it would not be severe.

In addition to intensity, duration of turbulence should be reported.

- Briefly: turbulence encountered for an extremely short period, usually only one or two jolts; /TB LGT /RM 2 MDT JOLTS could be reported /TB LGT BRFLY MDT).
- Occasional: less than ⅓ of the time.
- Intermittent: between ⅓ and ⅔ of the time.
- Continuous: more than ⅔ of the time.

Chapter 1 discusses how to use an SA to determine likely areas for turbulence. Recent and accurate PIREPs can verify or refute its presence. If reports or forecasts indicate turbulence, a pilot can minimize the hazard when encountered. Turbulence imposes gust loads that appear to be almost instantaneous. Gust loads increase with the speed of the aircraft and gust velocity.

If light or moderate turbulence is encountered or expected, avoid flight in the caution range, which is the yellow arc on the airspeed indicator. If severe turbulence is encountered or expected, slow to maneuvering speed, which is usually placarded in the vicinity of the airspeed indicator rather than printed on the indicator. Because the gust load factor decreases as wing loading increases, maneuvering speed increases with the aircraft's gross weight; therefore, it's often necessary to determine maneuvering speed based upon gross weight.

In moderate or greater turbulence, fly attitude rather than altitude. Disengage an autopilot's "altitude hold" feature. The aircraft should already be slowed to the turbulent-air penetration speed. Don't chase airspeed or altitude; it's comparable to riding a horse, go with it rather than fighting it. The objective is to avoid imposing additional abrupt maneuvering loads on the airframe. For the most part, ignore altitude unless terrain clearance becomes a problem. When VFR, try to avoid IFR cardinal altitudes (4,000, 5,000, 6,000, etc.) or opposite-direction VFR altitudes. When IFR, inform air traffic control of the problem and any altitude deviations required. ATC increases vertical separation during severe turbulence conditions.

ICING

As with turbulence, there is a tendency to overestimate icing intensity, especially by new or low-time pilots. A recently rated instrument pilot, after experiencing a second encounter with icing in a Cessna 172, reported severe icing. The encounter lasted about 30 minutes, and the pilot was unable to maintain altitude and forced to descend. This description, however, is only of moderate intensity. Icing intensities are formally classified for reporting purposes in the AIM; perhaps these unofficial definitions are more descriptive.

Trace. Ice becomes perceptible, and the rate of accumulation is slightly greater than the rate of sublimation. It is not hazardous, even though ice-protection equipment is not utilized, unless encountered for more than 1 hour. (Your passenger is admiring how pretty the ice looks on the wing. ATC has just instructed you to climb, and you advise them icing is probable and request descent. The controller calmly replies that in that case "you can declare an emergency or land." Shortly, you're handed off to the next controller, and you inquire about a lower altitude. "Is that Terry up there?" the new controller asks. It turned out to be a friend at L.A. Center, and the lower altitude was approved in about 15 minutes.)

Light. The rate of accumulation can create a problem if the flight continues for more than one hour. Occasional use of ice protection equipment removes or prevents accumulation. Ice should not present a problem if the protection equipment is used. (A student hasn't noticed the ice yet. Your pilot-friend in the backseat is beginning to wonder a bit. You're negotiating with ATC for a lower altitude, which can be approved in 15 miles; it takes only 8 minutes, but each minute seemed like 10.)

Moderate. The rate of accumulation, even for short periods, becomes potentially hazardous, and the use of ice protection equipment or a diversion becomes necessary. (A friend in an aircraft without ice-protection equipment survived moderate icing only because the terrain was lower than the freezing level.)

Severe. The rate of accumulation is such that ice protection equipment fails to reduce or control the hazard. Immediate diversion is necessary. Control is impossible. Certain pilots will report icing intensity as heavy; this is a misnomer because all ice is heavy.

The type of icing has also been classified for reporting purposes.

Rime ice. Rime ice is milky, opaque, and granular. It is normally formed when small supercooled water droplets instantaneously freeze upon impact with the aircraft. It is most frequently encountered in stratiform clouds at temperatures between $0°C$ and $-20°C$.

Clear ice. Clear ice is glossy, formed when large supercooled water droplets flow over the aircraft's surface after impact and freeze into a smooth sheet of solid ice. It is most frequently encountered in cumuliform clouds or freezing precipitation. Brief but severe accumulations occur at temperatures between $0°C$ and $-10°C$ with reduced intensities at colder temperatures and in cumulonimbus clouds down to as low as $-25°C$.

Mixed icing (rime and clear). Mixed ice is a hard, rough, irregular, whitish conglomerate formed when supercooled water droplets vary in size or are mixed with snow, ice pellets, or small hail. Deposits become blunt with rough bulges building out against the airflow.

Ice-protection equipment in the icing intensity definitions refers to aircraft and equipment certified for flight in known-icing conditions. Although many aircraft have limited ice protection equipment (pitot heat, prop anti-ice, alternate static source, etc.), it should never be construed as allowable for flight in icing. The heating capability is for emergency use only when icing is inadvertently encountered.

Icing potential exists anytime visible moisture exists (clouds or precipitation) at temperatures of 0°C or less. This contradicts the notion that icing can only occur in clouds. (Chapter 1 discusses the icing potential of snow and how freezing rain can be the most serious icing hazard.) Both phenomena can affect the VFR pilot. The greatest icing potential occurs between the freezing level and –10°C to –15°C, or within a layer approximately 5,000 and 7,500 feet deep. Icing has been encountered in convective clouds at altitudes of 30,000 to 40,000 feet in temperatures less than –40°C.

If the weather briefing indicates even a remote possibility of encountering ice, the pilot should accomplish several tasks. Ensure the pitot heat works with a very light touch during preflight. Check ice protection equipment for proper operation. Remember that a heated pitot is an anti-icing device to be turned on before encountering ice; deicing equipment usually requires ice buildup before activation. Improper operation can actually increase ice buildup and prevent its removal! Check alternate air or carburetor heat for an alternate source of air if the air filter ices over.

The object with ice is to minimize exposure. With temperatures 0°C or less, avoid flying in clouds or precipitation. If ice is encountered, immediately notify ATC and initiate a plan of action. The first consideration might be to climb to colder air or above the layer based upon PIREPs and the preflight weather briefing; ice will slowly sublimate (change from a solid directly into a gas) when on top. Or descend to warmer air based upon the actual freezing level that you noted during the climb. This requires a careful check of minimum enroute altitudes (MEA). Finally, if you have to, turn around because presumably you came from an ice-free area. The point is do something!

A nonturbocharged, nondeice Baron departed Reno, NV, for Southern California. Moderate icing and severe turbulence were forecast. The pilot elected to fly a direct course along the crest of the Sierra Nevada Mountains where the most intense icing and turbulence could be expected. The aircraft iced up resulting in a fatal accident.

The pilot had no way out because the MEA was the aircraft service ceiling. The terrain was well above the freezing level and the pilot failed to reverse course at the first sign of ice. What other options were available? The pilot could have crossed the mountains near Sacramento to minimize exposure to ice, and when over the Sierras the trip could have been successfully completed. The pilot could have flown toward Las Vegas where the weather was considerably better or simply waited for better weather conditions. The ice-covered airplane did not offer any safe options.

A Bonanza pilot departed the San Francisco Bay area on a flight to Los Angeles. Icing above 7,000 feet was forecast *and* reported. The pilot elected to fly at 11,000 feet. The last transmission was "I've iced up and stalled." The crash occurred in the San Joaquin Valley where the elevation was near sea level. Minimum altitudes in the vicinity of the crash were well below the freezing level. The pilot simply did nothing until control of the aircraft was lost.

If a descent through an icing layer is required, remember the objective is to minimize exposure. Under such circumstances, negotiate with ATC to obtain a continuous descent. Avoid, if possible, level flight in clouds. ATC is usually very responsive to such requests.

A word of caution. A popular notion in some aviation publications is that a pilot's mere mention of ice will constitute ATC handling as if an emergency had been declared. The icing might be an emergency, but remember that the controller's job is to separate aircraft within a finite amount of airspace. ATC might have to assign a higher altitude, but ATC cannot and should not be expected to fly the aircraft or assume the responsibility of pilot in command.

No one has any business flying in these conditions:

```
FAT UUA /OV FAT 090030.../FL160-240 /TP F18/IC SVR CLR
BFL UUA /OV PMD 330040.../FL100 /TP C402/IC SVR RIME/RM PUP 1 INCH
CANT SEE THRU WINDSHIELD
```

PIREPs are the pilot's only source of reported induction system icing: an iced-over air intake system and carburetor ice. Both result in loss of engine power, and left unchecked, complete power failure. The solution is to use the alternate air source on fuel-injected engines or carburetor heat on normally aspirated engines.

Induction system icing takes place anytime structural icing occurs. Symptoms are a gradual loss of power. On a flight from Van Nuys, CA, to San Francisco in a Cessna 172, I encountered light icing after an ATC request to climb. I periodically applied carburetor heat. Something unusual occurred. With carburetor heat on, the engine ran fine; off, the engine faltered. On the ramp at San Francisco, we parked next to a Navion that had also flown from the L.A. area. The Navion had flown at a higher altitude and encountered more ice. Sure enough, the Navion's air filter held a large chunk of ice. I realized that the carburetor heat in the 172 was functioning as an alternate air source. This is obvious now, but it wasn't at the time, which illustrates the hazards of learning by experience.

In addition to air-intake icing, normally aspirated engines can develop ice in the carburetor throat. Carburetor ice can form with outside air temperatures as high as 90°F. As air is accelerated through the carburetor and fuel evaporates, the temperature of the air can be lowered as much as 60°F. Whether ice will develop depends on the velocity of the fuel/air mixture, outside air temperature, humidity, and carburetor system. Conditions most favorable for carburetor ice are outside temperatures between 0°F and 60°F, high relative humidity, and low power settings.

Carburetor ice is detected in an engine with a fixed-pitch propeller by a loss of RPM; an engine with a constant-speed propeller will experience a loss of manifold pressure. At the first indication of carburetor ice (power loss or engine roughness), apply full carburetor heat. Realize that the application of carb heat will result in additional power loss and engine roughness because the engine must ingest the melted ice. Leave it on until the engine smooths out, which might take several minutes. Avoid using partial heat unless the aircraft is equipped with a carburetor air temperature gauge. It might be necessary to leave heat on for an extended period. A flight from Page, AZ, to Las Vegas, NV, required continuous use of heat to prevent carburetor icing in a Cessna 150. Don't forget to relean the mixture during extended use of carb heat.

Carbureted engines are more susceptible to icing during reduced-power operation. Some aircraft and engine manufacturers recommend the use of carburetor heat during all power reductions, others only when ice is suspected. Pilots should follow the aircraft manufacturer's recommendations. If full power is required, such as for a go-around, full carburetor heat and full power might cause early detonation or engine damage. It will certainly prevent the engine from developing full power, which might be crucial in an aircraft with low power operating in high density altitude. Know and follow the manufacturer's recommendations.

Pay attention. I had remained overnight in Amarillo, TX, because the previous afternoon a line of thunderstorms was approaching the area from the west, where I was headed. The Cessna 150 was parked into the wind when torrential rains moved through the area. The next morning was clear, with abundant surface moisture, temperature in the 60s, and nearly 100 percent relative humidity.

My first clue of trouble was the increased throttle setting required to obtain idle RPM. Engine runup also took more throttle than usual. I suspected carburetor ice and a water-saturated air filter because of the conditions. I had 13,000 feet of runway.

Full throttle only gave me about 2,200 RPM. The increased ground run to rotation speed, about 7,000 feet, should have been another clue. I was off the ground with no runway remaining and 200 feet of altitude when the engine started losing RPM. I applied carburetor heat. The very-rough-running engine was only producing about 1,700 RPM.

There was a tremendous psychological urge to reduce the heat and get that RPM back. I was preparing to crash straight ahead, but the engine was still producing power, and I decided to make a 180° turn and land on a taxiway. Only then did I inform a surprised tower controller of what happened; remember, a pilot's first job is to fly the airplane.

After running the engine for 20 minutes and trying one aborted takeoff, I launched into the air. The engine performed normally above the shallow, moist layer. It was a perfect example of having the clues and ignoring them. I was extremely fortunate.

REMARKS

Remarks amplify information or describe conditions not already reported:

SAC UUA /OV SAC/TM 1753/FL30/TP C172/TB MDT-SVR/RM LTGCG

The pilot observed lightning cloud to ground.

BFL UA /OV WJF-BFL/TM 1845/FL500AGL/TP UH60/SK 050 OVC/TB LGT OCNL MDT/RM THRU TSP PASS UNDER CLDS. OK FOR HELIO NOT SO HOT FOR FIXED WING
SNA UA /OV SNA-GMN/TM 2218/FL125/TP PA28/TA 00/WV 330015-020 /TB MDT/RM LIKE AN E TICKET AT DISNEYLAND

An E ticket got you on the big rides.

DEN UA /OV DEN 240060/TM 1715/FL350-390/TP H/DC8/TB MDT CAT /RM MDT CAT AT 373 SMTH BLO 367 [ZDV]

This report was filed by Denver Center's Weather Service Unit (ZDV) (CWSU). It amplifies the turbulence portion by indicating that moderate CAT was encountered at 37,300 and it was smooth below 36,700.

Remarks also describe low-level wind shear, which is shear that occurs within 2,000 feet of the surface. Because of its significance, pilot reports on wind shear are extremely important. Wind shear PIREPs include location, altitude, and airspeed changes. Low-level wind shear is different from turbulence in that airspeed changes in one direction (plus *or* minus) *with a sustained change* in vertical airspeed. Airspeed fluctuations (plus *and* minus) with no appreciable change in vertical speed should be reported as turbulence.

RNO UA/ OV RNO.../TP DC9 /RM LLWS 001-SFC +30 TO 40 KTS.

The pilot experienced a 30- to 40-knot increase in airspeed between 100 feet and the surface.

STS UUA/.../FL 030-020/TP C500/TM DURGD RY02 LLWS RESULTING IN 80 KTS CHG AIRSPEED.
RDD UA/.../FL010/TP PA31/RM LLWS +/– 20-25KT 010-SFC DRGD UNABLE TO LND RY16.

Interesting remarks abound:
/RM STRONGEST TURBC I HAVE EXPERIENCE IN 15 YEARS reported by a King Air pilot.
/RM 2 PASSENGERS INJURED...HAD TO TURN BACK reported by a Navy P3 pilot.
/RM IFR NOT RECOMMENDED DUE TO STRONG HEAD WINDS AND 2000 FT PER MIN UP AND DOWN DRAFTS WILL NEVER DO IT AGAIN reported by a Cessna 182 pilot.
/RM SOME REAL GOOD JOLTS PUT KNOT ON HEAD aircraft type missing.

/RM ONE LARGE JOLT, STEW FELL DOWN (SHE IS OK) LOTS OF DRINKS SPILLED reported by a Fairchild 27 pilot.

/RM UNA TO CONTROL HELICOPTER. RETURNED. NURSES KISSED THE GROUND.
/RM WIND O/G at FCH (Fresno Chandler Airport) IS 300-330 DEGREES AT 40 KTS AND ALL THE C150'S AND C152'S ARE INTMTLY FLYING ON THEIR CHAINS.
/RM SVR LLWS AFTER 3 APCHS UNABLE TO LAND reported by a Lear Jet.
/RM LOTS OF BAD TURBC. THIS ISN'T THE SMARTEST THING I'VE EVER DONE. HUGHES COPTER.
/RM ROUGHER THAN A COBB is the good old standby.
/OV AVX .../TP FBO/SK CLR/WX CLR/TB NONE/RM LET'S GO FLY. AVX is Catalina Island's Airport in the Sky. If you're in the L.A. area, try to get out there and order a Buffalo Burger at the restaurant.
LAX UA /OV SXC 213186/TM 1911/FL070/TP VYGR/SK BKN 040/TB LGT /RM VOYAGER 1 is an actual report filed by a Voyager pilot on the record-setting around-the-world flight.

A strong Santa Ana wind in Southern California was responsible for the conditions described in the following SA and PIREPs at Ontario, CA (ONT).

```
ONT SA 1450 CLR 50 46/01/0234G45/990/WIND 330V040
ONT UUA /OV ONT/TM 1425 /FL050/TP B727/TB MDT OCNL SVR
ONT UUA /OV ONT/TM 1445/FL050/TP B727/TB MDT OCNL SVR 050-SFC /RM
CRCLG FAP LNDG ONT
ONT UUA /OV ONT/TM 1450/FL020/TP B727/RM UNABLE TO LAND DUE TO
X-WINDS
```

PIREPs from air carriers, the military, and corporate aircraft tend to be more accurate because of the pilots' training and experience. Few student pilots fly DC-10s or F-14s. New and low-time pilots (inexperienced) tend to overestimate intensities of turbulence and icing; unfortunately the new pilot might not heed accurate reports or forecasts because he or she has a misconception about actual severe conditions. This is not to say PIREPs from pilots of Cessna 150s or Piper Tomahawks are never accurate and should be ignored.

Pilots must evaluate PIREPs within the context of surface reports, forecasts, and other PIREPs. A single report of severe turbulence from a Beech Sundowner under clear skies and light winds should be viewed with skepticism. On the other hand, a report of severe turbulence from a Cessna 172 should be taken very seriously with conditions favorable for a mountain wave and with appropriate advisories in effect. PIREPs that are not objective are worse than useless. Not only do they give a false impression to other pilots, but forecasters must take them as fact and issue advisories, which undermine forecast credibility.

We become observers every time we fly. Reports must be timely; some pilots have a tendency to wait until the latter portion or end of a flight to provide a report. A pilot on a flight from Seattle to Los Angeles contacted Oakland Flight Watch and reported conditions departing Seattle two and a half hours earlier. A somewhat overzealous briefer instructed a student to make a pilot report at the conclusion of the flight. The student calling the FSS the following day meekly apologized for failing to provide the report, then proceeded to recount in detail the conditions encountered. I'll bet that pilot doesn't forget on the next flight. Get into the habit of routinely providing timely reports because reports confirming the forecast are as important as those for unforecast weather.

The next time someone complains about the lack of weather information or forecast accuracy, ask if they routinely provide objective pilot reports. And the next time you fly and you "just don't get around to making a report," remember this: BFL UA /OV BFL /TM 1450 /FLUNK /TP ALL /RM WISH I HAD A TOP REPORT FROM BFL TO ONT.

FSS and NWS specialists sometimes tend to editorialize PIREPs, usually around the time of championship sporting events. Although unauthorized and unprofessional, pilots can expect to see these usually humorous reports from time to time.

Other comments contain personal, social, or political messages like the following example:

PILOT REPORTS (UA, PIREPs)

HWD UA /OV OAK 110007 TM/1600/FL060/TP BE33/SK CLR/WX FV99/TB NEG/RM DURC HWD NBND SVRL H LYRS AT FL015/FL042/FL060. SLANT VSBY 15-30MI. PTCHY ST OFSHR-THRU GOLDEN GATE OVR CITY OF OAKLAND. REPORTED BY A DERANGED BONANZA PILOT.

Yes, I was the deranged Bonanza pilot!

3
Forecasts

S IR WILLIAM NAPIER SHAW'S *MANUAL OF METEOROLOGY*, PUBLISHED IN THE late 1920s, nicely sums it up: "Every theory of the course of events in nature is necessarily based on some process of simplification of the phenomena and is to some extent therefore a fairy tale."

Aviation weather forecasts began with the Wright brothers' request for surface winds at Kitty Hawk, NC, in 1903. In the early 1920s, aviation forecast centers were established at Washington, Chicago, and San Francisco. Establishment of lighted airways in the mid-1920s encouraged the Weather Bureau to begin night forecasting. Several airlines had established their own forecasting systems by the mid-1930s. Because meteorologists could be held responsible for weather-related accidents, forecasts tended to be pessimistic.

Forecasts were limited by lack of observational data and the complexity of the atmosphere. Meteorologists were plagued with unexpected thunderstorms, the transient nature of icing and turbulence, and the unanticipated development of fog.

Advances were made in the late 1930s and 1940s with upper-air observations and facsimile transmission of weather charts. By the late 1950s, weather radar was added to the observational arsenal. Satellites and computers help produce forecasts today;

however, due to the lack of observational data and the complexity of the atmosphere, computer programs can only generate approximations.

Observational data comes from surface, upper-air, and radar reports, plus satellite imagery and PIREPs. Figure 3-1, the National Aviation Weather System chart, illustrates the importance of PIREPs and how they fit into the overall picture.

Analyzing and forecasting are accomplished at the National Meteorological Center (NMC) in Washington, D.C., the National Severe Storms Forecast Center (NSSFC), the National Aviation Weather Advisory Unit (NAWAU), and local weather service forecast offices (WSFO); NSSFC and NAWAU are located at Kansas City, MO.

National Aviation Weather System

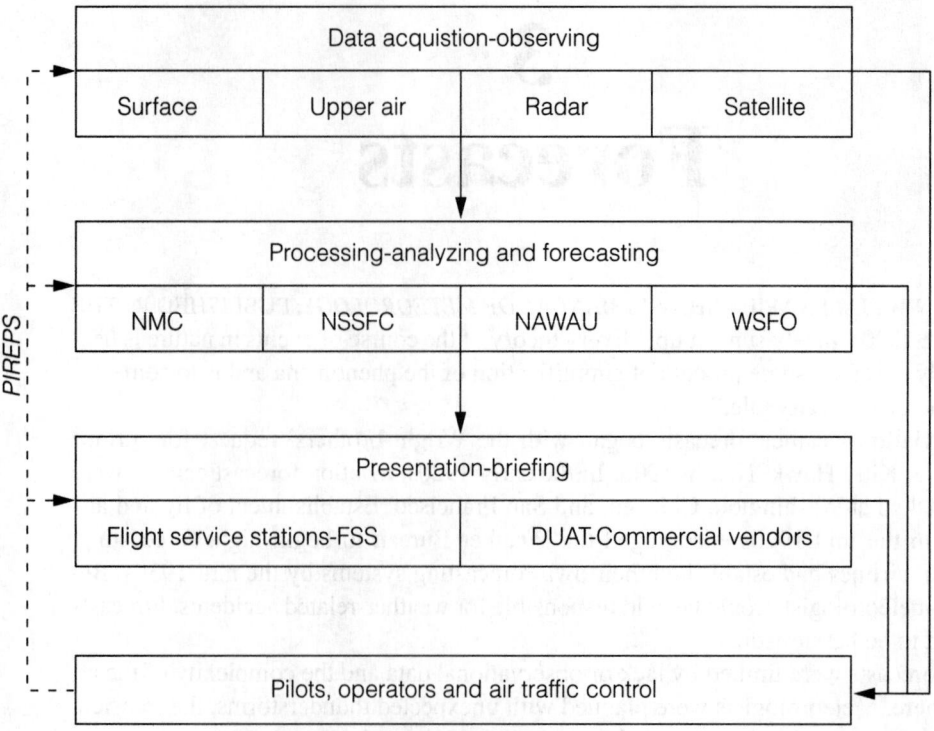

Fig. 3-1. *The National Aviation Weather System consists of many governmental offices and agencies and is dependent on the pilot's active participation.*

Meteorologists from the satellite section at the NWS Kansas City office discuss the day's weather (Fig. 3-2) using a satellite loop that provides a moving picture of weather systems. Meteorologists develop a forecast based on an equation:

Existing Weather + Weather Trend = Expected Weather

Fig. 3-2. *Meteorologists from the satellite section of the NWS's Kansas City office discuss the day's weather patterns using a satellite loop that provides a moving picture of weather systems.*

Trend can be the rate of change in the weather; advection and development create the weather trend. *Advection* is the movement of an atmospheric property from one location to another; fronts and upslope fog are examples of an atmospheric property moving from one location to another, which produces weather. Development is the growth of air mass thunderstorms, increase in afternoon thermal turbulence, or the dissipation of radiation fog.

ACCURACY

Are aviation forecasts accurate? The FAA admits that needs "cannot be met by an immediate application of existing technology. The need for accurate short-term forecasts exists in every phase of flight operations and is critical to an efficient, smoothly operating air traffic control system." To this end the FAA and National Weather Service established the en route flight advisory service, including the implementation of high-altitude Flight Watch and center weather service units (CWSU) at air route traffic control centers (ARTCC).

Forecast accuracy begins with observational data. (Or perhaps it "begins" to deteriorate with the data.) The limited number of observations hinders forecast accuracy.

The human observational network has dwindled, and automated observations with their own limitations have yet to be fully implemented. Upper-air observations, radar, and satellites help, but extensive areas remain outside the observational network.

Fully half of the forecast is based on existing weather.

Available data are computer-processed and analyzed. Computer equations represent the atmosphere at points approximately 125 miles apart, depending on the computer model, and at various heights. Large-scale (*synoptic*) weather systems are detected, but smaller scale (*mesoscale*) weather systems such as individual thunderstorms might not be detected. Additionally, because of computer model limitations, factors such as the interaction of water, ice, and local terrain cannot be adequately taken into account. These are major limitations to the second factor in the weather equation, which is the weather trend.

The National Weather Service monitors forecasts, but other than amendment criteria, no specific factors exist to determine forecast accuracy. Who defines accuracy? C. Donald Ahrens wrote in *Meteorology Today* (St. Paul: West Publishing Co., 1985), "At present, there is no clear-cut answer to the question of determining forecast accuracy." Few pilots and FSS briefers are aware of forecast limitations or amendment criteria; many have developed their own perceptions that are quite often erroneous due to misconceptions and misunderstandings.

SPECIFICITY

Each forecast is written for a specific purpose in accordance with specific criteria. Area forecasts cover entire states; TWEB route forecasts cover routes 50 miles wide, and terminal forecasts relate conditions within 5 nautical miles of an airport. Differences are to be expected due to scale, interpretation of the weather situation, issuance times, and starting conditions.

For example, localized areas of fog predicted in TWEB route or terminal forecasts might not appear in the area forecast. Or the area forecast might contain a prediction for thunderstorms that might not appear in individual terminal forecasts when the forecaster does not expect the phenomena to occur at that airport. Forecasters might legitimately differ in interpretation. The area forecast might predict frontal passage at one time and the terminal forecast at another time. Forecasts are issued at different times; therefore, available information used by forecasters to make predictions differs.

A thorough understanding of format, limitations, and amendment criteria is required to adequately apply a forecast to flight planning, especially using a self-briefing media. The FAA and NWS have said: "There probably is no better investment in personal safety, for the pilot as well as the safety of others, than the effort he (or she) spends to increase his (or her) knowledge of basic weather principles and to learn to interpret and use the products of the weather service." Then there's the legal requirement. Each pilot in command is required by regulations to become familiar with all available information concerning a flight. This includes "for a flight under IFR or a flight not in the vicinity of an airport, weather reports and forecasts . . . ." Even student pilots must receive instruction in the "use of aeronautical weather reports and forecasts" before venturing solo cross-country.

From Table 3-1, limitations on aviation weather forecasts, the following conclusions are apparent:

- Forecasts for good weather are more likely to be correct than forecasts for poor weather; this should be no surprise because good weather occurs more often than poor weather.
- Forecasts are most accurate during the first few hours of the period.
- Accuracy deteriorates below 80 percent beyond 4 hours when less than VFR conditions are forecast.
- Accurate forecasts of specific values beyond three hours are not yet possible.

Table 3-1. Limitations on aviation weather forecasts.

1. Forecasts 12 hours and beyond for good weather (ceiling 3,000 feet or more; visibility three miles or greater) are more likely to be correct than forecasts for poor weather (ceiling below 1,000 feet; visibility below one mile).
2. Poor weather forecast to occur within three or four hours has a better than 80 percent probability of occurrence.
3. Forecasts for poor weather within the first few hours of the period are most reliable with distinct weather system.
4. Errors occur with attempts to forecast a specific time poor weather will occur. Errors are less frequent forecasting poor weather within a time frame.
5. Surface visibility is more difficult to forecast than ceiling.

Forecasters can predict with 75 percent accuracy:

1. Within two hours, the passage of fast moving cold fronts or squall lines up to 10 hours in advance.
2. Within five hours, the passage of warm fronts or slow moving cold fronts up to 12 hours in advance.
3. Within one to two hours, the onset of thunderstorms, with radar available.
4. Within five hours, the time rain or snow will begin.

Forecasters cannot predict with an accuracy that satisfies operational requirements:

1. The time freezing rain will begin.
2. The location and occurrence of severe or extreme turbulence, or severe icing.
3. The location of the initial occurrence of a tornado or low-level wind shear.
4. Ceilings of 100 feet or zero before they exist.
5. The onset of a thunderstorm that has not yet formed.

Forecast issuance, valid times, and amendment criteria are often based on these limitations. Forecasts are most reliable for distinct weather systems: fast-moving cold fronts, squall lines, or strong high-pressure areas. Synoptic-scale systems are detected within the forecast models.

Certain phenomena are difficult or impossible to predict with accuracy: the time freezing rain will begin, severe or extreme turbulence, severe icing, the movement of a tornado, ceilings of 100 feet or zero before they exist, the onset of thunderstorms that have not yet formed, and low-level wind shear. These phenomena are often caused by mesoscale systems or are transitory and remain undetected within the normal observational system. Computer models are of limited use. The most hazardous weather is the most difficult to forecast.

Outlook forecasts for good weather are more likely to be correct than forecasts for poor weather. Errors in timing are more prevalent than errors of occurrence. One forecaster put it this way, "We're never wrong, our timing's just off sometimes." Or forecasts are 100 percent correct: 90 percent in the summer and 10 percent in the winter!

Unwarranted pessimism is a major forecast complaint. In a 1981 NASA study, pilots complained of canceling flights based on forecasts, and the weather turned out to be VFR. One pilot took off in spite of the forecast and completed the flight; however, the report did say, ". . . In this latter case, . . . the forecast was substantially correct and the pilot was fortunate enough to find breaks in the overcast . . . at (the) destination." Subsequent chapters reveal how the National Weather Service and its forecasters are making positive efforts to improve the products.

EXPECTATIONS

Pilots complain equally about pessimistic forecasts and unforecast weather. Pilots often expect too much from a forecast. Each situation is different with many variables and local factors. The limitations in Table 3-1 remain, and forecasts are going to be missed. The 1965 edition of *Aviation Weather* said it best: "The weatherwise pilot looks upon a forecast as professional advice rather than as the absolute truth."

It's essential to remember these concepts during the discussion of forecasts in subsequent chapters. Forecast issuance times, purpose, conditional terms, amendment criteria, and Federal Aviation Regulations reflect the limitations on aviation forecasts. Meteorologists sometimes produce a strategic forecast with values and conditional terms that will allow conditions to go from W0 X 0F to ceiling and visibility unlimited without a requirement to amend. That's exaggerated a bit, but it has occurred.

The National Weather Service's modernization/restructuring is well underway in the mid-1990s. Approximately 250 Weather Service forecast offices will be reduced to 115. A new computer system known as the *advanced weather interactive processing system* (AWIPS) will come on line. Automated weather observing systems (AWOS/ASOS) and *next generation weather radar* (NEXRAD) observations should expand the observational network. Terminal Doppler radars will detect microbursts and wind shear.

The *next generation weather satellite* (GOESNEXT) has been launched. The satellites are located at approximately 80°W and 135°W longitude and will provide enhanced pictures and other improvements. The *wind profiler*, a Doppler radar system, will replace radiosonde observations at the turn of the century for upper-air wind data. These technological advances promise to improve the accuracy and reliability of weather forecasts, but for now pilots will have to deal with today's system and its limitations.

Paraphrased from the FAA publication *The Weather Decision*: "At the weather briefing, keep in mind that meteorologists tend to be optimists." Say again, please?

4
Advisories

DURING THE LATTER PART OF THE 1950s THE WEATHER BUREAU ISSUED warnings of potentially hazardous or severe aviation weather in the form of "flash advisories." These were subsequently divided into AIRMETs (WA) and SIGMETs (WS). AIRMETs and SIGMETs alerted pilots that significant, previously unforecast, weather had developed. Twenty-one Weather Bureau offices routinely issued advisories, but by 1970 the number of flight-advisory weather service offices was reduced to nine. Although the *Weather Service Operations Manual* stated "the in-flight weather advisory program is intended to provide advance notice of potentially hazardous weather developments to en route aircraft," advisories were issued even when conditions were contained in the area forecast (FA). During this period there were AIRMETs by the number and SIGMETs by the score. It was not uncommon to have four or more AIRMETs continuously in effect, and to this end the *continuous AIRMET* (WAC) was developed. Weather advisories issued by adjacent offices were not always consistent and often overlapped. This led to confusion for briefers and pilots alike.

The area forecast was changed in 1978 to include a HAZARDs or flight precautions section to reduce the number of advisories. The FA was still issued by local NWS offices. When the San Francisco Weather Service Forecast Office (WSFO) issued the FA, there was invariably a flight precaution for occasional moderate turbulence within

5,000 feet of rough terrain. Pilots and briefers became equally disgusted with this generalization. Responsibility for issuing AIRMETs and SIGMETs for the 48 contiguous states was centralized in 1982 in Kansas City at the National Aviation Weather Advisory Unit (NAWAU).

In 1991, AIRMET criteria phenomena were removed from the area forecast and issued separately as the AIRMET Bulletin. Who initiated the change? The NWS Aviation Services Branch working with FAA headquarters, industry, and user groups instigated the change; to a somewhat limited extent, NWS and FAA field offices also provided input. The change occurred because of the redundancy of AIRMET/flight precautions and the disappearance of AIRMETs at the next FA issuance.

Gains:

- One advisory product instead of two (FA Flight Precautions/AIRMETs).
- More frequent AIRMET issuances, all at the same time.

Losses:

- Separate issuance times produce difficulty in composition and consistency.
- Combined 6-hour AIRMET, 6-hour outlooks are difficult to handle both from a composition and users' standpoint.
- There are more amendments mainly due to separate issuance times.
- The FA is no longer a stand-alone product.

SIGMETs were issued for convective activity until a DC-9 crashed in a severe thunderstorm near Atlanta in 1977. The *convective SIGMET* (WST) evolved from this accident. The National Aviation Weather Advisory Unit in Kansas City has WST responsibility. Specifically assigned meteorologists issue these advisories.

Center Weather Service Units (CWSUs) were established at air route air traffic control centers (ARTCCs) in 1980. The purpose of the CWSU is to assist controllers and flow control personnel, plus alert pilots of hazardous weather via a *center weather advisory* (CWA). As is often the case with government bureaucracy, the cart came before the horse. There were no instructions for ATC personnel when CWAs first appeared. Distribution went all the way from immediate broadcast to the trash can. The FAA took months to decide that the CWA had the weight of a SIGMET and apply identical distribution and broadcast procedures.

Public forecasts, alert weather watches (AWWs), and severe weather watch bulletins (WWs) are also produced. The severe local storms office (SELS) in Kansas City issues AWWs and WWs for severe thunderstorms and tornadoes.

The AIRMET bulletin, SIGMETs, convective SIGMETs, center weather advisories, and alert weather watches and bulletins are issued when phenomena reach specified criteria and, like urgent PIREPs (UUA), receive priority handling and distribution.

With the number of advisories, it would seem impossible to fly into an area of hazardous weather without warning, but this is not necessarily the case. An advisory cannot be issued for each thunderstorm, instance of turbulence or icing, mountain obscuration, or IFR condition. Severe weather can develop before an advisory is written and distributed. The absence of an advisory is no guarantee that hazardous weather does not exist or will not develop.

AIRMET BULLETIN (WA)

The AIRMET bulletin is issued regularly four times a day for each of the six FA forecast areas, which are depicted in appendix D. Issuance times are 0300Z, 0900Z, 1500Z, and 2100Z; during daylight savings time, the issuance times are 0200Z, 0800Z, 1400Z, and 2000Z. The AIRMET bulletin is valid for 6 hours with an additional 6-hour outlook. Each subsequent issuance is indicated by an update (UPDT) number, beginning with the second of the Zulu date "UPDT 1." If an amendment is issued between normal update times, it will be reflected in the header (SFOS WA 251730 AMD) and given the next update number.

The bulletin is divided into three sections designated AIRMET SIERRA, AIRMET TANGO, and AIRMET ZULU. AIRMET SIERRA describes areas of IFR conditions and mountain obscurement. AIRMET TANGO forecasts turbulence, low-level wind shear, and strong surface winds. AIRMET ZULU shows the location and intensity of icing and includes the freezing level.

Conditions must be widespread for inclusion as an AIRMET in the AIRMET bulletin—that is, occurring or forecast over an area of at least 3,000 square miles, approximately three times the size of Rhode Island. Localized occurrences do not warrant the issuance of an AIRMET.

What's local? According to Dick Williams, NAWAU forecaster, AIRMET bulletins that are "written on the scale of whole states do not endeavor to describe every single occurrence of IFR, icing, or turbulence. The forecaster, wishing to indicate that there (might) be isolated observations, pilot reports, or hazardous weather may use the term 'local.' No hard-and-fast rule exists for determining when 'local' becomes 'widespread.' The forecaster relies on observations, pilot reports, satellite imagery, and his or her own judgment in determining the extent of weather features."

Federal Aviation Regulations recognize the fact that every occurrence of adverse weather cannot be forecast. FAR 61 requires that even student pilots receive instruction in "The recognition of critical weather situations" and "estimating visibility while in flight . . .." In other words, the excuse that "They didn't tell me," is just that, an excuse, not a reason.

AIRMETs are issued when the following phenomena occur or are expected to develop.

- Moderate icing
- Moderate turbulence

- Sustained wind of 30 knots or more at the surface
- Ceilings less than 1,000 feet and/or visibility less than 3 miles affecting more than 50 percent of the area at any time
- Extensive mountain obscurement

AIRMET bulletins are issued based upon VOR locations to provide an overview and because hazards often affect more than one FA area; refer to the in-flight advisory plotting chart in appendix C, which also contains a list of location identifiers to decode chart locations. Since the plotting chart was developed, some VOR names have been changed if those VORs are not collocated with an airport of the same name. For example, the name of the Santa Barbara (SBA) VOR has been changed to San Marcos (RZS); however, the plotting chart identifier remains SBA. Affected areas start with the most northern location and continue clockwise.

From a forecast point of view, phenomena usually lie well within the delineated area. The phenomena might move through the area during the forecast period; therefore, the condition might only affect a portion of the area at any particular time. Details appear in the text of the bulletin; for example, an icing advisory might cross several FA boundaries. Specific altitudes might vary due to differing temperatures, humidity, and frontal locations. When phenomena are peculiar to a specific mountain range, coastal area, river basin, or valley, the geographical area might be included (SNAKE RIVER VALLEY or TEXAS WEST OF THE PECOS). (Appendix D contains geographical area designators.)

Conditional terms describe widely varying conditions over large areas. Area forecast conditional terms are defined in Table 4-1. The terms are essentially self-explanatory with the exception of occasional (OCNL), which describes a better than 50/50

Table 4-1. Weather advisory, area forecast conditional terms.

Term	Contraction	Definition
Occasional/	OCNL/	Greater than a 50% chance of occurrence/
Occasionally	OCNLY	occurring for less than ½ of the forecast period
Chance	CHC	30% to 50% probability
Slight chance	SLGT CHC	10% to 20% probability

The following terms refer to the aerial coverage of thunderstorms or precipitation.

Isolated	ISOLD	Single cells (no percentage)
Widely scattered	WDLY SCT	Less than 25% of area affected
Scattered	SCT	25% to 54% of area affected
Numerous	NMRS	More than 54% of area affected

chance of occurrence during less than half of the forecast period. OCNL often describes turbulence, icing, mountain obscurement, and IFR conditions. OCNL also reflects the transitory nature of these phenomena. A pilot might not encounter the condition flying through a forecast area of OCNL, but he or she has been warned. Look at it this way: If you run into it, they're right; if you don't run into it, they're still right! We'll discuss how to interpret and apply conditional terms throughout this chapter and chapter 5, which examines the area forecast.

IFR and mountain obscuration

AIRMET SIERRA describes the location of ceilings below 1,000 feet and visibilities less than 3 miles plus areas where the mountains are obscured by clouds or precipitation.

The AIRMET SIERRA portion of the AIRMET bulletin looks like this:

```
SFOS WA 261345
AIRMET SIERRA UPDT 2 FOR IFR AND MTN OBSCN VALID UNTIL 262000
.
AIRMET IFR . . . CA
FROM 70 WSW MFR TO UKI TO SNS TO 20W SNS TO 30W UKI TO FOT TO
70WSW MFR
OCNL CIG BLO 10 VSBY BLO 3F. CONDS ENDG 17Z-19Z.
.
AIRMET MTN OBSCN . . . CA
FROM 40 W SAC TO 20 NNW SBA TO 50WNW PSP TO 40E SAN TO 10S SAN
TO LAX TO 40W SBA TO 40NW SFO TO 40W SAC
CSTL MTNS OBSCD CLDS/FOG. CONDS ENDG CNTRL CA PTN AREA 19Z-
20Z..CONTG SRN CA BYD 20Z THRU 02Z.
```

This San Francisco AIRMET SIERRA bulletin (SFOS WA) was issued on the 26th day of the month at 1345Z. This is the third issuance for this Zulu day (UPDT 2). It is valid until the 26th at 2000Z.

IFR conditions are forecast for all or portions of California (AIRMET IFR . . . CA). Specifically, IFR is expected within the area from 70 nm west southwest of Medford, OR (MFR), to Ukiah, CA (UKI), to Salinas, CA (SNS), to 20 west of Salinas, to 30 west of Ukiah, to Fortuna, CA (FOT), and back to 70 west southwest of Medford.

Within the delineated area, occasional ceilings below 1,000 feet AGL and visibilities less than 3 miles in fog are expected. These conditions should end between 17Z and 19Z. The forecast is correct even if some localized areas of IFR remain beyond 19Z.

The mountain obscuration is expected to end in the Central California portion between 19Z and 20Z, but continue in the Southern California portion beyond 20Z through 02Z, which is the outlook period.

Turbulence and low-level wind shear

AIRMET TANGO describes the location, intensity, and height of turbulence that is not related to thunderstorms. Strong updrafts and downdrafts might also appear: STG UDDFS PSBL OMTNS (strong updrafts and downdrafts possible over the mountains). Isolated severe turbulence is forecast in AIRMET TANGO: ISOLD SVR TURBC. Localized or isolated moderate turbulence that does not require the issuance of an AIRMET also appears in this section (LGT or ISOLD MDT). When forecasted turbulence does not meet AIRMET criteria, only the general geographical area is indicated (WRN WA AND OR) Light or no turbulence might be indicated by a statement such as NO SGFNT TURBC EXPCTD or NO SGFNT TURBC OUTSIDE CNVTV ACTVTY.

Turbulence forecasts are based on wind flow, winds aloft, evaluation of terrain, and PIREPs. Moderate intensity is usually forecasted when the winds reach 25 to 30 knots, and severe turbulence is forecasted when winds exceed 40 knots. High-level turbulence is difficult to forecast.

Refer to the Chicago area forecast synopsis and AIRMET TANGO portion of the AIRMET bulletin below for the subsequent discussion.

```
CHIS FA 201040
SYNOPSIS VALID UNTIL 210500
COLD FNT ALG LN MOT-LBF-ABQ AT 11Z WL MOVE TO A DLH-DSM-ALI LN
BY 23Z AND DISIPT BY 210500Z.
CHIT WA 200745
AIRMET TANGO UPDT 1 FOR TURBC VALID UNTIL 201400
FROM ISN TO MQT TO COU TO ACT TO SJN TO FMN TO DEN TO ISN
OCNL MDT TURBC BLO 150 WI 100 ML OF COLD FNT. CONDS CONTG BYD
14Z THRU 20Z.
```

The turbulence forecast must cover the 6-hour forecast period 08Z through 14Z and the outlook period 14Z through 02Z. A cold front is expected to produce occasional moderate turbulence. The CHI FA synopsis and AIRMET TANGO illustrate how a hazard can move through the delineated area, only affecting certain portions of that area at any one time during the forecast period. Unfortunately some briefers fail to understand and consider this and issue the advisory even if it does not apply, which undermines the credibility of the forecast and briefing.

Frontal turbulence usually occurs below 15,000 feet MSL and caused by surface temperature differences exceeding 8°F within 50 miles of the front. Because temperature is the determining factor, speed or type of front is not involved in the extent of frontal turbulence; however, other types of turbulence, such as mechanical or wind shear, might also accompany a front. In fact, rapid changes in wind direction and speed below 3,000 feet AGL within 200 miles of an advancing front might produce low-level wind shear.

The probability of moderate turbulence is better than 50 percent; however, it is only expected to occur for less than half of the forecast period. A pilot could expect a greater probability in the vicinity of the frontal zone, with its approach, and shortly after passage. This is not inconsistent, but reflects the dynamic character of weather.

The following procedures can be used to avoid or minimize frontal turbulence; fly above the affected area or remain on the ground until frontal passage. Turbulence can be minimized by penetrating the front at a right angle, thus reducing exposure.

Thus far we've discussed thermal, evaporative cooling, mechanical, mountain wave, inversion, and frontal turbulence. We know that the delineated area in the AIRMET bulletin is usually larger than the affected area, and we must consider phenomena moving through the area during the forecast period. This explains a widely held misinterpretation. Turbulence will not necessarily occur at every location within an advisory area during the entire forecast period. Forecasts for mechanical turbulence often cover wide areas; however, the greatest intensity will occur in the vicinity of mountains, leaving valleys and coastal areas relatively smooth. The forecaster often reflects this by stating OCNL MDT TURBC VCNTY MTNS.

By understanding the causes of turbulence and the forecast, a pilot can determine the most likely areas and plan accordingly. This includes reducing to a turbulent-air penetration speed, securing objects, and briefing passengers before entering areas of probable turbulence.

Nonconvective (not related to thunderstorms) low-level wind shear will be included in a separate paragraph. The paragraph will state LLWS potential, location, and cause: LLWS POTENTIAL OVER MOST OF NEW ENGLAND AFTER 03Z DUE TO STG NWLY FLOW BHND CSTL LOW PRES SYS. There is a potential for LLWS over most of New England after 0300Z due to a strong northwesterly flow behind the low-pressure system over the coast. Nonconvective LLWS can be caused by fronts, low-level jet streams, terrain, valley effect, sea breezes, lee-side effect, inversions, or Santa Ana or similar foehnlike winds. It is difficult to predict the occurrence, exact location, and intensity of turbulence and LLWS. The forecaster must consider the often widespread and transitory nature of turbulence. Here again, PIREPs are the only means of validating the forecast.

AIRMET TANGO might include a paragraph for sustained surface winds in excess of 30 knots. The following is an example:

AIRMET STG SFC WINDS . . . CSTL WTRS
FROM 140W HQM TO 20SW HQM TO 140WSW UKI TO 140W FOT TO 140W HQM
18Z-20Z SUSTAINED SFC WINDS GTR THAN 30KNTS EPCD OVER THE WA OR AND NRN CA CSTL WTRS. CONDS CONTG BYD 20Z THRU 02Z.

"The wind bloweth where it listeth, and thou hearest the sound thereof, but canst not tell whence it cometh, and whither it goeth." John 3:8.

Icing and freezing level

AIRMET ZULU describes the location, intensity, and type (rime, clear, or mixed) of nonconvective icing. Layers where significant icing can be expected are expressed as specific values or ranges with bases and tops. This section includes forecasts for light or local moderate icing. Like AIRMET TANGO, light or local moderate icing will be described using geographical areas NE AZ AND NW NM; isolated severe icing is forecast in AIRMET ZULU. Trace or no icing is indicated by either NO SGFNT ICING EXPCD or NO SGFNT ICING EXPCD OUTSIDE CNVTV ACTVTY. A separate paragraph contains forecast freezing levels; terms such as *sloping* or *lowering* describe varying levels.

Icing is extremely difficult to forecast. Forecasters must determine which areas contain enough moisture to form clouds, which cloud areas will most likely contain supercooled droplets during the forecast period (6 hours with an additional 6-hour outlook), and the freezing level. Needless to say, this is not an easy task. Pilots should consider these as forecasts of icing potential to alert the pilot to the possibility of icing in clouds and precipitation within the areas and altitudes specified. Approximately 80 to 100 icing accidents occur each year; about half are from structural icing and the others involve induction-system and ground icing. Most are preventable.

Most of today's aircraft that are certified for flight into known-icing are quite capable of climbing through icing layers and flying well above potential icing areas. Considering this capability, icing PIREPs unfortunately are not as plentiful as briefers would like. PIREPs are the only means of validating the icing forecast. Some think that one reason for the lack of icing PIREPs is a pilot's fear of receiving a violation; unless an inspector is in the aircraft or emergency assistance is provided, violations are rare. I would hope these individuals would be as concerned about their fellow pilots and passengers—and our insurance rates. Chapter 2 reviewed two fatal icing accidents; the pilots would probably still be around if an accurate PIREP had been available or if they had confessed their fate and requested assistance at the first indication of ice.

This brings up the question: What is *known-icing*? You won't find it in FAR Part 1 definitions and abbreviations or the *Pilot/Controller Glossary*. Trying to determine an answer requires further consideration of numerous issues. The difficulty of forecasting icing and the transitory nature of icing have been established. Do we want a forecast of icing to forbid flight? Would a report of light icing above 8,500 feet, with bases at 8,000 and tops at 9,000 preclude flight for aircraft not certified for flight in known-icing? What if the terrain was at 7,800 and tops at 15,000? Do we really want a hard answer? If certain personnel in the FAA had their way, the definition would be any time there is visible moisture and a temperature of 5°C or less! This or a similar definition would certainly mark the end to many useful icing PIREPs. For now, "known icing" is a pilot report of icing.

A pilot is a candidate for a violation if icing is reported or forecast and an accident occurs, or if ATC provides emergency or special handling; however, more often than not, ATC is so busy and so happy to successfully usher the distressed pilot out of its airspace into safe flying conditions that no action is taken. No one should interpret this as meaning that I, the FAA, or ATC condones such actions. Regulations and common sense dictate that an aircraft not certified for flight in icing conditions must not do so. At present, the decision as to whether the flight can be made safely rests solely with the pilot; the responsibility to make that decision will stay with pilots until they prove that they are not worthy.

The following illustrates the decision-making process with forecast icing. The aircraft was a turbo Mooney on a flight from Bakersfield, CA, to Hayward, CA. The synopsis indicated that a moist but stable air mass was in the area. Bases along the route were reported around 5,000 and tops 9,000 to 11,000; the freezing level was 7,000. Except for the coastal mountains, terrain along the route was close to sea level. (The terrain is a very important factor in this instance.) The flight was planned at 12,000 because tops were relatively low and the aircraft had the performance to quickly climb through the potential icing layer.

Trace to light icing that was encountered during the climb quickly sublimated when on top. The pilot encountered some buildups that were above 12,000. Deviations to avoid these clouds were obtained from ATC. By circumnavigating the buildups, icing and turbulence were avoided. Should this be attempted in an airplane with lesser performance? Absolutely not! The bases and tops were known quantities. The airplane had the performance to quickly climb on top. If a quick climb had not been possible, the pilot had the option to return because the cloud bases at 5,000 feet were more than 4,000 feet above the surrounding terrain, and the bases were well below the freezing level.

Was the icing forecast correct? Yes, remember the definition of "occasional." If the pilot had required emergency assistance, he or she would have been a candidate for a violation. I do not intend to imply that this procedure is recommended. The decision rests solely with the pilot, based upon his or her training and experience and the capability of the aircraft.

The AIRMET ZULU portion of the AIRMET bulletin looks like this:

```
SFOZ WA 141345
AIRMET ZULU UPDT 2 FOR ICG AND FRZLVL VALID UNTIL 142000
AIRMET ICG . . . WA OR CA
FROM YQL TO GGW TO BFF TO ALS TO 120W SFO TO 120W FOT TO 120W
TOU TO YQL
OCNL MDT RIME/MXD ICGICIP BTWN 040 TO 140 WA BTWN AND 060 TO 160
OR/CA. CONDS CONTG BYD 20Z THRU 02Z.

FRZLVEL . . WA W OF CASCDS . . 045 LWRG BY 20Z TO 035.
WA CASCDS EWD . . AT/NEAR SFC WITH MULT FRZLVLS 30–35.
OR W OF CASCDS . . 55 NORTH TO 65 SOUTH. AFT 18Z 50 NORTH TO 75
SOUTH. OR CASCDS EWD..AT/NEAR SFC WITH MULT FRZLVLS TO 40–50
CA..AT/NEAR SFC SIERRAS AND NE PTN TIL 18Z. ELSW NEAR 70 NORTH
SLPG TO 90 SOUTH. AFT 18Z 70 NORTH SLPG TO 90 CNTRL AND 100 SOUTH.
```

The specific area extends from Lethbridge, Alberta (YQL), to Glasgow, MT (GGW), to Alamosa, CO (ALS), to 120 nm west of San Francisco (SFO) to 120 nm west of Fortuna, CA (FOT), to 120 nm west of Tatoosh, WA (TOU), and back to Lethbridge. This area includes the coastal waters.

Intensities, types, and altitudes might differ significantly over the areas affected. Occasional moderate rime or mixed icing in clouds and precipitation is forecast from 4,000 feet to 14,000 feet MSL over Washington, and from 6,000 feet to 16,000 feet MSL over Oregon and California. Conditions are expected to continue beyond the end of the forecast period 20Z, through the outlook period of 0200Z. Pilots planning flight beyond the forecast period of 20Z can expect this AIRMET to be in effect at least through 02Z.

It seems a bit redundant to include the base of the icing in the icing paragraph and freezing level in the subsequent paragraph. This came about as a result of a pilot obtaining a briefing with icing from the "freezing level." The base of the freezing level was not specified. Now the forecaster must enter a specific altitude in the icing paragraph. This might be in the form of "between" as in the previous example or "FRZLVL-150. FRZLVL 55-70." Pilots can expect the most significant icing within the layer specified in the icing paragraph. Trace or light icing can be expected between the freezing level and the altitude in the icing paragraph.

Next appears the freezing-level paragraph (FRZLVL). Areas are defined using geographical area designators that are described in appendix D. In Washington, west of the Cascades, the freezing level is expected to lower from 4,500 feet MSL to 3,500 feet MSL by 2000Z. East of the Cascades, the freezing level is at or near the surface with multiple freezing levels between the surface and 3,000 to 3,500 feet MSL. Multiple freezing levels are caused by overrunning warm air, such as with a warm front. Freezing rain occurs in these areas. Similar conditions are expected in Oregon, but at different levels and times.

In California, the freezing level is forecast at or near the surface in the Sierra Nevada Mountains and northeast portion (as described in the geographical designators map in appendix D) until 1800Z. Elsewhere, the freezing level is expected near 7,000 feet MSL in the north sloping to 9,000 feet MSL in the south. After 1800Z, the freezing level is forecast to remain around 7,000 in the north, and rise in the central and southern portions (described in appendix D) to 9,000 to 10,000 feet.

In this section, the forecaster has described an icing layer 6,000 to 10,000 feet deep that slopes upward about 2,000 feet from north to south. The type of ice that is forecasted (mixed) and depth of the anticipated icing layer indicate an unstable air mass. IFR pilots with aircraft that are certified for flight in icing conditions should have little trouble in these areas, assuming performance will allow them to climb out of the icing layer; however, the pilots must be prepared to contend with freezing rain east of the Cascades and icing to the surface in parts of California. For IFR pilots of aircraft without ice protection equipment, this forecast would be a very strong no-go indicator, especially east of the Cascades and the mountains of California. The VFR pilot flying in Washington or Oregon east of the Cascades will be just as susceptible to icing as an IFR pilot because of multiple freezing levels and possible freezing rain. But as we've seen, any flight decision cannot be based upon one forecast, especially taken out of context as in this case.

A pilot planning a flight below the freezing level normally should not expect to receive this advisory during an FSS preflight briefing because icing will not affect his or her proposed flight. Some briefers fail to understand and consider this and issue the advisory even though it is not a factor. This practice undermines the credibility of the forecast and briefing. A pilot planning a flight and briefed for low altitudes should keep this point in mind if he or she elects to or is instructed by ATC to climb to a higher altitude.

Pilots might well consider the advisability of accepting the clearance without additional information on icing and freezing levels. This point also applies to rerouting. If the pilot or ATC reroutes the aircraft, advisories that were not pertinent during the briefing might now apply. (Chapter 11 discusses various means to update weather and other pertinent information.)

Hawaiian AIRMETs use the same format as domestic AIRMETs. Alaskan AIRMETs are issued along with and at the same times as the area forecast; an example is contained in appendix E.

SIGMETs (WS)

SIGMETs often cover large areas, as AIRMETs do, because of the widely scattered and transitory nature of the reported phenomena; therefore, the term occasional (OCNL) frequently appears. Additionally, phenomena might move through the advisory area and affect only certain geographical districts. In the example of SIGMET NOVEMBER, strong updrafts and downdrafts are only expected in the vicinity of the mountains. Failure to completely read and understand an advisory has led many a pilot and briefer to unjustly criticize this product.

A SIGMET will only be issued when the phenomenon is widespread. Local occurrences of severe turbulence and icing will appear in an AIRMET bulletin. SIGMETs are issued when the following phenomena occur or are expected to develop:

- Severe icing
- Severe or greater nonconvective turbulence
- Moderate or greater clear air turbulence (CAT)
- Widespread dust storms, sandstorms, or volcanic ash that reduce visibility below 3 miles over an area at least 3,000 square miles

For a SIGMET, CAT is nonconvective turbulence occurring at or above 15,000 feet, although CAT usually refers to turbulence above 25,000 feet. Because of the difficulty to forecast this phenomenon, the reference *moderate or greater* (MOGR) indicates a threat of CAT. SIGMETs say severe or extreme only with actual reports.

In Alaska and Hawaii, the Weather Service Forecast Offices that are responsible for FAs also issue SIGMETs for these conditions. Additionally, these WSFOs issue SIGMETs for tornadoes, hail greater than ¾" in diameter, and embedded or lines of thunderstorms.

SIGMETs are identified by forecast area, alphabetic, and product designators. The forecast area specifies which FA the advisory applies to. Next appears the alphabetic designator for the phenomena being described. (To avoid confusion with international SIGMETs, domestic SIGMET names are NOVEMBER through YANKEE, excluding SIERRA, TANGO, and ZULU, which are reserved for AIRMETs.) The product designator (1, 2, 3 etc.) indicates the number of successive times the advisory has been issued. For example, a cold front causing severe turbulence might begin as SFO OSCAR

1. As the front moves into the Rocky Mountains, the report might become SLC OS-CAR 2, and into the Plains become CHI OSCAR 3. To assure continuity and alert pilots, briefers, and controllers that OSCAR 3 is the first CHI issuance, a referencing remark might be appended to the message (FOR PREVIOUS ISSUANCE SEE SLC OSCAR 2). Updates often contain changes; they must be reviewed for affected area, altitudes, and times. It's important to note both phenomena and product designators.

```
SFON WS 152130
SIGMET NOVEMBER 3 VALID UNTIL 160130
CA
FROM FOT TO 50NW RNO TO 50NE BFL TO SBA TO 40W SBA TO 30W SFO
TO FOT
OCNL SVR TURBC BLO 100 XCP BLO 150 VCNTY SIERRAS. STG UDDFS
VCNTY MTNS AND LLWS POTENTIAL BLO 20 AGL. CONDS CONTG BYD
0130Z.
```

In the above example the FA designator is San Francisco ("SFO"N). The issuance date-time group follows the product designator (WS). This SIGMET was issued on the 15th day of the month at 2130Z (152130). The alphabetic phenomena designator is N (SFO"N"). SIGMET NOVEMBER is spelled out on the following line along with the product designator (3). WSs are valid for 4 hours as indicated by the VALID UNTIL time, the 16th day of the month at 0130Z (160130). San Francisco SIGMET NOVEMBER 3 affects all or part of California. Like the AIRMET bulletin, specific geographical areas are described using VORs on the in-flight advisory plotting chart that is in appendix C, starting with the most northern and continuing clockwise. Phenomena will usually lie well within the delineated area.

Figure 4-1 shows SIGMET NOVEMBER 3 laid out (the gray area) on an in-flight advisory plotting chart. Inset in Fig. 4-1 is a portion of the common geographical area designators map from appendix D. The gray area again shows the area covered by the SIGMET. The advisory affects a small portion of Northern California and most of Central California. Plotting might be required to determine the extent of the advisory, which is extremely helpful when trying to visualize affected areas. Occasional severe turbulence is expected below 10,000 MSL, except below 15,000 MSL in the vicinity of the Sierra Nevada mountains. Strong updrafts and downdrafts in the vicinity of the mountains plus low-level wind shear are anticipated. Conditions are expected to continue beyond the end of the advisory period (0130Z). This means SIGMET NOVEMBER 4 should be issued prior to 0130Z; however, if factors such as data-transmission difficulties delay the updated advisory, pilots should consider the advisory still in effect unless a cancellation message is received.

CONVECTIVE SIGMETs (WST)

Convective SIGMETs provide detailed and specific forecasts for thunderstorm-related phenomena. NAWAU's WST unit makes extensive use of radar data to analyze thunderstorm systems. Meteorologists compare radar data with satellite imagery, lightning

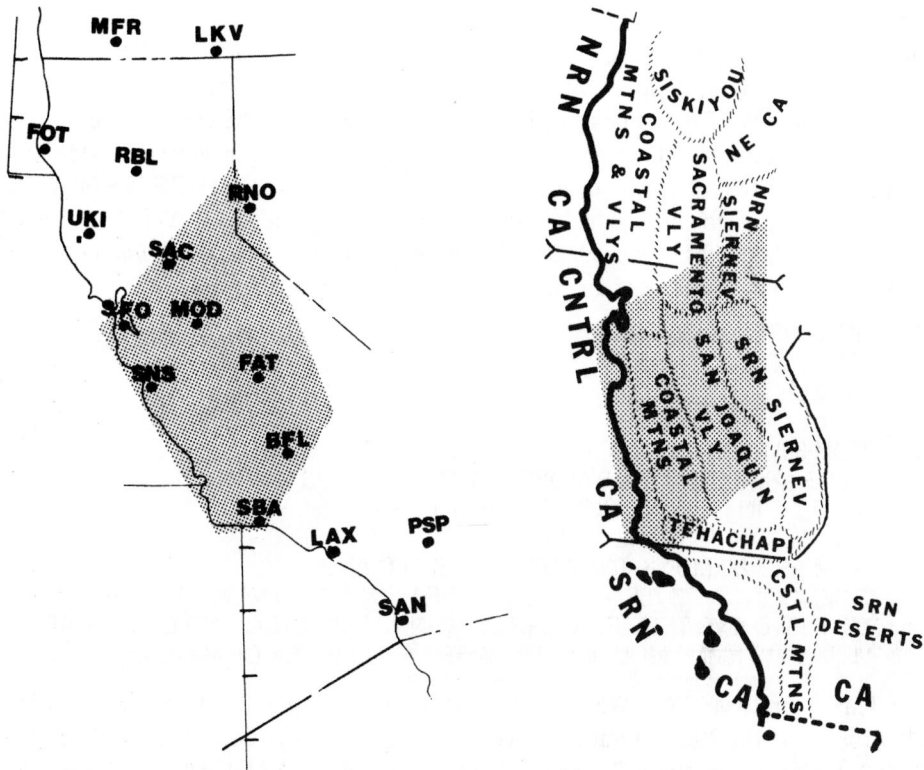

Fig. 4-1. *SIGMET NOVEMBER 3 laid out on an in-flight advisory plotting chart is compared to the same area on the geographical area designators map. Weather advisories do not necessarily pertain to all of the delineated area. To apply a weather advisory, pilots must compare the delineated area with terrain and the text of the advisory.*

information, and other conventional sources to determine the need for a SIGMET. WSTs are issued when the following phenomena occur or are expected to develop:

- Tornadoes
- Hail greater than or equal to ¾" diameter
- Isolated severe thunderstorms
- Embedded thunderstorms
- A line of thunderstorms
- Thunderstorms (VIP level 4 or greater) affecting 40 percent or more of an area of at least 3,000 square miles

Regarding these reportable WST phenomena, a severe thunderstorm produces hail at least ¾" in diameter, or surface winds of 50 knots; an embedded thunderstorm occurs within nonconvective precipitation; and Table 9-1 contains a description of radar precipitation intensity VIP levels.

When tornadoes, hail, severe or embedded thunderstorms, or a line of thunderstorms occur, a WST might be issued regardless of the size of the affected area.

The 48 contiguous states are divided into three areas for convective SIGMET issuance: west (MKCW WST), central (MKCC WST), and east (MKCE WST). These areas are shown on the in-flight advisory plotting chart in appendix C. WSTs are issued on an unscheduled basis as needed beginning with Number 1 at 0000Z and contain a forecast for up to 2 hours and an outlook from 2 to 6 hours.

```
MKCC WST 191155
CONVECTIVE SIGMET 20C
VALID UNTIL 1355Z
TX
FROM 30W LBB-70SW SPS-ABI-90WSW SJT-30W LBB
DVLPG AREA SVR TSTMS MOVG FROM 2330. TOPS ABV 450.
HAIL TO 2 IN . . . WIND GUSTS TO 50 KT PSBL.
OUTLOOK VALID UNTIL 1755Z
FROM ARG-MEM-TXK-100WSW SJT-LBB-SPS-ARG
TSTMS CONT TO DVLP ALG DRYLN OVR WRN TX AS PVA AND CD AIR ALOFT
MOVE EWD OVR THE AREA. UPR LVL WINDS RMN STRG AND DIFFLUENT
WHILE AMS RMNS MDLY UNSTBL WITH LIFTED INDEX OF MINUS 6.
```

This central (MKC"C") WST was issued on the 19th day of the month at 1155Z (191155). It is the 20th central issuance for this ZULU day (CONVECTIVE SIGMET "20"C). It affects portions of Texas and is valid for 2 hours (VALID UNTIL 1355Z).

Specific areas are described using the VORs on the in-flight advisory plotting chart. In this case, from 30 west of Lubbock, to 70 southwest of Wichita Falls, to Abilene, to 90 west southwest of San Angelo, to 30 west of Lubbock. The advisory warns of a developing area of severe thunderstorms moving from 230° at 30 knots. The continuous line in Fig. 4-2 encloses this area; it's nice to have the chart to visualize locations. Tops above 45,000 feet, hail 2" in diameter and wind gusts to 50 knots are possible.

The WST outlook was designed primarily for preflight planning and aircraft dispatch. It normally includes a meteorological discussion of factors considered by the forecaster. The outlook is supplemental information not required for weather avoidance, but useful to CWSU and FSS specialists for analysis and background information. Normally the outlook will not be included in broadcasts nor provided during a briefing.

The example outlook is valid for an additional 4 hours and covers an area from Walnut Ridge, AR, to Memphis, TN, to Texarkana, AR, to 100 west southwest of San Angelo, TX, to Lubbock, TX, to Wichita Falls, TX, and back to Walnut Ridge. The broken line in Fig. 4-2 encloses this area.

The outlook translates: Thunderstorms are expected to continue to develop along a dry line over western Texas as positive vorticity advection and cold air aloft move

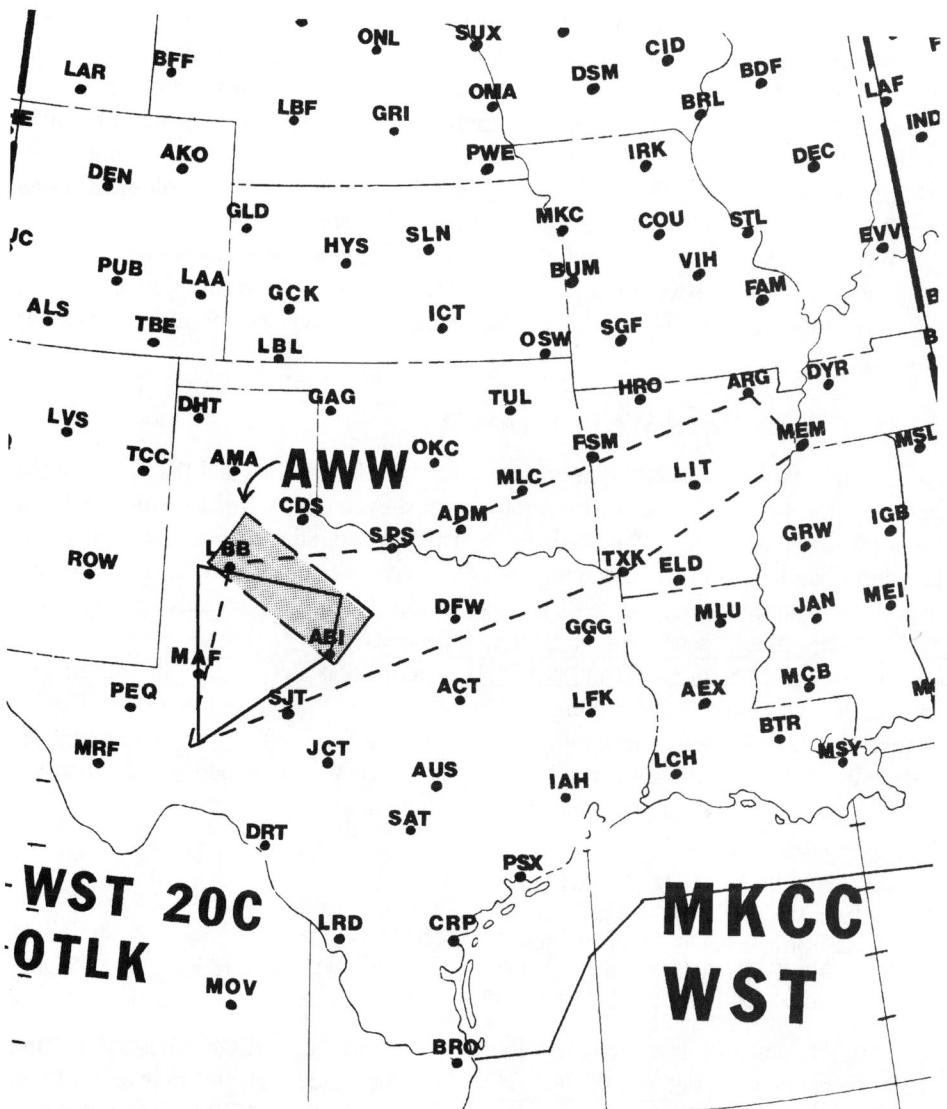

Fig. 4-2. *Because convective SIGMETs and alert weather watches are issued and valid for different times and serve different purposes, aerial coverage might not coincide.*

eastward over the area. Upper-level winds remain strong and diffluent while the air mass remains moderately unstable with a lifted index of minus six. (Standard contractions are contained in appendix A.)

The outlook interpretation is that thunderstorms are going to develop along a dry line (or temperature-dew point front), which is the boundary between dry air from the southwestern deserts and moist air from the Gulf of Mexico. Positive vorticity advection, which is the upward-moving air that serves as a lifting mechanism, and cold air (instability) aloft move eastward. The upper-level winds remain strong and diffluent (a divergent flow activates and perpetuates thunderstorms), while the air mass remains moderately unstable. The three elements needed for thunderstorm development are present: moisture, instability, and a lifting mechanism. (Most meteorological terms used in the outlook discussion are contained in the glossary.)

When thunderstorms or related phenomena are the purpose of an advisory, severe or greater turbulence, severe icing, and low-level wind shear are implied; therefore, these conditions will not be specifically addressed in the advisory nor during a weather briefing.

ALERT WEATHER WATCH (AWW)

Alert weather watches (AWWs) alert forecasters, briefers, pilots, and the public to the potential for severe thunderstorms or tornadoes. A *severe weather watch* bulletin (WW) is issued after the AWW notification. The WW contains details on the phenomena described in the AWW. (An example of a severe weather watch bulletin is in appendix E.) These unscheduled bulletins are primarily a public forecast, whereas the WST is a combined observation and aviation forecast.

Although SELS (Fig. 4-3) and NAWAU meteorologists coordinate their products, criteria and timeframes differ; therefore, aerial coverage might not coincide. The issuance of an AWW might precede or coincide with a WST. AWWs are numbered sequentially beginning on January 1. The following AWW was issued just prior to the WST in the previous example. (Refer to the shaded area in Fig. 4-2.)

```
MKC AWW 191147
WW 99 SEVERE TSTM TX 191215Z–191800Z
AXIS..70 STATUTE MILES EITHER SIDE OF LINE . .
20N LBB/LUBBOCK TX/-50NE ABI/ABILENE TX/
HAIL SURFACE AND ALOFT . . 2 INCHES. WIND GUSTS 60 KNOTS. MAX TOPS
TO 550. MEAN WIND VECTOR 22035.
```

When the area is described with locations not on the in-flight advisory plotting chart, a separate line titled AVIATION COORDS will be added. The *mean wind vector* is the direction and magnitude of the mean winds from 5,000 feet AGL to the tropopause. The vector may be used to estimate cell movement—in the example, 220° at 35 knots.

The WST describes a developing area of interest to aviation. Expect subsequent WSTs to cover the area toward the northeast, into the AWW, and along the area described in the WST outlook.

CENTER WEATHER ADVISORIES (CWA)

CWAs are unscheduled in-flight advisories that are issued when conditions are expected to significantly affect IFR operations. CWAs are intended to help IFR pilots avoid hazardous weather. The advisories update or expand AIRMETs, SIGMETs, or

Fig. 4-3. *Forecasters at the severe local storms section of the Kansas City NWS office prepare alert weather watches and bulletins.*

the FA, and are issued when conditions meet advisory criteria. In such cases, the CWSU will coordinate with NAWAU forecasters for the issuance of the appropriate advisory. CWAs are also issued when local hazardous conditions develop that do not warrant other advisories. Because the center weather advisories often report localized phenomena, the area might be described using locations other than those on the geographical area designators map, or VORs on the in-flight advisory plotting chart.

The CWA numbering system is somewhat complex, but facilitates computer processing and distribution. Each CWA will have a phenomenon number: 1 through 6. A separate phenomenon number will be assigned each distinct condition: turbulence, icing, thunderstorms, and the like.

```
ZDV1 CWA 01 132050–132250
CAUTION FOR MDT-SVR TURBC/MTN WAVE ACTVTY PSBL ALL FLGT LVLS
OVER AND NEAR MTNS IN WY AND CO. STRONG CLD FNT MOVG THRU
AREA THIS AFTN AND EVE WITH PSBL SFC WND GUSTS TO 60KT. HIGH WND
WARNINGS ARE IN EFFECT FOR FNT RANGE AREA BEGINNING 14/0000Z.
THIS ADVRY SUPPLEMENTS SIGMET NOVEMBER 2.
```

This Denver Center CWA describes phenomenon Number 1 (ZDV"1"). It is the first issuance for this phenomenon (ZDV1 CWA "01"), valid from the 13th day of the month at 2050Z until 2250Z. If the advisory requires updating at 2250Z, it will become ZDV1

CWA 02. The CWA advises caution for moderate to severe turbulence associated with mountain wave activity for all flight levels over and near the mountains of Wyoming and Colorado. Additionally, a strong cold front moving through the area during the afternoon and evening might bring surface winds with gusts to 60 knots. High-wind warnings are in effect for the Front Range area beginning on the 14th at 0000Z. This advisory supplements SIGMET NOVEMBER 2.

This example illustrates many uses of the CWA. The advisory expands on SIGMET NOVEMBER 2 by indicating phenomena is due to mountain wave activity and mechanical turbulence. The CWA also mentions high-wind warnings for the Front Range, which is a local geographical reference to the mountains on the western fringes of Denver that extend from Fort Collins (north of Denver) to about 40 miles southwest of Denver.

```
ZHU1 CWA 01 061355–061455
FM A BPT to 40SE LFT LN . . S 150 MI INTO GULF . . . AREA SCT LVL 3-5
TSTMS MOVG N 15 KTS. NMRS TOPS ABV 450.
```

The Houston Center CWSU has issued this advisory for an area of scattered VIP-level 3 to 5 thunderstorms (see Table 9-1) moving north at 15 knots with numerous tops to above 45,000 feet MSL. The area extends along a line from Beaumont, TX, to 40 miles southeast of Lafayette, LA, and south 150 miles into the Gulf of Mexico. These locations are not on the in-flight advisory plotting chart. The condition has not yet met the criteria for a WST.

CWSUs also issue *meteorological impact statements* (MIS). Strictly an in-house product, the MIS alerts controllers of weather that might affect the flow of IFR traffic. The MIS describes conditions already contained in other advisories and forecasts. From time to time, overzealous FSS briefers might refer to an MIS, or tower controllers might record it on the ATIS.

DISSEMINATION

Advisories are routinely provided during FSS standard briefings and are offered during abbreviated briefings. (FSS weather briefings are discussed in detail in chapter 11.) During routine FSS radio contacts, advisories within 150 miles will be offered when they affect the pilot's route. It's important to note the SIGMET series and number to ensure receipt of the latest information.

The *hazardous in-flight weather advisory service* (HIWAS) provides a continuous broadcast of advisories and urgent PIREPs over selected VORs. HIWAS outlets are identified on aeronautical charts by a small black square in the lower right corner of the NAVAID box and in the NAVAID data section of the *Airport/Facility Directory*. (HIWAS is fully examined in chapter 12.) Jeppesen charts list "HIWAS" next to the NAVAID providing the service.

When a WA, WS, WST, AWW, or CWA affects an area within 150 miles of a HIWAS outlet or an ARTCC sector's jurisdiction, an alert is broadcast once on all frequencies except flight watch and emergency. Approach controls and towers also broadcast an alert, but it might be limited to phenomena within 50 miles of the terminal. When the advisory affects operations within the terminal area, an alert message will be placed on the ATIS. Here again, overzealous controllers have been known to place SIGMET alerts on the ATIS for conditions hundreds of miles away.

In spite of criticism that advisories cover too much area, issuance of advisories has become more conservative. Ironically, some pilots and briefers now criticize the forecast for not containing enough precautions. Virtually all criticism, however, is due to misconceptions and misunderstanding the product.

The existence of an advisory, or lack thereof, does not relieve the pilot from using good judgment and applying personal limitations. Like all pilots, I have had on occasion parked my Cessna 150 and hopped on an airliner. These instances lend credence to the aviation axiom: "When you have time to spare, go by air; more time yet, take a jet." When you don't have the equipment or qualifications to handle the weather, don't go! This doesn't mean cancel every time that you hear an advisory, but take a close look at all available information.

For a flight from Las Vegas to Van Nuys, CA, I was told by the briefer: "Well, you aren't going today!" My jaw locked up and I replied: "Oh yes I am!" I hadn't looked at the weather yet; my statement was a gut reaction to this individual's horrible technique. Advisories for turbulence, mountain obscurement, and rain showers were in effect. As is often the case in that part of the country, a direct flight was out. But by choosing a course over lower terrain, VFR is frequently possible. My decision to make the trip was based upon my experience, knowledge of terrain, a thorough review of all available weather reports and forecasts, and always having an out should the weather ahead become impassable. I proceeded down the Colorado River to Needles, then to Daggett to avoid the higher terrain, and finally over Palmdale to Van Nuys.

The lack of an advisory does not guarantee the absence of hazardous weather. An unfortunate pilot learned this lesson the hard way. The synopsis described a moist unstable air mass. Thunderstorms were not forecast for the time of flight, but were expected to develop and were already being reported along the route. The airplane did not have storm detection equipment and encountered extreme turbulence when the pilot inadvertently entered a cell. The pilot filed an IFR flight plan based on the fact that there were no advisories. About a half hour into the flight, according to the pilot's statements to the FAA and NTSB, "We noticed a heavy layer of clouds at and below our altitude and some 20 miles ahead . . .. The layer in front of us seemed to be light cumulus with a heavier layer behind it (not ominous looking)."

After the encounter the pilot could not understand why no "precaution or warning regarding that system" was provided. The pilot said the accident "would not have happened" if information about the weather had been provided. No advisories were in effect because at the time of the briefing none were warranted. The pilot had the clues of a moist and unstable air plus thunderstorms that were already reported, but the pilot put complete trust in a forecast that included no flight precautions or advisories.

The preceding examples illustrate two "go" decisions: one a routine flight, the other almost fatal. The intent wasn't to brag about personal skills or criticize another pilot. The intent was to show decision making based upon available information, a knowledge of weather products, and limitations of pilot and aircraft.

All too often, briefers hear pilots flying aircraft without storm detection equipment say "Thunderstorms? Ahhh, well, I'd better go IFR." Not for me, thanks. I want to be clear of clouds where I can see and avoid convective activity. Dennis Newton, author of *Severe Weather Flying*, said, "If you are instrument-rated and are flying without radar or a Stormscope, and if you fly long enough in IFR conditions favorable to air-mass thunderstorms, sooner or later you are going to get caught."

If we wish to be accorded and exercise the privileges of pilot-in-command, we have to understand the system and its limitations. We must evaluate all available information as required by FARs and make a flight decision based upon our knowledge and limitations and that of the aircraft and its equipment.

A popular aviation saying: Aviation in itself is not inherently dangerous, but to an even greater degree than the sea, it is terribly unforgiving of any carelessness, incapacity or neglect.

5
Area forecasts (FA)

PILOTS, BRIEFERS, AND METEOROLOGISTS SHARE MISCONCEPTIONS AND misinterpret the purpose and scope of the area forecast (FA). A forecast for the desert portion of a fuel proficiency air race predicted scattered thunderstorms and rain showers with wind gusts to 35 knots (SCT TRW G35). Pilots criticized the forecast, complaining they didn't encounter any gusts; pilots had remained clear of the winds by avoiding the thunderstorms. The thunderstorms were there, and you can bet gusty winds could be found in the vicinity of the cells. The forecast was perfectly correct.

Others criticize an FA for being too lengthy, and in the next ironic breath they criticize an FA for not containing enough detail. A West Coast FSS manager was quoted in an April 1986 *Pacific Flyer* article by Lance Stalker: "Before, you had people that were familiar with the local conditions and put that [knowledge] into their forecasting." That's still true. Local NWS offices issue TWEB route and terminal forecasts (FT). The area forecast has never been intended to cover every single condition.

The area forecast predicts conditions over an area the size of several states. Due to limitations on size, computer storage, and communications equipment, the forecast cannot be divided into smaller segments and cannot provide the detail available in TWEB route forecasts or terminal forecasts (FT). Widely varying conditions over relatively large areas must be included; therefore, small-scale events are often described using the conditional terms occasional, chance, and isolated. The FA provides a forecast for the en route portion of a flight and destination weather for locations without FTs. This contradicts a widely held notion that without an FT there is no destination forecast. Conditions are forecast from the surface to 70 millibars (approximately 63,000 feet).

FAs in the 1960s were issued every 6 hours, valid for 12 hours, with a 12-hour outlook. This was time-consuming for the forecaster and, therefore, expensive. FAs in the 1970s were issued twice a day, valid for 18 hours with an additional 12-hour categorical outlook (IFR, MVFR, or VFR). Today they are issued three times a day and are valid for 12 hours with a 6-hour outlook. The increased number of issuances and reduced valid times directly reflect forecast limitations.

Area forecasts for the 48 contiguous states were reduced to six in 1982. Rather than being issued by local offices, responsibility was transferred to the National Aviation Weather Advisory Unit (NAWAU) in Kansas City. (Appendix D depicts FA coverage areas.) During this period, the area forecast was divided into five sections: hazards, synopsis, icing and freezing level, turbulence and low-level wind shear, and significant clouds and weather.

In 1991, with the advent of the AIRMET bulletin, the FA was reduced to the "synopsis" and "VFR clouds and weather." Alaska local NWS offices in Anchorage (ANC), Fairbanks (FAI), and Juneau (JNU) issue FAs three times a day; the Honolulu (HNL) WSFO issues FAs four times a day. Alaskan and Hawaiian FA areas are depicted in appendix D and use a similar format as those in the contiguous states. The Miami WSFO issues a Gulf of Mexico FA for the area west of 85°W and north of 27°N, which includes the coastal plains and waters from Apalachicola, FL, to Brownsville, TX. (An example is contained in appendix E.)

Table 5-1 contains FA issuance times. Note that UTC or zulu issuance times within the 48 contiguous states change twice a year with daylight saving time. Because the SFO and SLC FAs are issued together, and the Mountain time zone is one hour ahead of Pacific, the SLC FA becomes available one hour earlier local, in that time zone. Forecasts become valid on the hour following issuance: issued 1040, valid 1100.

The area forecast begins with a heading describing the coverage of the product and valid times.

```
DFWC FA 280945
SYNOPSIS AND VFR CLDS/WX
SYNOPSIS VALID UNTIL 290400
CLDS/WX VALID UNTIL 282200...OTLK VALID 282200-290400
OK TX AR TN LA MS AL AND CSTL WTRS
```

This is the Dallas-Fort Worth FA issued on the 28th day of the month at 0945Z (DFWC FA 280945). (The C in DFW"C" is left over from the FAs prior to 1991. In those days, DFWC was the "significant clouds-and-weather section" of the FA.) The next line identifies this as the synopsis and VFR clouds and weather. The synopsis is valid until the 29th at 0400Z (18 hours). The clouds-and-weather section is valid until the 28th at 2200Z (12 hours), with an outlook from the 28th at 2200Z until the 29th at

Table 5-1. Area forecast issuance times.

SFO & SLC UTC	CHI & DFW UTC	BOS & MIA UTC	LOCAL/SLC FA
1040/1140	0940/1040	0840/0940	4:40 a.m./5:40 a.m.
1940/2040	1840/1940	1740/1840	12:40 p.m./1:40 p.m.
0240/0340	0140/0240	0040/0140	8:40 p.m./9:40 p.m.

ANC & FAI		JNU		HNL	
UTC	LOCAL*	UTC	LOCAL*	UTC	LOCAL**
1440	5:40 a.m.	1340	4:40 a.m.	1540	5:40 a.m.
2240	1:40 p.m.	2240	1:40 p.m.	2140	11:40 a.m.
0640	9:40 p.m.	0640	9:40 p.m.	0340	5:40 p.m.
				0940	11:40 p.m.

* Alaskan Standard Time
** Hawaiian Standard Time

0400Z (six hours). This FA covers the states of Oklahoma, Texas, Arkansas, Louisiana, Mississippi, and Alabama, and includes the adjacent coastal waters (OK TX AR TN LA MS AL AND CSTL WTRS).

Following the heading is the disclaimer paragraph.

SEE AIRMET SIERRA FOR IFR CONDS AND MTN OBSCN.
TSTMS IMPLY SVR OR GTR TURBC SVR ICG LLWS AND IFR CONDS.
NON MSL HETS DENOTED BY AGL OR CIG.

The first sentence refers the user to AIRMET SIERRA for IFR conditions and areas of mountain obscuration. This has led to confusion for pilots and briefers alike. Often when IFR is forecast, the body of the FA will only contain a tops forecast. For example, AIRMET SIERRA might forecast ceilings and visibilities below 1,000 feet and 3 miles. The FA might only contain ST TOPS 030 (stratus tops 3,000 feet). The bases of the clouds and visibilities are contained in the AIRMET bulletin.

From *Aviation Weather* (AC 00-6): "A thunderstorm packs just about every weather hazard known to aviation into one vicious bundle." To eliminate redundancy and serve as a "we-told-you-so," the following statement appears on every FA: TSTMS IMPLY SVR OR GTR TURBC SVR ICG LLWS AND IFR CONDS.

Thunderstorms imply possible severe or greater turbulence, severe icing, low-level wind shear, and IFR conditions; this shouldn't be news to anyone. A report or forecast of thunderstorms implies these and other hazards associated with thunderstorms: hail, lightning, gusty winds, and altimeter errors. The body of the FA will not contain specific precautions, nor will briefers normally include this statement. The fact that thunderstorms are reported or forecast infers all thunderstorm-associated hazards.

NON MSL HGTS DENOTED BY AGL OR CIG.

This statement simply means all heights are mean sea level (MSL), unless noted as above ground level (AGL 30 SCT-BKN) or ceiling (CIGS 15-20 OVC). This distinction can be significant, especially in mountainous areas. Forecasts for western and Appalachian states with mountains will normally reference cloud bases to MSL; forecasts for flat terrain in the remainder of the United States will normally reference bases to AGL. This differentiation must be considered when comparing the FA with SAs.

One evening several FSS specialists brought the "east of the Cascades" portion of the SFO FA to my attention. They contended the observations and the forecast had nothing in common; however, after converting the SAs to MSL heights, the observations and forecast were perfectly consistent. This example emphasizes the point that a pilot or briefer must have a thorough knowledge of terrain to apply a forecast.

SYNOPSIS

The synopsis is usually a brief and generalized statement that describes the location and movement of pressure systems, fronts, and weather patterns. The following example is quite detailed.

> COLD UPR SYS OFF WA CST AT 19Z INVOF 48N 127W MOVG EWD AT ABT
> 10-15 KTS. PVA INDUCED/ENHANCED CNVTV CLDS WERE ROTG ONSHR
> FROM NRN CA THRU SWRN WA. AMS THIS RGN APPRS QUITE MOIST AND
> UNSTBL AND WL SPRD ACRS NRN HLF OF FCST AREA THRU THE PD.

A cold upper-level low-pressure system is off the Washington coast at 11 a.m. PST about 100 miles west of Seattle moving east at about 10 to 15 knots. Positive vorticity advection is inducing and enhancing convective clouds that were rotating onshore from northern California through southwestern Washington. The air mass in this region appears quite moist and unstable and will spread across the northern half of the forecast area through the period.

I prefer a detailed synopsis because my training and experience allows extra insight regarding the weather situation. This is especially important if the weather improves or deteriorates more rapidly than forecast. This synopsis would normally be summarized during FSS briefings and on broadcasts as "An upper-level low off the Pacific Northwest is bringing moist unstable air over Washington, Oregon, and Northern California." Translating and summarizing in this manner is a prime function of FSS weather briefers. Pilots using DUATS will have to decode, translate, and interpret the synopsis on their own.

From this example, a conclusion might be that on a flight from northern California to Washington, the best weather lies to the east, ahead of the system. The example also contradicts a widely held misconception that fronts are the only weather-producing systems.

A synopsis describes the cause of the weather; therefore, language and detail will depend upon the situation. The synopsis will vary from the lengthy detail in the previous example to HI PRES OVR THE ERN GLFALSK WL MOV OVR THE PNHDL AND WKN, which translates into "High pressure over the eastern Gulf of Alaska will move over the panhandle and weaken."

The importance of a synopsis cannot be overemphasized. For example, the significance of a forecast for IFR conditions will depend on whether IFR is due to a coastal marine layer, upslope fog covering several states, a frontal system, or a tropical storm. Here's another example.

```
CHIS FA 300940
SYNOPSIS VALID UNTIL 310400
AT 10Z CDFNT FROM LS SWWD THRU NWRN IA INTO NWRN KS THEN WWD
THRU CO. HI PRES OVR OH VLY AND MT. THE CDFNT WL CONT EWD AND BY
00Z WL EXTEND FROM LWR MI SWWD INTO SRN KS AS HI PRES BLDS OVR
DKTS.MRNG FOG/ST OVR ERN GRTLKS WL IPV BY 16Z. AFTN/EVE TSTMS
MOST ACTV ALG FNT FROM IL NEWD THRU MI...WILLIAMS...
```

This Chicago FA synopsis (CHIS FA) was issued on the 30th day of the month at 0940Z (300945), valid until the 31st at 0400Z. The synopsis covers the entire 18-hour forecast period. At 1000Z a cold front extended from Lake Superior southwestward through northwestern Iowa into northwestern Kansas, then westward through Colorado. High pressure dominates the Ohio Valley and Montana. The cold front will continue eastward and by 0000Z will extend from the lower portion of Michigan southwestward into southern Kansas as high pressure builds over the Dakotas. The morning fog and stratus over the eastern Great Lakes will improve by 1600Z. Afternoon and evening thunderstorms will be most active along the front from Illinois northeastward through Michigan. This forecast was prepared by Williams.

The CHIS synopsis has been plotted in Fig. 5-1. Notice the ease of visualizing conditions. A VFR flight from La Crosse, WI (LSE), to Denver (DEN) might be well advised to delay until after frontal passage. An IFR flight might be planned directly to Minneapolis, MN (MSP), to penetrate the front at a right angle and minimize exposure to its weather.

VFR CLOUDS AND WEATHER

The VFR clouds-and-weather section includes sky condition, non-IFR cloud heights and visibilities, obstructions to visibility, weather, surface winds, and a 6-hour categorical outlook.

Sky condition contains cloud height, amount, and tops. Heights are normally mean sea level, with above ground level and ceilings generally limited to layers within 4,000 feet of the surface. Because tops of building cumulus, towering cumulus, and cumulonimbus are quite variable, only upper limits appear: TOPS TO 350 or CB TOPS 300, and the like. When multiple or merging layers are forecast, which would not permit VFR flight between layers, only the top of the highest layer appears: 80-100 BKN-OVC LYRD TO 200 or MEGG/NMRS LYRS TO 180, and the like. Because of its scope, FA tops cannot be more precise. TWEB route forecasts might contain more detail.

Usually combined with weather, surface visibility and obstructions are forecast when expected to be 5 miles or less: 3-5R-F (visibility 3 to 5 miles in light rain and fog) or VSBY 3-5RW-WDLY SCT TRW- (visibility 3 to 5 miles in light rain showers and

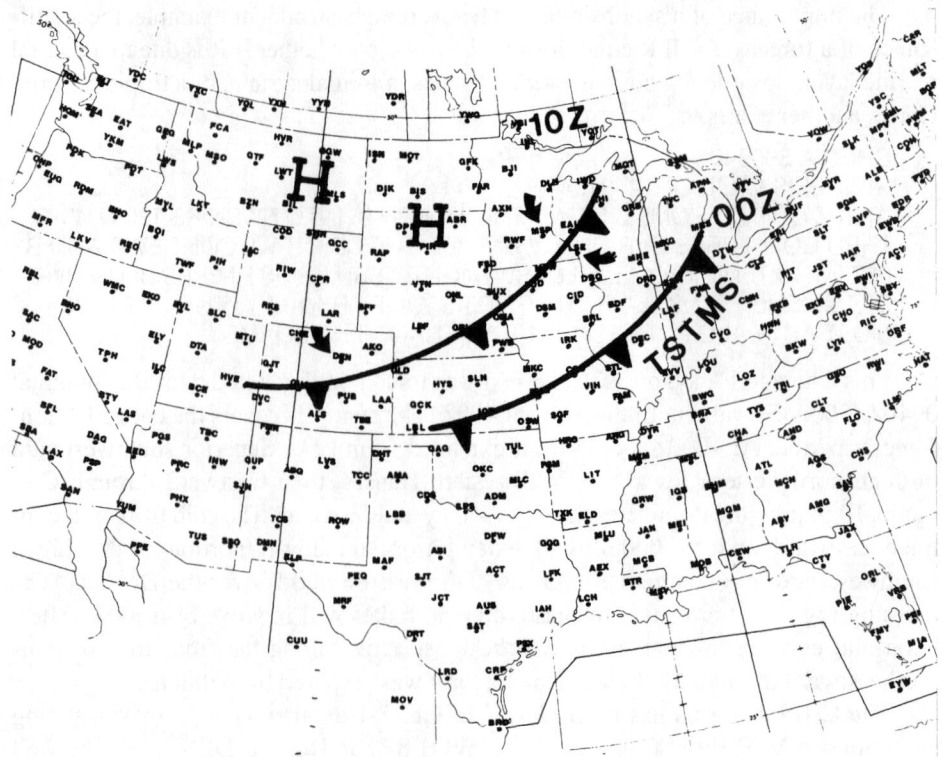

Fig. 5-1. *The locations and expected movements of fronts and weather systems can be visualized by plotting the synopsis.*

widely scattered light rain showers and thunderstorms). Table 4-1 defines widely scattered as less than 25 percent of the area. Because of the scope of this product, the forecast cannot be more precise. A pilot can interpret this forecast to mean the thunderstorms should be circumnavigable.

The absence of a visibility forecast only implies general visibilities greater than 5 miles. Because they are not within the scope of this product, widespread visibilities of 6 miles, or local conditions less than 5 miles, might exist and not be included in the FA. TWEB route and terminal forecasts might contain greater detail.

Widespread areas of surface winds that are expected to be operationally significant appear in the forecast: 20G30. Direction is true, referenced to the eight points of the compass (north, northeast, east, southeast, etc.). The lack of a wind forecast only implies widespread sustained speeds less than 30 knots. TWEB route and terminal forecasts can often be used to determine winds of lesser speeds and local conditions. Often associated with convective activity, TRW G40 translates to wind gusts of 40 knots expected to accompany thunderstorms.

The forecaster divides the FA area using standard geographical designators. (The most common designators can be found in appendix D.) The extent and detail will depend on the weather situation. The example below illustrates a standard division of the Dallas-Fort Worth FA.

```
SWRN TX
WEST OF PECOS RVR...CLR OR SCT CI.
EAST OF PECOS RVR...AGL 30 SCT 100 SCT. ISOLD RW-/TRW-.
```

This portion of the DFW FA covers southwestern Texas (SWRN TX). The forecaster has further divided the area into west and east of the Pecos River. The regional reference is a common feature in this FA; if the Pecos River ever dries up, I don't think they'll be able to write a DFW FA.

In this next example of a VFR clouds-and-weather section, the synopsis indicates midlevel moisture with a stable air mass in the valleys. A typical situation during the winter months for northern and central regions of California.

```
NRN/CNTRL CA
CNTRL VLYS AND CSTL VLYS SFO NWD..ST TOPS 020-025 CNTRL VLYS AND
010-020 CSTL VLYS. CONDS IPVG CSTL PTN 18-21Z.
ELSW..150 SCT CI ABV. OCNL 100 BKN 150 NRN CSTL WTRS AND CSTLN
WITH CHC R-.
```

In this forecast, the meteorologist has divided California, specifying this portion for northern and central sections. The forecaster will further divide the area within the text of the FA. Figure 5-2 pictorially displays the forecast; a portion of the written forecast is below each excerpt from the geographical area designators map. The shaded area (gray tint) represents the affected area. The forecast for the central valleys (Sacramento and San Joaquin) and coastal valleys (San Francisco northward) (ceilings below 1,000 feet, visibilities below 3 miles in fog) is contained in AIRMET SIERRA; therefore, that forecast is omitted from the FA. Stratus tops are expected between 2,000 to 2,500 feet in the central valleys, and 1,000 to 2,000 feet in coastal valleys (tops are always mean sea level); coastal valleys are forecast to improve between 18Z and 21Z. Notice how the forecaster specifies a period (18Z-21Z), rather than an exact time, which is another reflection on the limitations of forecasts.

Radiation fog has formed from the moisture of previous storms trapped in valleys and cooled at night under stable air. The relatively shallow layer in the coastal valleys is expected to improve by the middle of the day. Conditions in the central valleys will continue. VFR flights will have to be delayed until afternoon in the coastal valleys, and most probably will not be possible at any time in the central valleys. IFR operations will be possible, assuming the pilot has takeoff, landing, and alternate minimums. TWEB route and terminal forecasts should be consulted because the lowest values normally found in the FA are ceilings less than 1,000 feet and visibilities less than 3 miles. It is not within the scope of this product to provide more detail.

A pilot planning IFR into an airport without an FT must specify an alternate. Additionally, airports within this area without an FT do not satisfy alternate requirements. A pilot must specify either an airport with an FT that is forecasting alternate minimums or an airport out of the affected area. VFR operations above the valley fog will not be

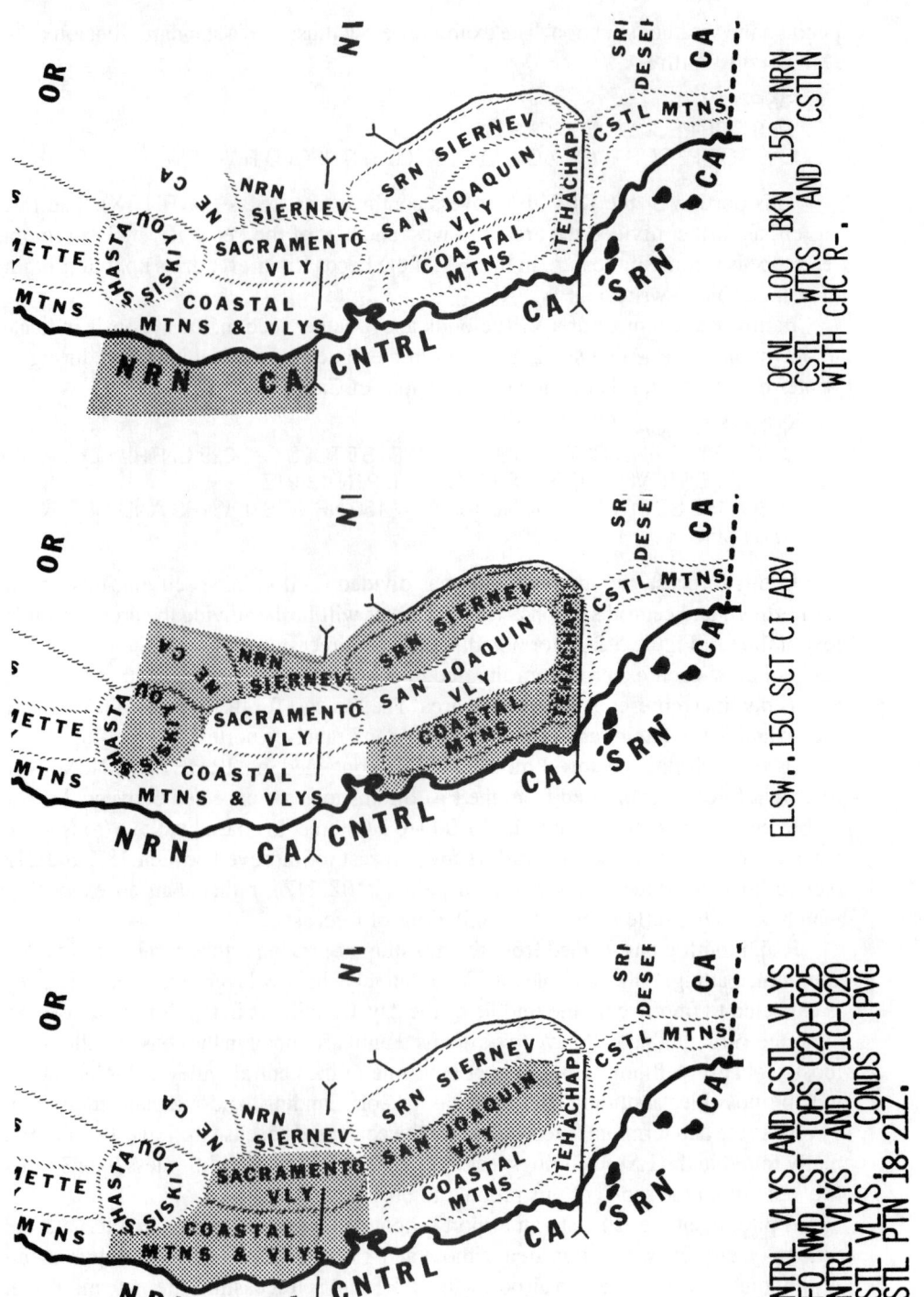

Fig. 5-2. *With DUAT, the pilot will have to determine which parts of the forecast apply to his or her intended route. A copy of the geographical area designators map will be extremely helpful.*

restricted; therefore, it is important to realize that VFR *and* IFR pilots might be flying above extensive areas of zero-zero conditions. These conditions will pose a risk to single-engine operations in case of engine failure; pilots should weigh this danger very carefully.

Elsewhere, the forecast for the coastal mountains (including northern coastal mountains above the stratus), northern mountains, Sierra Nevada mountains, and the coast south of San Francisco is 15,000 MSL scattered with cirrus above. There will be no restrictions to either VFR or IFR operations in these areas. Landing fields in these areas would be suitable IFR alternates for valley airports. Additionally, over the northern coastal waters and coastline, due to sufficient moisture at the middle levels, the forecast calls for occasional 10,000 MSL broken and tops 15,000 MSL with a chance of light rain.

OUTLOOK

A 6-hour categorical outlook (OTLK) appears at the end of each 12-hour VFR clouds-and-weather statement. The outlook consists of the following categories: IFR, MVFR (marginal visual flight rules), and VFR; these categories do not necessarily correspond to FAR definitions.

Refer to Table 5-2. References to IFR or MVFR explain the phenomena causing the condition. For example:

- IFR CIG Ceiling less than 1,000 feet.
- IFR CIG F Ceiling less than 1,000 feet and visibility less than 3 miles.
- MVFR HK Visibilities between 3 and 5 miles in haze and smoke.
- VFR WIND Ceiling greater than 3,000 feet and visibility greater than 5 miles with sustained wind 30 knots or greater.

Table 5-2. Area forecast outlook categories.

Category	Ceiling (feet)		Visibility (miles)
IFR:	less than 1,000	and/or	less than three
MVFR:	1,000 to 3,000	and/or	three to five
VFR:	more than 3,000	and	more than five
WIND:	Sustained winds or gusts of 30 knots or greater.		

A "categorical outlook" is another direct reflection on the limitations of aviation forecasts. The outlook is based on synoptic scale events and might not contain local conditions.

Pilots must carefully consider an outlook. VFR cannot be interpreted as clear, although conditions might actually be clear. VFR translates into ceiling greater than 3,000 feet AGL and visibility greater than 5 miles; how much "greater" is not specified. Often, mountain obscuration is not considered. VFR might be expected in valleys, while VFR flight through mountainous areas might not be possible. This category is an indicator that airports within the area will not require an IFR alternate. MVFR cannot necessarily be interpreted as allowing VFR flight.

If conditions are at the lower limit of MVFR, IFR most often would be required. This category is an indicator that airports will be above instrument minimums, but require an alternate. IFR indicates that VFR flight is out of the question; however, the IFR pilot cannot interpret this category as indicating an airport will be above instrument minimums. The category indicates that an IFR alternate will be required.

A word of caution to the IFR pilot: Outlooks are merely indicators. Categorical outlooks can never be used to determine IFR alternate requirements or suitable IFR alternate airports. FAR requirements must be met based upon the latest forecasts prior to departure.

AREA FORECAST AMENDMENT CRITERIA

Weather advisories (AIRMETs, SIGMETs, CWAs, etc.) automatically amend the area forecast. The synopsis will normally be amended with a significant change in the synoptic pattern. Other sections are amended whenever the weather improves or deteriorates.

This is a broad statement and pilots must remember the FA describes conditions over large areas. We should not expect amendments for local or localized changes, which might be reflected in TWEB routes or FTs or their amendments. When wide-scale changes do occur, the FA will be amended. The outlook will be amended when a change from IFR to VFR or VFR to IFR is expected. A change in the wind outlook alone, however, is not sufficient for an amendment.

Like the AIRMET bulletin, FA amendments are indicated by AMD in the header (CHIC FA 041535 AMD) and UPDT in the body (ND...UPDT). The Chicago FA was amended at 1535Z; the state of North Dakota (ND) contains the changes.

A DAY ON THE FA DESK

Figure 5-3 shows the FA position at NAWAU in Kansas City. The following article by NAWAU meteorologist Paul Smith appeared in the FAA's June 1988 *Air Traffic Bulletin*.

"A typical day shift begins at 6:45 a.m. with a briefing from the midnight forecaster. We keep this as brief as possible, explaining current conditions and any expected trouble spots for the upcoming day.

"Three forecasters start the shift together, one (forecaster) for the east, central, and west. Each forecaster is responsible for two FAs: east has Boston and Miami, central has Chicago and Dallas-Fort Worth, and west has Salt Lake City and San Francisco. The three forecasters' work areas are adjacent to one another to allow easy coordination.

"Several things are routine on every shift. For instance, surface maps are analyzed every 2 hours to keep up with current weather conditions over the area. PIREP collectives 'alarm' (a system to automatically alert the forecaster to the availability of new information) at our consoles twice an hour and are displayed in text form and graphically on a computer-generated map. (This emphasizes the importance of PIREPs.) Surface and upper-level guidance material is received from NMC in Washington.

Fig. 5-3. *Meteorologists of the National Aviation Weather Advisory Unit in Kansas City prepare area forecasts for the contiguous United States.*

"Many other things are not routine and occur as weather conditions warrant. When SIGMETs are in effect or are being considered, coordination with the CWSUs occurs regularly. AIRMETs and amendments must be issued if forecast conditions go sour. When things of this type occur on a shift, the forecaster might become rushed to meet product deadlines.

"Product composition is accomplished on computer terminals. When composed and checked, our products are transmitted to WMSC (the switching center in Kansas City) for nationwide dissemination. While we do not have backup procedures in case of computer failure, delayed FAs can and do occur due to computer or communications failure.

"Each forecaster has a different routine in preparing a forecast. Usually, the meteorologist spends the first hour or so on shift analyzing the current weather situation and reviewing computer guidance concerning the evolution of weather systems over the following 6 to 24 hours. SIGMETs, AIRMETs, and forecast amendments have high priority and are issued as needed to keep briefers up to date.

"Approximately 3 hours prior to FA issuance, the forecasters begin work in earnest on the two FAs being written. Most forecasters draw up tentative outlines for their flight precaution areas, then consult with the person writing the adjacent forecast to develop a common set of VOR points to describe the entire area. Transmission times are staggered; the BOS and MIA FAs are transmitted first, followed an hour later by CHI/DFW, and finally SLC/SF0.

"The issue-time differences make early coordination a necessity; thus, for flight precaution areas extending from the Rockies to the East Coast, the west forecaster must make decisions very early on the aerial outline of expected weather conditions. IFR areas, because of their changeable nature, are often the last VOR outline to be 'nailed down.' Negotiation and point changes are made right up to our transmission deadline.

"The significant clouds-and-weather (now VFR clouds-and-weather) portion of the FA is usually the last section to be composed. The forecaster makes use of many separate sources of information to develop the FA including SAs, PIREPs, FTs, TWEBs, satellite imagery, radar, radiosonde data, prognostic charts, standard-level charts, and CSIS (the NWS's) interactive computer system. During periods of extensive adverse weather conditions with multiple flight precautions and frequent conversations with CWSU meteorologists, the FA forecaster is particularly busy as the transmission deadline approaches."

USING THE AREA FORECAST

Below is an example of a Boston AIRMET bulletin and Boston area forecast used for the following discussion. It might be helpful to refer to Fig. 5-4 to better visualize conditions.

```
BOSZ WA 301345
AIRMET ZULU UPDT 3 FOR ICG AND FRZLVL VALID UNTIL 302000

.

ME CSTL WTRS
LGT TO LCL MDT RIME ICGIC ABV FRZLVL BTW 080 AND 150 OVER ME AND
NEW ENG CSTL WTRS.
OTRW...NO SGFNT ICING EXPCD.

.

FRZLVL...50-80 ME. 80-100 NH VT MA RI CT NY LO LE OH. 100-110 NJ PA WV
MD DC DE VA.

BOST WA 301345
AIRMET TANGO UPDT 3 FOR TURBC VALID UNTIL 022000

.

AIRMET TURBC...NY PA OH LE LO WV MD
FROM MQT TO YOW TO BGM TO EKN TO LOZ TO ARG TO MCW TO MQT
MDT TO LCL SVR TURBC 200-400. CONDS DMSHG BY 20Z.

.

BOSS WA 301345
AIRMET SIERRA UPDT 3 FOR IFR AND MTN OBSCN VALID UNTIL 302000

.

AIRMET IFR...ME NH MA RI AND CSTL WTRS
FROM MLT TO YSJ TO 150SE ACK TO ORF TO PVD TO CON TO MLT
```

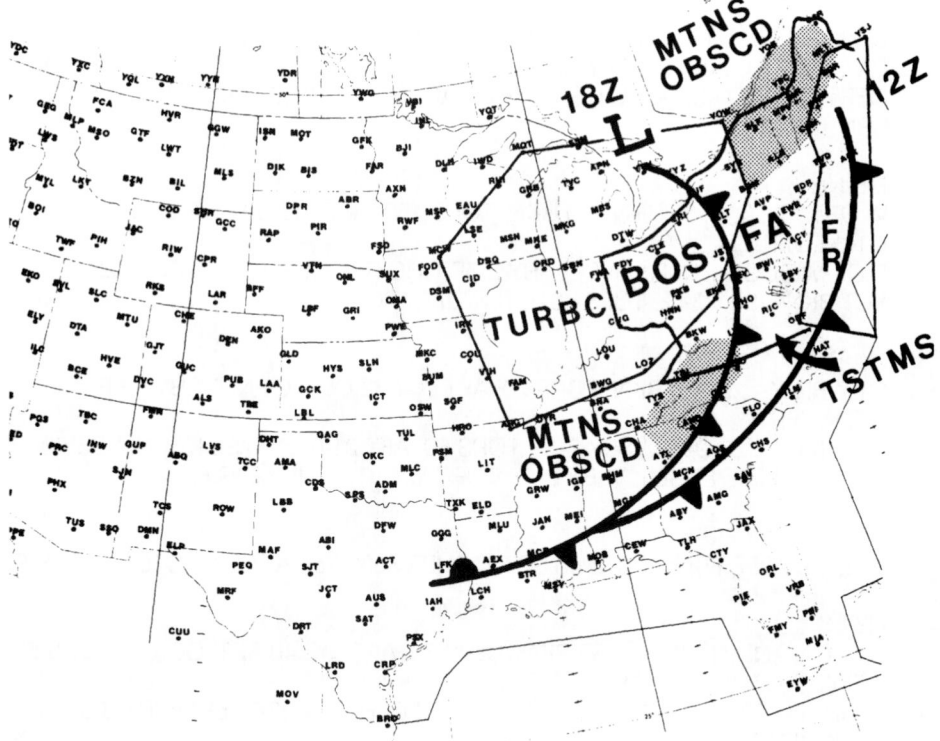

Fig. 5-4. *Plotting area forecasts, adverse conditions, and synopses are the next best things for DUAT and FSS users without access to weather graphics.*

OCNL CIG BLO 10 VSBY BLO 3F. CONDS CONTG NEW ENG AND CSTL WTRS
BYD 20Z THRU 02Z.

.
AIRMET MTN OBSCN...ME NH VT MA NY
FROM CAR TO CON TO 40SE ALB TO BGM TO YOW TO CAR
MTNS OSBCD IN CLDS. CONDS CONTG BYD 20Z IPVG BY 01Z.

.
OTLK VALID 2000-0200Z...MTN OBSCN WV VA
AFT 00Z MTNS OBSCD IN CLDS AND PCPN. CONTG THRU 02Z.

.
BOSC FA 301745
SYNOPSIS AND VFR CLDS WX
SYNOPSIS VALID UNTIL 011200
CLDS/WX VALID UNTIL 010600...OTLK VALID 010600-011200
ME NH VT MA RI CT NY LO NJ PA OH LE WV MD DC DE VA AND CSTL WTRS

.
SEE AIRMET SIERRA FOR IFR CONDS AND MTN OBSCN.
TSTMS IMPLY SVR OR GTR TURBC SVR ICG LLWS AND IFR CONDS.
NON MSL HGTS DENOTED BY AGL OR CIG.

.

SYNOPSIS...AT 18Z LOW WAS OVER LH WITH CDFNT CURVG THRU LO ERN PA CNTRL NC NRN AL BCMG STNRY TO SRN TX. BY 12Z CDFNT WILL MOVE OFF NRN AND MID ATLC CST CURVG ACRS SC NRN GA BCMG STNRY TO SRN TX.

.

ME NH
W OF MLT-CON...30 BKN 100 BKN. TOPS 150. AFT 01Z 50 SCT/BKN. OTLK...VFR.
E OF MLT-CON...10-20 OVC. VSBY 3-5F. CHC R-. TOPS 150. AFT 00Z 20 SCT/BKN. VSBY 3-5F. TOPS 40. OTLK...IFR CIG F.

.

MA RI CT
CT WRN MA...20-30 BKN. TOPS 60. AFT 00-03Z CLR. OCNL VSBY 3-5F. OTLK...IFR F.
RI ERN MA...20 OVC. VSBY 3-5F. TOPS 50. AFT 21-00Z 20 SCT/BKN. VSBY 3-5F. AFT 00-03Z 20 SCT. VSBY 3-5F. TOPS 40. OTLK...IFR CIG F.

.

VT
30 BKN. TOPS 80. AFT 01Z 40 SCT. OCNL VSBY 3-5F AFT 03Z. OTLK...MVFR F.

.

NY LO
LO WRN AND NERN NY...30 BKN. TOPS 60. AFT 01Z 40 SCT. OCNL VSBY 3-5F AFT 02Z. OTLK...MVFR F.
SERN NY...30 SCT/BKN. TOPS 60. AFT 22Z CLR. AFT 02Z CLR. VSBY 3-5F. OTLK...IFR CIG F.

.

NJ PA
40 SCT/BKN. TOPS 60. ISOLD TRW SRN NJ SERN PA TIL 00Z. CB TOPS 350. VSBY 3-5F. AFT 00Z CLR. VSBY 3-5F. OTLK...MVFR F.

.

OH LE
40 SCT. AFT 00Z CLR. CONL 100 SCT/BKN WRN AND SRN PTNS OH. TOPS 150. OTLK...VFR.

.

WV
NRN 2/3...50 SCT. AFT 00Z CLR. AFT 03Z 100 SCT/BKN. VSBY 3-5F. TOPS 150. OTLK...VFR BCMG MVFR CIG F BY 10Z.
SRN 1/3...50 SCT/BKN. AFT 00-03Z 30-40 BKN 100 BKN. VSBY 3-5F. WDLY SCT RW/TRW. TOPS 180. CB TOPS 350. OTLK...IFR CIG TRW F.

.

MD DC DE VA
MD DC DE NRN VA...40-50 SCT/BKN. WDLY SCT RW/TRW. OCNL 20 BKN/OVC CSTL SXNS. CB TOPS 350. OTLK...MVFR CIG F.
SRN VA...40-50 BKN. WDLY SCT TRW. OCNL 20 OVC WDLY SCT TRW ERN PTN WITH CHC SVR TSTMS. AFT 00-03Z 20-30 BKN. WDLY SCT RW/TRW. CB TOPS 450. OTLK...IFR CIG TRW F.

.

CSTL WTRS
N OF ACK...10-20 OVC. VSBY 3-5F. CHC R-. TOPS 150. AFT 00-03Z 10-20 OVC.
VSBY 3-5F. TOPS 40. OTLK...IFR CIG F.
S OF ACK...20 OVC. VSBY 3-5F. WDLY SCT RW/TRW. AFT 00-03Z 20-30 BKN.
TOPS 60. OTLK...MVFR CIG.

A pilot's first task is to determine the valid time of the forecast. This Boston FA was issued on the 30th day of the month at 1745Z, becomes effective at 1800Z, and is valid until the 1st at 0600Z. (Old forecasts occasionally remain in the system due to late revisions or computer trouble.) The pilot then determines if the product covers the proposed flight. Notice that the Boston FA includes Lake Erie (LE), Lake Ontario (LO), and the District of Columbia (DC). (The forecast area has been outlined in Fig. 5-4.) The synopsis paragraph follows the standard reference to AIRMET SIERRA, thunderstorm, and height statements. (To visualize conditions, the synopsis has been plotted in Fig. 5-4.) The main weather feature is a cold front forecast to move through the FA area. Notice the 18-hour valid time through the outlook period.

Refer to AIRMET ZULU. Light to locally moderate rime icing is forecast in clouds from 8,000 to 15,000 MSL over Maine and New England coastal waters. The forecaster does not expect extensive moderate icing. This is an example of icing that does not meet "AIRMET criteria"; otherwise, no significant icing is expected (this does not rule out trace icing). The freezing level paragraph indicates that the freezing level is expected to slope from around 5,000 feet in northern Maine to 8,000 feet in the south. This condition would be a definite consideration for aircraft without ice protection equipment. Freezing levels are always included.

Refer to AIRMET TANGO. Moderate to locally severe turbulence is forecast between 20,000 and 40,000 feet within the delineated area (Fig. 5-4). Notice that only portions of New York and Pennsylvania are affected, with turbulence expected to diminish by 2000Z. Because severe turbulence will be localized, a SIGMET has not been issued. If a SIGMET is in effect, a statement following the paragraph will refer to the advisory: SEE SIGMET OSCAR 1 FOR SVR TURBC. Is this the only area where significant turbulence can be expected? No, remember that thunderstorms imply moderate or greater turbulence and low-level wind shear, which are not addressed separately.

Refer to AIRMET SIERRA. Occasional ceiling below 1,000 and visibilities below 3 miles in fog are forecast. (The outlined section in Fig. 5-4 labeled IFR over New England represents this area.) Conditions will continue beyond 2000Z through 0200Z. (The gray shaded areas of Fig. 5-4 depict mountain obscurement.) The mountains of New York and New England are expected to be obscured in clouds; conditions are expected to improve by 0100Z. A second mountain obscuration paragraph provides details on conditions in West Virginia and Virginia. Mountains are expected to become obscured in clouds and precipitation after 0100Z and continue through 0200Z, which is the end of the outlook period. This is an example of a condition developing during the period.

Generally, advisories for IFR and mountain obscurement are synonymous. They represent low ceilings and visibilities that will prevent VFR flight within all or part of the affected area. Mountain obscurement refers to mountainous areas, and IFR refers to flat lands. Pilots, especially those who are accustomed to flying over flat terrain, must use caution with a mountain obscurement forecast. Weather reports from valley stations might indicate good VFR flying conditions, but VFR flights through passes and over the mountains might be impossible. Again, it's imperative for the pilot to have a sound knowledge of terrain.

When the forecast indicates the phenomena will be "occasional," for example the IFR paragraph in the Boston AIRMET bulletin, the probability of occurrence is greater than 50 percent, but the occurrence is expected to last for less than half the forecast period. A forecast for "occasional" should not in itself warrant the cancellation of VFR; however, all available information must be considered and caution exercised with suitable alternates available. The absence of a conditional term indicates the phenomena will be widespread, for example the BOS AIRMET SIERRA mountain obscurement paragraphs. This means a greater than 50 percent probability of occurring for more than half the forecast period. VFR flight probably will not be possible, but there could be areas where the phenomenon does not exist. This is not an error, but a limitation of the forecast product.

VFR clouds-and-weather paragraphs describe expected weather for the forecast area. Areas are usually described using the geographical area designators in appendix D. In Maine and New Hampshire west of a line from MLT to CON (the mountainous area), 30 BKN 100 BKN, TOPS 150 are expected from 1800Z until after 0100Z (heights are all above mean sea level). The mountain peaks in the 3,500- to 5,000-foot range will be obscured as advertised in AIRMET SIERRA; icing in clouds can be expected where minimum en route altitudes are above 5,000. After 0100Z, conditions are forecast to improve to 5,000 scattered to broken, which is consistent with AIRMET SIERRA. General visibilities through the period will be 5 miles or greater, with no significant precipitation. The outlook is VFR, but remember its definition.

East of the line from MLT to CON (the coastal plain), 10-20 OVC, VSBY 3-5F, CHC R-, with TOPS 150 (MSL), with occasional IFR conditions (AIRMET SIERRA). The forecaster expects conditions to begin to improve, but not until after 0000Z. VFR will be iffy. MEAs are lower, and ice should not be a factor for flights below 5,000. After 0000Z, conditions will improve somewhat, but remain marginal-to-IFR. Tops will lower significantly in the coastal areas. The outlook indicates that the ceilings below 1,000 and the visibilities below 3 miles in fog will persist.

Refer to the New Jersey and Pennsylvania paragraph. Notice that isolated thunderstorms and rain showers are forecast for southern New Jersey and southeastern Pennsylvania. Because the thunderstorms are expected to be isolated, circumnavigation should be possible; we wouldn't want to be poking around in clouds without storm detection equipment. Under the circumstances, climbing above the lower layer where visual separation from convective activity can be maintained would be a prudent procedure. What about turbulence and icing? They're implied with any forecast of thunderstorms. The outlook indicates that visibilities will be between 3 and 5 miles in fog. The ceiling? Well, greater than 3,000 feet.

Refer to the Maryland, District of Columbia, Delaware, and Virginia paragraph. Notice for southern Virginia that the forecast indicates OCNL 20 OVC WDLY SCT TRW ERN PTN WITH CHC SVR TSTMS. Over the eastern portion of southern Virginia, a chance of severe thunderstorms is expected. The outlook indicates ceiling less than 1,000 feet, visibilities less than 3 in fog, rain showers, and thunderstorms.

The final paragraph forecasts conditions for the coastal waters. The area is divided into north and south of Nantucket, Massachusetts.

This FA describes a distinct weather system. Notice how the AIRMET bulletin, synopsis, and VFR clouds-and-weather are all tied together. FAs will not always be this detailed or consistent, but the example has provided most variables used in this product.

Both the area forecast and AIRMET bulletin must be carefully reviewed to determine conditions en route.

THE HIGH-ALTITUDE SIGNIFICANT WEATHER PROG

The high-altitude significant weather prognosis provides a pictorial view of forecast weather from 400 to 70 millibars (approximately 24,000 through 63,000 feet). All heights are pressure altitudes (flight levels). The prog is manually prepared at the National Meteorological Center. The chart depicts cumulonimbus clouds, tropical cyclones, severe squall lines, moderate or severe turbulence, widespread sandstorms or dust storms, surface fronts, tropopause height, and jet streams. Charts are valid at 0000Z, 0600, 1200Z, and 1800Z.

Scalloped lines enclose areas of forecast cumulonimbus development divided into three categories:

- Isolated (ISOL) less than ⅛ coverage
- Occasional (OCNL) ⅛ to 4/8 coverage
- Frequent (FRQ) 5/8 to 8/8 coverage

Cumulonimbus clouds imply hail and moderate or greater turbulence and icing. Bases and tops are shown by numerical figures below and above a short horizontal line. Bases below 24,000 feet are indicated by XXX below the line. For example, along the southwestern coast of Mexico in Fig. 5-5, the forecast indicates an area of isolated (less than ⅛ coverage), embedded cumulonimbus with tops to 40,000 feet, and bases below 24,000 feet. Clear air turbulence that is not associated with cumulonimbus clouds is indicated by a dashed line containing a turbulence intensity symbol and forecast height. Figure 5-5 shows that over the central United States, moderate turbulence is expected between 25,000 and 37,000 feet. Off New England, severe turbulence is forecast between 25,000 and 34,000 feet.

The expected location of jet streams is indicated by long black lines. Figure 5-5 shows the jet stream through the central United States at 33,000 feet (FL 330), with a core speed of 100 knots indicated by two flags; each flag represents 50 knots, and each

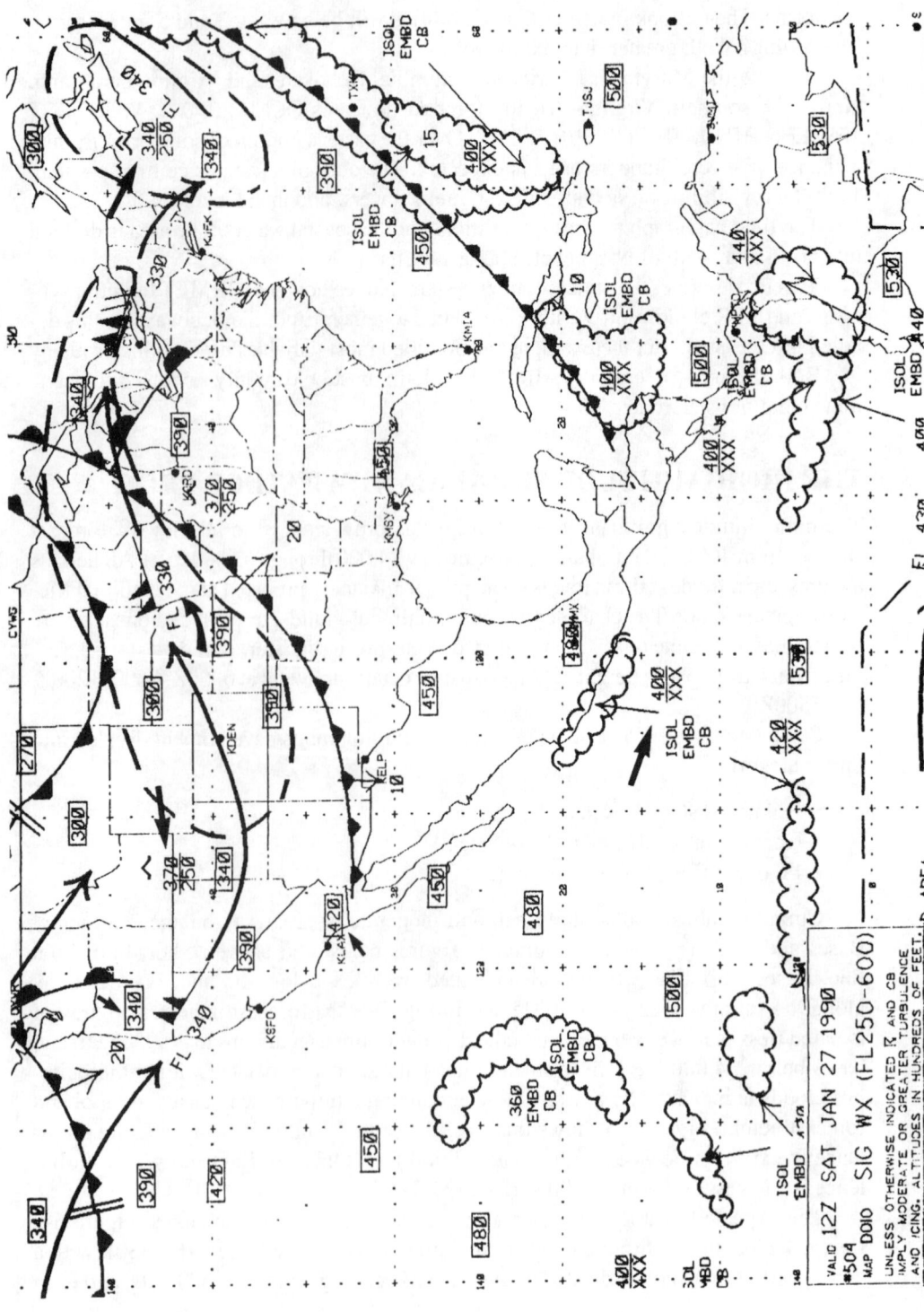

Fig. 5-5. *The high-altitude significant weather prog pictorially displays weather above 24,000 feet. Because the chart is not amended, the area forecast and weather advisories must be consulted to complete the forecast weather picture.*

barb represents 10 knots. The chart also contains the expected height of the tropopause, which you'll recall is the boundary between the troposphere and stratosphere. The height of the tropopause over the West Coast is indicated by numerals enclosed in a box and forecast around 42,000 feet in Southern California, lowering to 34,000 feet in Oregon (Fig. 5-5). A five-sided polygon indicates areas of high or low tropopause heights.

Significant frontal boundaries are depicted using standard symbols contained in Table 10-2; movement is indicated by an arrow and speed is printed by the arrowhead. For example, the cold front in the central United States (Fig. 5-5) is expected to move to the southeast at 20 knots.

The names of tropical cyclones, when relevant, are entered next to the symbol. Severe squall lines are depicted within areas of CB by the symbology - V - V - V. Widespread sandstorms or dust storms are enclosed by scalloped lines and labeled with the appropriate symbol contained in Table 10-1 and vertical extent.

Volcanic activity is indicated by a trapezoidal figure depicting an eruption. The legend of the chart will include the eruption symbol and any pertinent information, such as the volcano's name, latitude and longitude, date and time of the first eruption, and a reminder to check SIGMETs.

USING THE HIGH-ALTITUDE SIGNIFICANT WEATHER PROG

The chart is often an excellent place to begin planning a high-altitude flight (above FL 240). For example, a pilot flying from San Francisco to New York might plan to follow the jet stream over the Great Lakes to take advantage of tailwinds. If turbulence were a significant factor, for example on a medical lifeguard flight, the pilot might plan a route through the Texas panhandle and then to New York to avoid wind shear turbulence. A flight from east to west might also select this route to avoid maximum jet stream headwinds.

The chart can also be used to determine areas of cumulonimbus activity, cloud tops, and tropopause heights. If the lack of storm detection equipment or aircraft performance does not allow the pilot to climb above thunderstorms, he or she might wish to avoid areas where coverage is expected to be occasional or frequent. The pilot should note the expected height of the tropopause; if thunderstorms develop, the isothermal layer of the tropopause caps all except severe activity.

Pilots, like FSS briefers, can use the chart for the en route portion of a high-altitude weather briefing, but pilots as well as briefers must not substitute the chart for the FA and weather advisories. The chart is not amended, and severe weather might develop that is not depicted. The FA is also needed to provide weather details for climbout and descent.

With the pilot becoming more and more responsible for interpretation, as well as weather acquisition, he or she will need both the in-flight advisory plotting chart and geographical area designator maps to determine affected areas. We've seen the ease of visualizing conditions using the weather log. When any item is not completely understood, a flight service station specialist should be consulted.

Forecasters do make mistakes. This usually takes the form of omitted paragraphs or forecast areas, or data from the previous FA that is inadvertently repeated. Forecasters occasionally use incorrect contractions. The DUAT-decode function made the synopsis of one FA read "upper-level Lake Ontario over northern Baja California." The forecaster used the contraction "LO" for LOW, and LO decodes "Lake Ontario." With the NWS virtually out of the pilot briefing business, a pilot's only access to the forecaster will be through an FSS.

Weather advisories automatically amend the FA. And pilots must understand that the FA will only be amended when a significant change occurs. With few exceptions, hazardous weather must be extensive to warrant an advisory or amendment.

FAs are issued at 6- to 8-hour intervals, and pilots should become familiar with these times. Because forecasts are only valid for 12 hours, pilots checking with flight service or using DUATS just prior to the next issuance will only have 5 to 6 hours of valid forecast, excluding the outlook. Pilots interested in conditions for the following day should call for or request the forecast after 9 p.m. This will usually ensure receipt of the latest information.

There's no question that the FA has improved. The NWS's decision to consolidate in Kansas City and the commitment of NAWAU forecasters has resulted in standardization and consistency. The FA is a valuable and useful tool as long as we understand its purpose, format, conditional terms, and limitations.

Recall from chapter 1: Never rely on a single piece of data. The FA is only one product, one piece of the weather picture. Localized conditions might not be included. Flight decisions should never be based solely upon this or any other product. The FA must be used along with all available information. Forecast limitations cannot be overemphasized along with using all available sources.

> Whatever may be the progress of the sciences, never will observers who are trustworthy and careful of their reputations venture to forecast the state of the weather.
> Dominique Argo, French astronomer and physicist (1786–1853)

Dick Williams obviously hasn't heard this, as you might recall he puts his name on the forecast!

6
TWEB forecasts

TWEB AVIATION FORECASTS WERE DEVELOPED AS SCRIPTS FOR THE FAA's transcribed weather broadcast (TWEB) and pilots automatic telephone weather answering service (PATWAS). TWEB may also be used on the telephone information briefing service (TIBS), available at automated flight service stations (AFSSs).

Forecasts are available by phone from PATWAS and TIBS and broadcast continuously over selected low- and medium-frequency radio beacons. Telephone numbers can be found in the telephone numbers section of the *Airport/Facility Directory*; frequencies are most easily found on aeronautical charts and are also available in the A/FD, although not separately listed. (Refer to chapter 11 regarding obtaining information.) TWEB route forecasts may also be used during preflight, in-flight, and flight watch briefings. (TWEB route numbers and locations are contained in appendix B.)

TWEB forecasts were revised in February 1988, and issuance and valid times were standardized. TWEBs are issued three times a day and valid for 15 hours. Time references transmitted with the product remain UTC, but they are issued for local times in all time zones and are normally available just prior to valid time: 5 a.m.–8 p.m. (morning); 2 p.m.–5 a.m. (afternoon); and 9 p.m.–noon (evening).

403 TWEB 200419 RNO-BOI. ALL HGTS MSL XCP CIGS. GENLY CLR.

TWEB Route Number 403 forecasts conditions from Reno, NV, to Boise, ID (RNO-BOI). The forecast for the evening issuance (9 p.m.–noon) was valid on the 20th day of the month from 04Z until 19Z the following day (200419). TWEB routes describe a corridor 50 nm wide, which is 25 miles either side of the route. The 403-route forecast covers conditions 25 miles either side of a line from Reno to Boise. It is important not to extrapolate the forecast beyond its designated area.

The loss of reporting stations has caused revisions to the TWEB network resulting in the issuance of certain routes on a part-time basis.

> 426 TWEB 250520 TSP MTNS SOLEDAD-CAJON-BNG PASSES AND ADJ MTNS. DLAD.

The evening issuance of Route 426 for the Tehachapi Mountains, Soledad, Cajon, Banning Passes, and adjacent mountains will be delayed due to lack of observations. Additionally, amendments might not be available for selected routes as noted by a specific remark: NO AMDTS AFT 04Z. Significant changes can occur without an amendment, so this serves the same purpose as NO SPL on SAs. It's a warning that caution must be exercised, especially during marginal and deteriorating conditions.

Pilots, FSS specialists, and forecasters have been instrumental in revising TWEB routes. For example, the 420 route at one time forecast conditions from Concord to Arcata to Crescent City, CA; however, with the loss of 24-hour observations at Crescent City and only part-time observations at Concord, the forecast was suspended during the night. By changing the end points on the route to Oakland and Arcata with 24-hour observations, the route once more became available full-time. Other routes in the area have also been revised as a direct result of pilot input. This is an excellent example of how pilots, briefers, and forecasters can improve the system by working together.

SYNOPSIS

In addition to route forecasts, NWS offices responsible for TWEBs prepare a synopsis. The synopsis contains a brief description of fronts, pressure systems, and local climate or terrain factors affecting the routes, valid for the same period as the route forecasts. The TWEB synopsis often contains more detail than the FA and might be of more value describing local conditions. More detail is possible because the TWEB synopsis only covers about one-fifth the area of the FA. (Refer to appendix B for TWEB synopsis locations.)

> LAX SYNS 191203 UPR RDG ALG W CST WITH NLY FLO ALF. SFC HI CONTS TO BLD OVER GT BASIN FOR CONTG SANTA ANA CONDS.

This Los Angeles synopsis covers the routes issued by the Los Angeles WSFO. An upper-level ridge is along the West Coast with a northerly flow aloft. Surface high-pressure continues to build over the Great Basin, which is the high plateau generally including Nevada, Utah, and southern Idaho, for continuing Santa Ana conditions, which are the strong, sometimes warm, northeasterly winds that occur in Southern California.

The detail of a TWEB synopsis depends on the weather pattern.

ATL SYNS 301809 STNRY FNT NR MYR-ABY-PAM. HI PRES CNTRD OH DRFTG SEWD.

This Atlanta synopsis describes a stationary front from Myrtle Beach, SC (MYR), to Albany, GA (ABY), to Panama City, FL (PAM). High pressure that is centered over Ohio is drifting southeastward.

A more complex weather system is reflected in a longer synopsis.

IND SYNS 041809 CD FNT XTNDS FRM NRN LK ERIE THRU CNTRL OH INTO NRN AL. WK TROF EXTNDS FRM SFC LO SW LWR MI THRU W CNTRL IN. UPPR LVL LOW OVER NRN IL WL BRNG SCT TRW MAINLY OVR WRN AND NRN RTES THIS AFTN AND EVE. PRZYBYLINSKI

In this Indianapolis synopsis, a cold front extends from northern Lake Erie through central Ohio into northern Alabama. A weak trough extends from a surface low in southwest Lower Michigan through west central Indiana. The upper-level low over northern Illinois will bring scattered rain showers and thunderstorms mainly over the western and northern routes this afternoon and evening. The synopsis was prepared by Przybylinski.

Pilots using DUATs will have to contact an FSS for clarification of contractions or phrases that they do not understand. One evening I had a pilot call the FSS with such a question. He didn't understand the last word of the synopsis. I explained that it was the name of the meteorologist who wrote the forecast.

Earlier it was mentioned how personal comments sometimes get into weather products.

SFO SYNS 091203 SFC HI PRES OFSHR WITH LWR PRES OVER INTR CA. THIS IS MY LAST SET OF TWEBS.....EVER. SHORTY THOMAS.

Then every once in a while something gets transmitted that shouldn't have.

SFO SYNS 032121 THE QUICK BROWN FOX JUMPED OVER THE LAZY DOG BECAUSE THERE WAS NO WEATHER OF ANY MERIT TO TALK ABOUT.

I'll leave the names off of this one.

SIGNIFICANT CLOUDS AND WEATHER

TWEBs use the same terminology and contractions as the FA. Forecasts contain significant clouds and weather, with "significant" defined as phenomena affecting at least 10 percent of a route or phenomena that are important to flight planning; the latter is subject to the forecaster's judgment. Forecasts consist of sky condition, visibility, weather including mountain obscurement, strong surface winds, and nonconvective low-level wind shear. Included cloud types might be cumulus, stratocumulus, and cirrus.

Cloud heights are based on a standard reference: ALL HGTS MSL XCP CIGS (all heights mean sea level except ceilings) or ALL HGTS AGL XCP TOPS (all heights above ground level except tops). The use of AGL and CIGS is normally limited to layers within 4,000 feet of the surface. TWEBs were basically a low-altitude forecast: below 18,000 feet. That no longer necessarily applies, but forecasters emphasize cloud conditions below 12,000 feet. With no clouds or scattered clouds above approximately 12,000 feet, the forecaster might use the phrase NO SGFNT CLDS/WX (no significant clouds or weather); any clouds present should be easily circumnavigable.

Mountain obscurement has been added (MTNS OBSCD ABV, MTN RDGS OBSCD, or ALL PASSES OBSCD), and geographical features might be specified: (TSP MTNS (Tehachapi Mountains), CSTL MTNS). Because of the local nature of the forecast, not all locations appear on the geographical area designators map. Pilots unfamiliar with locations on TWEB forecasts will have to consult an FSS.

When forecast, tops will normally only be included for layers with bases below 12,000 feet. Like the FA, for multiple or merging layers that would not permit VFR flight between layers, only the top of the highest layer appears; however, because of the local nature of the product, tops are often more specific and useful. Cloud tops might also be indirectly specified: CIG BLO 10 OBSCG TRRN BLO 15 (cloud tops are expected to be 1,500 feet).

Surface visibilities are forecast when expected to be 6 miles or less; however, to prevent misunderstanding, 6+ might be used to indicate unrestricted visibility. Weather and obstructions to vision appear as necessary.

Sustained surface winds are normally forecast when expected to be 25 knots or greater. Notice the TWEB threshold for winds is 5 knots less than the FA. Direction is referenced to true north and given as cardinal headings or in degrees: respectively, LO LVL WNDS W TO SWLY LCLY TO 25 KT. or 30–3415–25 (the latter translates to "wind 300° to 340° at 15 to 25 knots"). Gusts might also appear: WNDS G40. The lack of a wind forecast only implies sustained speeds less than 25 knots over 90 percent of the forecast area.

TWEB conditional terms describe variability; refer to Table 6-1, which lists the conditional terms used on TWEB forecasts. Conditional terms have been intentionally changed to coincide with the FA. Also used in briefings, it is important to note interchangeable terms:

- "Widely scattered" and "locally"
- "Scattered" and "areas"
- "Numerous" and "widespread"

These conditional terms can lead to confusion.

414 TWEB 312213 RDD-SAC-FAT-BFL...AFT 02Z-05Z BCMG 30-50 SCT SCT 30 BKN RW- SRN SAN JOAQUIN VLY....

This would seem to say that after the period 02Z to 05Z conditions will become 3,000 to 5,000 feet scattered, scattered 3,000 feet broken with light rain showers in the southern San Joaquin Valley. The forecaster's intent is to describe conditions becoming 3,000 to 5,000 feet scattered, "areas" (25 percent to 54 percent of the route; the condi-

Table 6-1. Conditional terms used on TWEB forecasts.

Term	Contraction	Definition
Isolated	ISOLD	Single cells or localized conditions (no percentage). Implies circumnavigable.
Widely scattered or local	WDLY SCT LCL	Less than 25 percent of area/route affected.
Scattered or areas	SCT AREA	25 percent to 54 percent of area/route affected.
Numerous or widespread	NMRS WDSPRD	More than 54 percent of area/route affected.

tional term scattered) 3,000 feet broken with light rain showers. In this case, the forecaster could have written BCMG 30-50 SCT AREAS 30 BKN RW- to prevent any confusion. In spite of this example, forecasters do attempt to avoid ambiguity when preparing forecasts.

As an FSS supervisor, I counsel specialists to keep briefings clear and concise. One area of excess verbiage is "high thin cirroform clouds"; all cirroform clouds are high. But leave it to the National Weather Service when sure enough one day I find 426 TWEB...HI THIN CIFM CLDS. What's a parent to do?

The length and detail of a TWEB forecast varies widely. Beyond examples of one-liners, this is an example of the other extreme:

> 418 TWEB 250520 OAK-SAC-RNO. ALL HGTS MSL EXP CIGS. VCNTY OAK 40-60 SCT. PTCHY FOG FRMG WITH AREAS CIG/VSBY BLO 10/3F. PCPN SPRDG TO AREA AFT 18Z WITH CIGS 10-20 AND SCT CIG/VSBY BLO 10/3R-F. MTNS BCMG OBSCD. SAC VLY CIG 30-50 VSBY 3- 5F WITH SCT CIG/VSBY BLO 10/3F. FOG BCMG MORE XTSV AFT ABT 08Z WITH WDSPRED CIG/VSBY BLO 10/3F. PCPN SPRDG TO AREA AFT 18Z WITH CIGS 10-20 AND SCT CIG/VSBY BLO 10/3R-F. OVR SIERNEV TO RNO 100-200 SCT-BKN WITH SCT SW-. LATE IN PD PCPN SPRDG ACRS MTNS WITH SCT CIG/VSBY BLO 10/3R-/S-F. MTNS OCNL OBSCD BCMG OBSCD.

Well, try getting *that* and a half-dozen more like it on a 3-minute tape for the transcribed weather broadcast and you get some idea why FSS specialists speak so fast. This example does show, however, how detailed a TWEB can be and how it can relate to a geographical area. It begins with conditions in the vicinity of Oakland, through the Sacramento Valley, and then over the Sierra Nevada Mountains to Reno.

LOCAL VICINITY FORECASTS

Local vicinity forecasts have been developed to cover metropolitan environs. The forecasts normally cover a radius of 50 nm. (Locations are listed in appendix B.)

> 358 TWEB 200419 SEA-PGTSND LCL VCNTY. ALL HGTS MSL XCP CIG.20 SCT 30-50 SCT-BKN...AFT 10Z CONDS BCMG 15 SCT-BKN 20-30 BKN...AFT 17Z CONDS BCMG 30 SCT.

This local vicinity forecast covers the Seattle and Puget Sound area. The forecast indicates scattered to broken clouds deteriorating after 1000Z and then improving after 1700Z.

Exceptions to the 50-mile radius occur in Southern California where large, irregular, and homogeneous geographical regions exist (Fig. 6-1). The 431 TWEB covers the Los Angeles Basin, from the San Fernando Valley along the San Gabriel and San Bernardino Mountains to Hemet, and back to Santa Ana. The 426 route forecasts conditions over the Tehachapi Mountains and mountains and passes north and east of Los Angeles, including a specific forecast for the Soledad, Cajon, and Banning passes. The 429 route for the Southern California deserts, south of a Palm Springs-Needles line, is illustrated by the shaded portion of the inset in Fig. 6-1.

Compare the 431 TWEB for the Los Angeles Basin route with the San Francisco area forecast for the same time and location.

> 431 TWEB 071203 LAX BASIN. ALL HGTS MSL XCP CIGS. 20-50BKN-SCT. ISOLD SHWRS. CIGS LCLLY AOB 10...VSBY LCLLY AOB 3HF. HIR INLD MTNS OBSCD. AFT 18Z INCRG SHWRS WITH LCL EMBDD TSTMS.
> CA
> CST...15-25 SCT-BKN 50-70 SCT-BKN 140-160 OVC 180 WITH SCT VSBYS 3-5RW- AND WDLY SCT TRW-. CB TOPS TO 300.

Sky conditions are comparable. The TWEB includes local ceilings and visibilities below 1,000 feet and 3 miles; these conditions are too localized for inclusion in the FA. The TWEB predicts increasing showers and thunderstorms after 18Z; the FA includes the same forecast, but without the time reference. The TWEB indicates mountains obscured; the AIRMET bulletin, which is not shown, stated mountains occasionally obscured. This is consistent, given that the FA and AIRMET bulletin must cover conditions for all of California's coastal sections and the TWEB covers only the Los Angeles Basin. The FA might contain more detail on tops, as in this case stating "tops 18,000 feet with CBs to 30,000 feet." Both forecasts are consistent within the scope of each product.

AMENDMENT CRITERIA

Table 6-2 contains TWEB amendment criteria. An amendment will be issued when phenomena occur, are expected to develop, or are no longer anticipated. Amendments are also required for thunderstorms and low-level wind shear. Routes will also normally be amended when trends indicate, in the forecaster's judgment, that the forecast will be substantially in error or unrepresentative. FSS specialists make inquiries when discrepancies develop or significant differences occur between FA and TWEB forecasts. Amendments can originate in this manner.

Forecasters consider issuance times in the decision to amend. Because it takes time to write and distribute an amendment, if the next issuance is in less than 2 hours, the forecaster might elect to delay and issue the new forecast, rather than amend. As can be seen from Table 6-2, amendments are directed at low altitude changes. If these criteria are not met, the forecast is considered accurate and will not be amended.

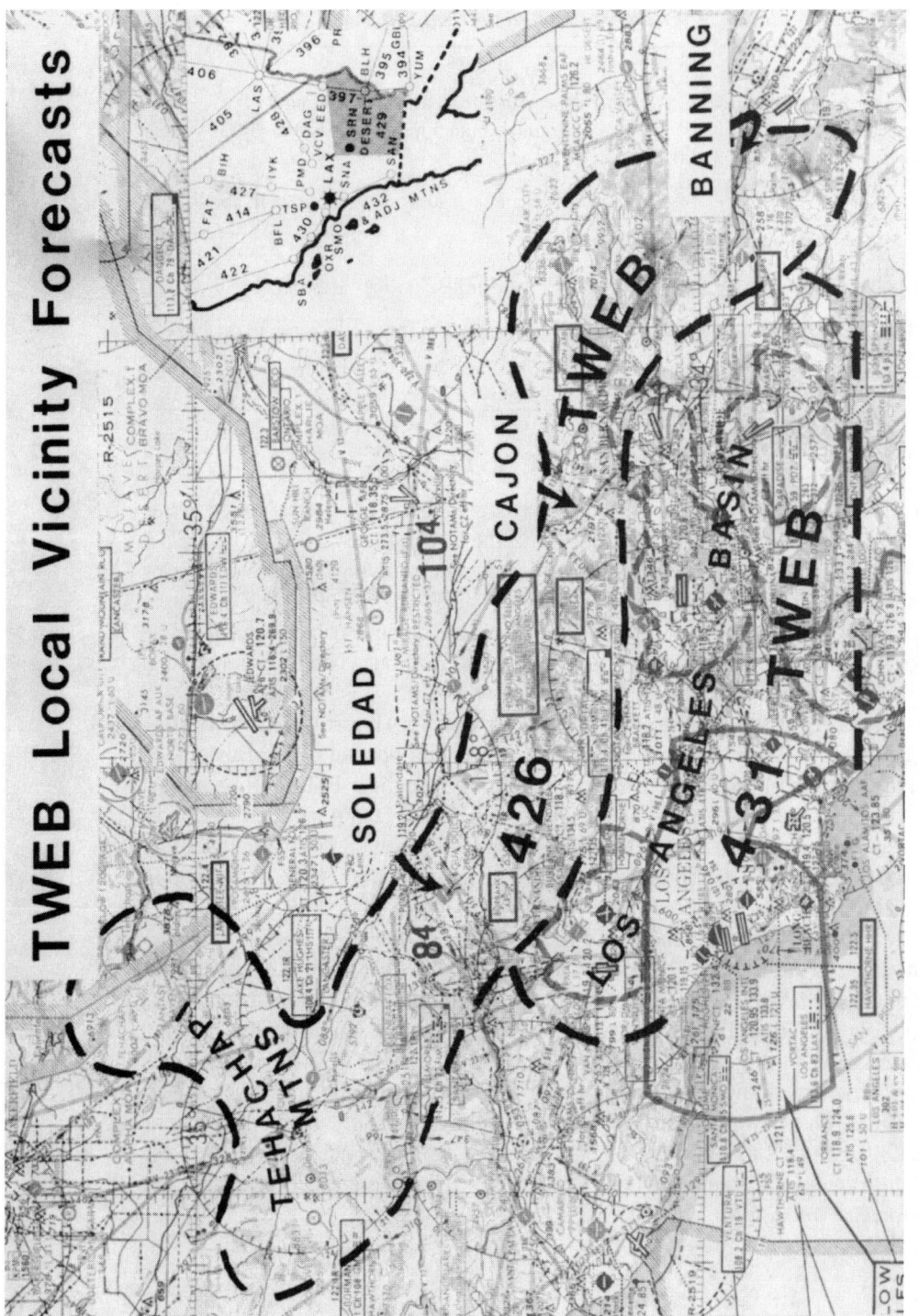

Fig. 6-1. *TWEB local vicinity forecasts cover a 50-mile radius of large metropolitan areas, except in Southern California where large, irregular, homogeneous geographical areas exist. (Background chart is only intended to help illustrate the boundaries of forecast areas and is not intended to illustrate aeronautical-chart symbology.)*

Table 6-2. TWEB forecast amendment criteria.

Ceiling with a forecast of:	amend if:	Visibility with a forecast of:	amend if:
ceiling more than 3,000	less than 3,000*	more than 5 miles	less than 5*
		5 miles or less	increase to above 5
3,000 or less	increase to or above 3,000		
3,000 to 1,000	less than 1,000	3 to 5 miles	less than 3
less than 1,000	increase to or above 1,000	less than 3 miles	increase to or above 3

* Or, in the forecaster's judgment is not representative.

USING THE TWEB ROUTE FORECAST

Figure 6-2 compares and contrasts the area forecast with a TWEB route. Notice in the NERN OH portion of the 072 TWEB that AFT 16Z the forecast becomes 30 OVC VSBY 6+. VSBY 6+ is included to prevent confusion or misunderstanding. The previous portion of the forecast includes VSBY 3-5RF...AREAS...VSBY BLO 3R-F.... The inclusion of 6+ indicates that after 16Z, visibilities are expected to be unrestricted. The absence of 6+ would be interpreted as ceiling improving after 16Z, but visibilities remaining predominately 3 to 5 miles, with some locales below 3 miles.

Let's plan a flight from Pittsburgh's Allegheny County Airport (AGC) to Harrisburg's Capital City Airport (CXY) using the Pennsylvania portion of the Boston FA and the Pittsburgh to Harrisburg portion of the 072 TWEB CLE-PIT-HAR. A pilot's first task is to determine the valid time of the forecast. The BOS FA significant clouds and weather section was issued on the 29th day of the month at 0845Z, valid at 0900Z until 2100Z. The TWEB, also issued on the 29th, is valid from 0900Z until 2400Z. (Only used to indicate the *end* of a time period, 2400Z might be written as 2359Z. Almost always, when 2400Z is intended, it will be written as 0000Z.)

With valid time nailed down, a pilot's next job is to ensure that the forecast covers the intended route. The FA divides the area into northwest and southeast of a Johnstown, PA, to Saranac Lake, NY, line (JST-SLK); the line is drawn on the excerpt from the in-flight advisory plotting chart (Fig. 6-2). The PIT-HAR portion of the 072 TWEB is divided into two areas: (1) southwestern Pennsylvania west of the ridges and (2) the ridges themselves and east of the ridges. Notice from the excerpts of the geographical area designators map and TWEB routes map in Fig. 6-2 that a JST-SLK line coincides with the west slopes of the Allegheny Mountains. The FA and TWEB cover the same geographical area.

The forecast in Boston's AIRMET SIERRA:

OCNL CIG BLO 10 OVC AND VSBY BLO 3RW-F. MTNS OCNLY OBSCD.
BOSC FA
NW JST-SLN LN

BOSTON AIRMET SIERRA FORECASTS OCNL CIG BLO 10 OVC AND VSBY
BLO 3RW-F. MTNS OCNLY OBSCD.

BOSC FA 290845
CLDS/WX VALID UNTIL 292100...

NY LO NJ PA
NW OF JST-SLK LN,.,20-30 SCT-BKN 40-60 OVC 120 RW-.
18Z...30-50 BKN-OVC 100 BKN-OVC 120.
SE OF JST-SLK LN,.,15-25 SCT-BKN 40-60 BKN-OVC 150. WDLY
SCT EMBDD TRW. TSTMS BCMG SCT AND PSBLY SVR AFT 18Z. CB
TOPS ABV 450.

072 TWEB 290924 CLE-PIT-HAR. ALL HGTS AGL XCP TOPS. NERN
OH...CIGS 20-40 OVC VSBY 3-5R-F...AREAS CIGS 12 OVC VSBY BLO 3R-
F...WDLY SCT CIGS BLO 10 OVC. AFT 16Z CIGS 30 OVC VSBY 6+...WDLY
SCT R-. AFT 19Z 45 SCT. SMRN PA W OF RDGS...CIGS BLO 10 OVC VSBY
BLO 3RW/TRW F... AFT 14Z CIGS 20 BKN VSBY 5RW-. AREAS 45 BKN
VSBY BLO 3RWF...WDLY SCT TRW. RDGS AND E...45 OVC VSBY 5R-.
AREAS 12 OVC VSBY BLO 3RF. AFT 14Z CIGS 10-20 OVC...AREAS BLO
10 OVC...WDLY SCT TRW. RDGS OBSCD.

Fig. 6-2. TWEB forecasts often contain more detail than is possible in the FA, although they are usually perfectly consistent within the scope of each product.

20-30 SCT-BKN 40-60 OVC 120 RW-. 18Z...30-50 BKN-OVC 100 BKN-OVC 120.
072 TWEB
SWRN PA W OF RDGS
CIGS BLO 10 OVC VSBY FLO 3RW/TRW F...
AFT 14Z CIGS 20 BKN VSBY 5RW-...AREAS 45 BKN VSBY BLO 3RWF...WDLY
SCT TRW.

Ceilings below 1,000 feet and visibilities below 3 miles are expected to prevail over the TWEB route during the beginning of the forecast period, whereas the FA only forecasts occasional IFR. VFR flight will be doubtful based on the TWEB forecast. The TWEB expects improvement after 1400Z, the FA around 1800Z; therefore, a VFR pilot can expect a delay, although VFR flight should be possible after 1400Z to 1800Z, probably with circumnavigation of scattered areas of lingering IFR conditions. The FA is of more value to the IFR pilot because it indicates layered clouds with tops to 12,000 feet MSL; however, the widely scattered thunderstorms predicted in the TWEB must be considered. For an aircraft without storm detection equipment, the only option is to fly between layers or on top to visually avoid convective activity.

After 1400Z, ceiling 2,000 broken, visibilities of 5 miles in light rain showers are expected to prevail over the route according to the TWEB; however, one-quarter to one-half of the route (AREAS) is forecasted to have 4,500 feet broken AGL, visibilities below 3 miles in rain showers and fog. The higher clouds and lower visibilities would not necessarily coincide. Given that terrain elevations are around 1,000 feet, FA and TWEB cloud heights are perfectly consistent. This illustrates the importance of determining whether forecast heights are MSL or AGL. Finally, less than one-quarter of the route (WDLY SCT) is expected to be affected by rain showers and thunderstorms. The TWEB forecaster covers the possibility of thunderstorms, whereas the FA forecaster does not; this is a matter of professional judgment.

BOSC FA
SE OF JST-SLK LN
15-25 SCT-BKN 40-60 BKN-OVC 150 RW-. WDLY SCT EMBDD TRW. TSTMS
BCMG SCT AND PSBLY SVR AFT 18Z. CB TOPS ABV 450.

072 TWEB
RDGS AND E
45 OVC VSBY 5R-...AREAS 12 OVC VSBY BLO 3RF.
AFT 14Z CIGS 10-20 OVC...AREAS BLO 10 OVC...WDLY SCT TRW. RDGS
OBSCD

The FA forecasts occasional ceilings and visibilities below 1,000 feet and 3 miles, skies obscured in light rain showers and fog, with scattered-to-broken clouds between 1,500 and 2,500 feet MSL, and broken-to-overcast clouds between 4,000 and 6,000 feet MSL. Terrain varies from about 500 feet around Harrisburg and in the valleys to about 2,500 feet along ridges in the Johnstown area. Remembering that TWEB heights are AGL, the TWEB forecasts predominately 5,000 and 7,000 feet MSL (converted to MSL) overcast, with scattered AREAS of 2,000 to 3,500 MSL overcast (approximately). After 1400Z, TWEB cloud heights become 2,000 to 4,000 MSL, with "areas" below 1,000 AGL.

Here is a comparison of FA/TWEB heights converted to MSL:

- FA/AIRMET 15-25 SCT-BKN 40-60 BKN-OVC OCNL CIG BLO 10 OVC
- TWEB 50-70 OVC AREAS 20-35 OVC...14Z. 20-40 OVC AREAS BLO 10 OVC

In general, both forecasts predict occasional ceilings below 1,000 feet AGL, with higher clouds at about 2,000 to 3,000 feet MSL, and another layer between about 4,000 to 7,000 feet MSL. For people who deal with exact numbers, this might not look close, but from a meteorological forecast point of view, these forecasts are perfectly consistent given the scope of each product. The TWEB can usually be more specific about the probability and coverage of weather. The FA, covering a larger area, forecasts thunderstorms increasing to scattered and possibly severe after 1800Z; however, the TWEB expects coverage to remain widely scattered.

VFR flight will be marginal, with the mountains occasionally obscured in clouds, rain, and possible thunderstorms. Other factors, primarily current conditions, would have to be considered to determine the advisability of a VFR flight.

For an IFR flight, the major concern again will be convective activity. The FA indicates widely scattered embedded thunderstorms becoming scattered and possibly severe after 1800Z; the TWEB only contains widely scattered thunderstorms. This is again consistent given that the FA must cover a larger area. This can be interpreted as thunderstorm activity increasing and possibly becoming severe somewhere within the FA-delineated area, which includes portions of New York and New Jersey, as well as Pennsylvania, but not within the area covered by the TWEB route. The smaller aerial coverage is a distinct advantage of TWEB forecasts. TWEBs can often be used to determine where specific phenomena (thunderstorms) will occur within an FA area. The advisability of an IFR flight would depend upon current radar reports and overall current weather conditions. A flight decision can never be made solely with the limited information in this example. Weather advisories, synopses, and current conditions must be considered.

TWEBs do not directly forecast turbulence or icing; however, turbulence and strong updrafts and downdrafts are implied by weather associated with these phenomena. For example, strong winds and mountain wave activity indicate turbulence. Icing can be expected above the freezing level where visible moisture exists, or in areas of freezing precipitation. And like the FA, thunderstorms imply severe or greater turbulence, icing, and low-level wind shear. Specific forecasts for turbulence and icing must be obtained from weather advisories. The TWEB route forecast is a supplement to weather advisories, not a substitute.

TWEB routes and synopses often provide more precise timing and detail than is possible in the FA; they provide additional specific information on visibility, surface winds, and local conditions that is not possible in the FA. But like other forecast products, TWEBs cannot cover every instance of hazardous weather. Although the number of TWEB routes is extensive, many areas are not covered. Never extrapolate nor extend the forecast beyond its defined area.

Revised TWEB standards were adopted to standardize the product; however, as is often the case, interpretation of the new instructions was not consistent. Just after implementation, a message appeared from one weather service office: TWEB ROUTES DELAYED DUE TO RIDICULOUS NEW FORMAT.

7
Terminal forecast (FT)

GILBERT, PERHAPS THE CENTURY'S STRONGEST HURRICANE TO THAT DATE to strike the United States, was approaching the Texas Gulf Coast on September 16, 1988. With the hurricane's arrival, several San Francisco-area oil companies planned to fly to Brownsville, presumably to check the fate of their oil rigs. Similar to the weather, no one could change the fate of the materiel; the rigs would be OK, damaged, destroyed, or perhaps blown away. The following terminal forecast (in the format used at that time) for Brownsville (BRO) was written just prior to Gilbert's arrival. This is decoded, translated, and interpreted at the end of this chapter:

BRO FT 160808 20 SCT C60 OVC 4RW 0220G30 OCNL C20 BKN 2RW+ CHC C5 X 1/2TRW+ Q50. 00Z 3 SCT C12 OVC 2RW+ 0860Q75 OCNL C3 X 1/4 T+RW+ G120. 02Z LIFR CIG TRW WND.

Some think the only place a pilot will actually see a terminal forecast is on an FAA exam. This might have been true in the past, but with FSS automation and DUAT, responsibility for decoding, translating, and interpreting this product will rest with the pilot. Incorrect interpretation could lead to anything from an embarrassing chat with an inspector from the local flight standards district office to an aircraft accident.

A terminal forecast is prepared by local weather service forecast offices, such as the Redwood City, CA, office (now in Monterey, CA) in Fig. 7-1; the forecasts contain

Fig. 7-1. *Meteorologists at local NWS offices prepare and monitor terminal forecasts. This forecaster at Redwood City, California, is responsible for FTs in northern and central California.*

specific information for individual airports. FTs forecast cloud heights and amounts, visibility and obstructions to visibility, weather, wind, and low-level wind shear. (FT locations are listed in appendix B.)

The meteorologist normally must have at least two consecutive SAs before issuing a terminal forecast. As a minimum, these observations usually contain sky condition, visibility and obstructions to visibility, weather, temperature, dew point, wind, and altimeter setting. Part-time observations normally require 3 hours after the first observation for the FT to be written, distributed, and become available. The FT might not be issued if an element is missing. For example, with a missing dew point, pilots could expect to see XYZ FT 071919 FT NOT AVBL DUE MSG DEW PT.

FT time references are UTC, with specific intervals specified: 18Z, 23Z, and the like. Terminal forecasts are valid for 24 hours.

FTs are issued in the contiguous United States three times a day. Times vary with time zone, based upon operational and other considerations. Refer to Table 7-1 for FT issuance times.

Table 7-1. Terminal forecast issuance and valid times.

Time zone	Issued/UTC	Local
Eastern	0730/0830	3:30 a.m.
	1630/1730	12:30 a.m.
	2330/0030	7:30 p.m.
Central	0730/0830	2:30 a.m.
	1630/1730	11:30 a.m.
	2330/0030	6:30 p.m.
Mountain	0030/0130	6:30 p.m.
	0830/0930	2:30 a.m.
	1730/1830	11:30 a.m.
Pacific	0130/0230	6:30 p.m.
	0930/1030	2:30 a.m.
	1830/1930	11:30 a.m.

Note: UTC issuance times change one hour with daylight-saving time in the contiguous states.

Alaska (most locations)	0340	6:40 p.m.
	1040	1:40 a.m.
	1540	6:40 a.m.
	2040	11:40 a.m.
Hawaii	0540	7:40 p.m.
	1140	1:40 a.m.
	1740	7:40 a.m.
	2340	1:40 p.m.

FTs have been criticized for not being written more often and not being valid for a longer period. The National Weather Service, after receiving input from the FAA and other aviation users, has revised FT preparation procedures, format, and amendment criteria. This new policy addresses many of these complaints. Some changes have been made in anticipation of full conversion to *terminal aerodrome forecasts* (TAF) in 1996.

```
TUL FT 190909 250 -SCT 0410.
15Z 60 SCT C150 OVC 0515.
23Z 30 SCT C80 BKN 0712 OCNL C30 BKN CHC C10 OVC 1TRW.
03Z C10 BKN.
```

This Tulsa, OK, FT was issued on the 19th day of the month, valid from 0900Z until 0900Z on the 20th (190909). The first forecast section (250 -SCT 0410) is valid from 0900Z until the next forecast section becomes effective at 1500Z. The FT now provides a specific 24-hour forecast (09Z-09Z). FT categorical outlooks were deleted in 1994. The FT forecasts conditions within 5 nautical miles of the center of the runway complex. In areas where thunderstorms, showers, or fog regularly develop beyond 5 miles but directly affect operations, the term "vicinity" (VCNTY) is used, which indicates phenomena occurring beyond 5 miles, but within 25 miles of the airport.

BODY AND REMARKS

The forecast group is divided into body and remarks. Remarks amplify or describe conditions that differ from that in the body of the forecast group. Conditional terms describe variability. Table 7-2 defines terms used on the FT.

Table 7-2. Terminal forecast conditional terms.

No conditional term	50% or greater probability of occurrence during more than one half of the forecast period.
OCNL (occasional) (TEMPO on TAF)	50% or greater probability but only expected to persist for one half or less of the forecast period.
CHC (chance) (PROB40 on TAF)	30% to 49% probability.
VCNTY (vicinity)	Phenomena expected to occur beyond 5 nm and up to 25 nm from the airport.
VC (vicinity) on TAF	Phenomena expected to occur beyond 5 sm and up to 10 sm from the airport.

The body contains sky cover (CLR, 250 -BKN, C50 OVC, etc.). FTs are similar to SAs in that they report cloud heights above ground level. This is important when using an FT product or comparing it with observations, PIREPs, or other forecasts. The letter "C" designates the ceiling; its definition is the same as used on SAs. FTs are specifically intended for arriving and departing aircraft; therefore, cloud layers above 15,000 feet might not be included when a lower ceiling appears. Visibility will be forecast when prevailing visibility is expected to be 6 miles or less; the omission of a visibility forecast implies visibilities greater than 6 miles.

Visibilities greater than 6 miles are often referred to as unrestricted by FSS briefers. The symbol 6+ indicates that unrestricted visibility is intended, which prevents misunderstanding. For example, C10 OVC 2RW OCNL 10 SCT will be written C10 OVC 2RW OCNL 10 SCT 6+ to clearly indicate that visibility is expected to be occasionally unrestricted. Weather and obstructions to vision appear with a visibility forecast: 3HK, 5TRW, and the like. Weather and obstructions to vision use the same contractions and definitions as SAs.

A popular aviation magazine reported that wind conditions received from towers were miles per hour and FT wind conditions were knots. Let's settle this speed matter and two other wind matters once and for all:

- Is wind direction *true* or *magnetic* north?
- Is it blowing *from* or *toward* the reported direction?
- Is the speed *knots* or *miles per hour*?

All official aviation wind observations and forecasts are reported in relation to ***true north***, given as the direction *from* which the wind is blowing, with speed in ***knots***. *The only time that a pilot will receive winds in relation to magnetic north is from a tower, AWOS broadcast, or as part of a local airport advisory because* ***runway headings are magnetic.***

Omission of a wind forecast implies speed less than 6 knots; otherwise, wind direction and speed will appear on the FT. Gustiness is a forecast for rapid fluctuations of 10 knots or more indicated by the G following the wind group: 3425G40 (wind speed 25 knots with gusts to 40 knots), or G50 (wind gusts to 50 knots). Unlike the FA or TWEB, the FT does provide a specific wind forecast.

Low-level wind shear (LLWS) on terminal forecasts refers to nonthunderstorm shear. Wind direction, speed, and height might be included. LLWS appears when PIREPs report an airspeed gain or loss of 20 knots or more; LLWS also appears when vertical shears of 10 knots or more per 100 feet are expected or reported.

Refer to the 2300Z portion of the TUL FT. The body forecasts 30 SCT C80 BKN 0712 and the remarks state OCNL C30 BKN CHC C10 OVC 1TRW. The body decodes as 3,000 scattered, ceiling 8,000 broken, and wind 070° at 12 knots. The remarks indicate that the 3,000-foot AGL scattered layer might be "occasionally" broken (50 percent or greater probability but expected for 50 percent or less of the period), and a "chance" exists (30–49 percent probability) for ceiling 1,000 overcast and visibility 1 mile in moderate rain showers and thunderstorms.

A word of caution, it's dangerous to extrapolate the FT beyond 5 miles. Over flat terrain, such as the Midwest, chances are that nearby airports might have similar conditions; however, this does not relieve the pilot from checking appropriate FA or TWEB forecasts. In hilly or mountainous terrain, this practice can be disastrous. For example, afternoon surface winds at San Francisco International Airport are routinely strong and gusty during the summer, whereas winds at other Bay Area airports remain relatively calm due to wind direction and terrain.

Conversely, an airport in the middle of a valley might have benign surface winds, but surface winds can prevent landings altogether at a nearby airport that is below a canyon. This is not uncommon in the Los Angeles Basin when Santa Ana winds blow. Winds at the Ontario, CA, Airport might be out of the west at less than 5 knots, but at Rialto, about 10 miles away, winds might be out of the north gusting to more than 40 knots. Rialto lies just below the Cajon Pass.

From the discussion thus far, let's decode, translate, and interpret the following FT.

18Z 35 SCT C80 BKN OCNL C35 BKN 2312 CHC RW/TRW.
23Z CFP C25 BKN 80 OVC 0808 CHC C15 OVC 1TRW.
02Z 45 SCT 100 SCT 1713 TRW VCNTY OCNL C45 BKN 100 BKN.

At 18Z, 50 percent or greater probability, during more than one-half the period (18Z–23Z), exists for scattered clouds at 3,500 feet, ceiling 8,000 feet broken; however, there is a 50 percent or greater probability for one-half or less of the period that the scattered clouds at 3,500 feet will be broken, winds 230° at 12 knots, and a 30 to 49 percent probability of rain showers, or rain showers and thunderstorms. The winds would not necessarily coincide with broken clouds. Conditions are expected to deteriorate around 2300Z, so expect the remarks portion of the forecast to be more prevalent during the later portion of the period.

Cold front passage is expected around 2300Z. But conditions will not instantly change, which you already understand based upon the discussion thus far. Table 3-1, which notes the limitations on aviation weather forecasts, reveals the best that could be expected would be frontal passage within 2 hours, sometime between 2100Z–0100Z. At that time, conditions are forecasted to deteriorate to a "50 percent or greater probability, during more than one-half the period from 2300Z to 0200Z" of ceilings 2,500 feet broken, 8,000 feet overcast, and wind 080° at 8 knots. Remarks indicate a 30–49-percent probability of ceilings 1,500 feet overcast and visibility 1 mile in rain showers and thunderstorms. Here we could reasonably expect conditions described in the remarks (CHC C15 OVC 1TRW) to be more prevalent during the beginning of the period, associated with frontal passage.

By 0200Z, 50 percent or greater probability, during more than one-half of the period, exists for conditions to improve to 4,500 scattered, 10,000 scattered, wind 170° at 13. Rain showers and thunderstorms are expected beyond 5 miles, but within 25 miles of the airport. The scattered layers are expected to be broken for one half or less of the forecast period.

It's extremely important to understand the synoptic situation and obtain frequent updates. Remember the limitations on aviation forecasts in Table 3-1. Timing is a major problem, and if observations reveal that frontal passage occurred at 2100Z, we could reasonably expect the conditions described at 2300Z to prevail. Conversely, if frontal passage occurs at 01Z, expect the improvement advertised in the 0200Z portion of the forecast to be delayed. Within the scope of this product, frontal passage within 2 hours is considered accurate and under most circumstances will not generate an amendment.

The lowest conditions forecast (CHC C15 OVC 1TRW) are associated with frontal passage. Even under these conditions, special VFR should be possible if available; however, based solely upon the forecast, VFR flight should be planned either before 2100Z or after 0100Z to be safe. The thunderstorm activity would be of concern to VFR and IFR flights. Convective activity should be circumnavigable, but might cause arrival and departure delays; fuel reserves and alternatives for VFR and IFR flight plans should be planned accordingly.

Remarks must be considered when determining IFR alternatives; therefore, a forecast of CLR 5F CHC C5 X 2L-F would require an alternate. Additionally, this airport would not qualify as a legal alternate because the ceiling is forecast for less than 600 feet. Forecasters are aware that remarks as well as the body must be considered when determining legal destinations, alternates, and fuel loads. This makes the remarks operationally more significant when they describe lower conditions. Forecasters will use remarks sparingly and be as concise as possible to allow the use of an airport as an alternate.

Provisions of FAR Part 135, regarding air taxis, and FAR Part 121, regarding air carriers, are also considered. Both regulations state: "No person may . . . begin an IFR . . . operation unless the latest weather report or forecast, or any combination of them, indicate that weather conditions at the ETA at the next airport of intended landing will be at or above authorized IFR landing minimums." Forecasters could effectively close airports to certain operations.

Here is an example of how forecasters deal with this situation. The SAs and FT are for Monterey, CA, based upon observations taken by a controller at the part-time tower.

```
MRY SA 1250 W3 X 1/4F
MRY SA 1345 W0 X 1/4F
MRY SA 1449 W0 X 1/8F
MRY SP 1522 W1 X 1/2F
MRY FT RTD 061410 1420Z C0 X 1/4F OCNL C3 X 1F. 17Z C4 X 2F OCNL 5 SCT
6F. 19Z...
```

Monterey's first report was observed at 1250Z; therefore, the FT is routinely delayed (RTD). When two consecutive observations became available, the FT was written, valid at 1420Z; however, due to transmission and distribution time, a pilot wouldn't expect the FT to be available until some time after 1420Z. Notice the 1420Z forecast

C0 X 1/4F OCNL C3 X 1F, even though W0 X 1/4F is reported. When actual conditions improve to landing minimums (ceiling 200 and visibility ½ mile) FAR 135 and 121 operations can legally land. The forecaster said that "between 1420Z and 1700Z conditions will improve to IFR landing minimums (and) between 1700Z and 1900Z the weather will become VFR."

Pilots continually demand to know *exactly* when an airport will improve to IFR landing minimums or VFR conditions. Accuracy simply is not possible. FAR alternate airport and fuel reserve minimums take these limitations into consideration; however, departing with appropriate alternates and adequate fuel does not relieve the pilot of updating weather information en route.

With the implementation of automated observations, the FT forecast will not change in appearance; however, all the cautions that apply to automated SAs must be considered when using an FT that is based upon those observations. Like the FAA, with respect to operational requirements, the National Weather Service has concluded that remarks are not a factor in forecasting.

AMENDMENT CRITERIA

The NWS forecast manual states: "Forecasters should strive to amend FTs prior to the occurrence of expected changes which meet these criteria. If this cannot be done, amendments to the body and/or remarks portion of FTs shall be issued promptly whenever conditions meeting the criteria occur, and, in the forecasters' judgment, these conditions will persist."

Amendment criteria have changed significantly in the 1990s. Table 7-3 describes terminal forecast amendment criteria; you can probably see the reason for some of the criticism of this product. Ceilings above 3,000 feet and visibilities more than 6 miles must decrease below these values before an amendment is required. For example, amendments are not required for a forecast ceiling of 10,000 decreasing to 5,000 feet, although the forecaster has the authority to amend the forecast whenever he or she considers it operationally significant. Amendment criteria intervals decrease when conditions lower and become more significant. The forecast is considered accurate as long as the criteria in Table 7-3 are met.

The new amendment criteria reflect operations needs. Note that a decrease or increase from basic VFR (1,000 and 3), nonprecision approach minimums (600 and 2), and precision approach minimums (200 and ½) require an amendment. As we will see, however, this does not mean that an amendment will be issued at the moment these criteria are met.

According to Table 7-3, changes in wind speed less than 12 knots do not require an amendment. At speeds of 12 knots or greater, the speed must differ by 10 knots or more; therefore, a wind forecast of 20 knots would have to decrease to 10 or increase to 30 before an amendment is required. Basically, FT wind speeds are considered accurate when within 10 knots of forecast. A pilot competent enough to handle 20 knots must realize

Table 7-3. Terminal forecast amendment criteria.

Ceiling

Forecast	Amend if:
more than 3,000	3,000 or less
2,000 to 3,000	more than 3,000–less than 2,000
1,000 to 1,900	more than 1,900–less than 1,000
600 to 900	more than 900–less than 600
200 to 500	more than 500–less than 200
100 or less	200 or more

Visibility

Forecast	Amend if:
6 or more	5 or less
3 to 5	more than 5–less than 2
2	more than 2–less than 2
1 to 1½	more than 1½–less than 1
½ to ¾	more than ¾–less than ½
less than ½	½ or more

Wind

Forecast	Amend if:
Speed 12 knots or more	Direction differs by 30°; speed differs by 10 knots
Gusts with mean speed of 12 knots or more	Gust 10 knots above forecast

Weather

Forecast	Amend if:
Thunderstorms, freezing precipitation, or ice pellets	Does not occur or no longer expected
No thunderstorms, freezing precipitation, or ice pellets	Occurs or is expected
LLWS	No LLWS expected
No LLWS	LLWS occurs or is expected

that with a forecast of 20 knots, actual winds could increase to almost 30 knots without a requirement to amend. Within these parameters, the forecast is considered correct.

The only weather phenomena requiring amendments are thunderstorms, freezing precipitation, ice pellets, and LLWS. The unforecast occurrence or ending of rain or snow does not require an amendment. This fact has led many pilots to erroneously criticize the forecast.

An amendment is required when phenomena described in the "vicinity" move or are expected to move within 5 miles of the airport. The phenomena may appear in the body or remarks of the amendment.

At locations with part-time observations, the remark NO AMDTS AFT (TIME)Z will appear. As used in TWEBs, this serves as a warning that significant changes can occur without an amendment. Extra caution must be exercised operating into these airports, especially during marginal or deteriorating conditions.

A forecast of landing minimums, VFR or IFR, is no guarantee those conditions will exist at an estimated time of arrival. An IFR alternate affected by the same weather pattern as the destination might satisfy FARs, but leave a pilot out on the proverbial limb with a busted forecast. For example, if a pilot's destination was Fresno in California's Central Valley, Bakersfield might qualify as a legal alternate; however, with "tule fog"—a colloquialism for a condition of extensive wintertime fog—in California's Central Valley, a viable alternate airport that is unaffected by tule fog might be along the coastline. This would also apply to airports affected by upslope fog or frontal systems. If at all possible, a pilot should select an alternate not affected by the weather pattern that is affecting the destination. At the risk of being redundant, we're back to knowing the synopsis and continually updating weather en route.

As of this writing, amendments are indicated by the contraction AMD: SFO FT AMD 131310. Numbering amendments (AMD 1, AMD 2, etc.) has been discontinued.

TERMINAL AERODROME FORECASTS (TAF)

The United States and Canada will issue domestic terminal forecasts in the international terminal aerodrome forecast (TAF) format beginning in 1996; METAR formatting is also part of the conversion (see chapter 1). According to sources within the National Weather Service, "a few of the changes seem to defy logic; in fact, they are illogical." At first glance, the new format appears confusing, but the basic forecast actually remains the same.

The National Weather Service already issues TAF-coded reports for airports served by long, usually transoceanic international flights; TAF is also used in Mexico. TAFs are issued four times a day at 0000Z, 0600Z, 1200Z, and 1800Z and are valid for 24 hours. TAFs cover a 5-sm radius of the airport. Forecasts using TAF codes are also issued by local base weather offices for many military locations. Military TAFs have different criteria than those issued by the NWS, therefore, some differences will occur.

TAFs are issued in the following format:

- Type
- Location
- Issuance time
- Valid time
- Forecast

The two types of TAF issuance are a routine forecast (TAF) and an amended forecast (TAF AMD). TAF and domestic FT amendment criteria are the same as shown in Table 7-3. Like domestic FTs, TAFs might be corrected (COR) or routinely delayed (RTD). TAF location is identified by the four-letter ICAO station identifier; KLAX is Los Angeles. Issuance date and time are a six-digit group. The first two digits represent the day of the month, and the last four digits represent UTC issuance time. The valid period is a four-digit group, usually 24 hours, in UTC.

The body of the TAF uses the following format:

- Wind
- Visibility
- Weather
- Sky Condition

Like the METAR code, wind is forecast as a five- or six-digit group when considered significant to aviation. The contraction KT follows the wind forecast and denotes the units as knots. Gusts are noted by the letter G. The current practice of not forecasting a wind speed that is less than 6 knots will probably continue.

Prevailing visibility up to and including 6 statute miles is forecast. Visibility greater than 6 miles is indicated by the letter P for plus: P6SM (visibility greater than 6 statute miles). The military and many international TAFs forecast visibility in meters. A conversion of meters to miles is found in Table 7-4.

Weather and obstructions to vision use the same format and codes as the METAR report discussed in chapter 1 and illustrated in Table 1-3. Vicinity (VC) refers to an area between 5 and 10 sm from the airport. With no significant weather expected, the weather group is omitted. When significant weather is forecast, then expected to change in the future to no significant weather, the contraction NSW (no significant weather) appears. NSW will not appear when any of the following phenomena are expected to occur:

- Freezing precipitation
- Moderate or heavy precipitation
- Drifting or blowing dust, sand, or snow
- Dust storms or sandstorms
- Thunderstorms
- Squalls
- Tornadoes
- Phenomena expected to cause a significant change in visibility

Sky condition uses the same format and contractions as METAR: amount, height, cloud type, or vertical visibility. When cumulonimbus clouds are expected, CB is appended to the cloud layer. CB is the only cloud type forecast in TAFs. Total obscurations and vertical visibility are forecast; like METAR, there are no provisions for

Table 7-4. Terminal aerodrome forecasts (TAF).

Temperature: TT_1T_1/tt

T – Temperature group

T_1T_1 – Temperature in Celsius

tt – Time UTC

Turbulence: 5ihhhd

5 – Turbulence group

i – Turbulence intensity

hhh – Base height hundreds of feet

d – Thickness in thousands of feet

Turbulence intensity:

0 None

1 Light turbulence

2 Moderate turbulence in clear air, infrequent

3 Moderate turbulence in clear air, frequent

4 Moderate turbulence in cloud, infrequent

5 Moderate turbulence in cloud, frequent

6 Severe turbulence in clear air, infrequent

7 Severe turbulence in clear air, frequent

8 Severe turbulence in cloud, infrequent

9 Severe turbulence in cloud, frequent

Icing: 6ihhhd

6 – Icing group

i – Icing intensity

hhh – Base height hundreds of feet

d – Thickness in thousands of feet

(0 indicates to top of clouds)

Icing intensity:

0 No icing

1 Light icing

2 Light icing in cloud

3 Light icing in precipitation

4 Moderate icing

5 Moderate icing in cloud

6 Moderate icing in precipitation

7 Severe icing

8 Severe icing in cloud

9 Severe icing in precipitation

Visibility in meters:

Meters	Statute mile	Meters	Statute mile	Meters	Statute mile	Meters	Statute mile
0000	0	1000	⅝	2400	1½	4800	3
0200	¹⁄₁₆	1200	¾	2600	1⅝	6000	4
0300	⅛	1400	⅞	2800	1¾	8000	5
0400	³⁄₁₆	1600	1	3000	1⅞	9000	6
0500	⁵⁄₁₆	1800	1⅛	3200	2	9999	6+
0600	⅜	2000	1¼	3600	2¼		
0800	½	2200	1⅜	4000	2½		

partial obscurations. (Not used on domestic TAFs, CAVOK might replace visibility, weather, and cloud groups on international TAFs when visibility is 10 kilometers or more, the criteria for NSW exist, and no clouds below 1,500 meters, which is 5,500 feet, are forecast.)

Conditional terms

TAF conditional terms consist of probability (PROB) forecasts and temporary (TEMPO) conditions.

A PROB group indicates the probability of occurrence of thunderstorms or other precipitation events. When the probability is in the 30 to 49 percent range, PROB40 is used. This is followed by a four-digit time group giving beginning and ending times. For example, PROB40 2102 1/2SM +TSRA OVC005CB indicates that there is a probability, between 2100Z and 0200Z, of visibility one-half mile in thunderstorms with heavy rain (not severe thunderstorms) and a ceiling of 500 overcast with cumulonimbus clouds. The PROB40 contraction is the equivalent to CHC on domestic FTs.

TEMPO indicates that temporary conditions are expected to occur during the forecast period. TEMPO describes any condition expected to last for generally less than an hour at a time. The time during which the condition is expected to occur is indicated with a four-digit group giving beginning and ending time UTC. For example, SCT030 TEMPO 1923 BKN030 means 3,000 scattered temporarily between 1900Z and 2300Z, ceilings 3,000 broken; ceilings of 3,000 broken are expected to exist for periods of less than 1 hour during the 1900Z to 2300Z timeframe. TEMPO is equivalent to OCNL on domestic FTs.

Forecast change groups

Forecast change groups consist of from, followed by a time group (FMtt), and becoming, followed by a time group (BECMG TTtt). The FMtt group is used when a rapid change is expected, usually less than 1 hour. For example, BKN020 FM16 SKC translates into "before 1600Z, ceiling 2,000 broken is forecast, and around 1600Z the sky condition will change to clear." The BECMG TTtt group indicates a more gradual change in conditions over a longer period of time, usually 2 hours. For example, 5SM HZ BKN 030 BECMG 0507 3SM BR OVC 020 translates into "before 0500Z, visibility 5 miles in haze and ceiling 3,000 broken, then during the period of 0500Z to 0700Z, conditions changing to visibility 3 miles in fog and ceiling 2,000 overcast."

Below is an example of a TAF for Seattle Tacoma International Airport.

TAF
KSEA 041045Z 1212 00000KT P6SM SCT015 OVC025 TEMPO 1214 -RA
BECMG 1314 24007KT SCT015 BKN025 TEMPO 1418 -SHRA BECMG 1718
SCT020 BKN035 PROB40 1801 -TSRA BECMG 0001 22006KT SCT015 BKN035
TEMPO 0112 -SHRA=

This is the TAF for Seattle Tacoma issued on the 4th day of the month at 1045Z (041045). The forecast is valid from the fourth at 1200Z until the fifth at 1200Z (1212). At 12Z, the surface winds are expected to be calm (00000KT); this might not appear with winds less than 6 knots, or VRB05KT, which means variable at 5 knots, might be used. The visibility forecast is more than 6 statute miles (P6SM). The sky-condition forecast is 1,500 scattered with ceiling 2,500 overcast. Occasionally (TEMPO) between 1200Z and 1400Z (1214), light rain is expected. Prevailing conditions are expected to become (BECMG) between 1300Z and 1400Z (1314) wind 240° at 7, visibility is implied to be more than 6 miles, with clouds 1,500 scattered and a ceiling 2,500 broken. Occasionally between 1400Z and 1800Z, light rain showers are expected. Between 1700Z and 1800Z (wind is implied to be 240° at 7 knots, and visibility is more than 6 miles), 2,000 scattered with ceilings of 3,500 broken are forecast. There is a chance (PROB40) of thunderstorms with light rain. Between 0000Z and 0100Z, wind is forecast to be 220° at 6, visibility should be more than 6, clouds are expected to be 1,500 scattered with a ceiling 3,500 broken, and occasionally between 0100Z and 1200Z light rain showers are forecast.

Now compare the TAF with the domestic SEA FT for the same period. Keep in mind the parameters and limitations discussed thus far.

```
SEA FT 041010...12Z 15 SCT C25 OVC OCNL R-.
14Z 15 SCT C25 BKN 2407 OCNL RW-.
18Z 20 SCT C35 BKN 2407 CHC TRW-.
01Z 15 SCT C35 BKN 2206 OCNL RW-...
```

Well, what do you know, they're identical. This shouldn't be a surprise because both were written by the same forecaster. The TAF, however, directly reflects the limitations on aviation forecasts through the use of weather change indicators and variability terms. The lack of TAF indicators and terms on domestic FTs does not mean the TAF forecast is more accurate. As we've seen, the indicators and terms are implied, and must be considered when applying the forecast.

Pilots can also expect to see TAF code used on military forecasts. Below is an example of the McCord Air Force Base (Tacoma, WA) forecast for the same time period as the previous examples.

```
TCM 1212 20009KT 9999 SCT010 BKN020 OVC050 T12/13 540109 610703
QNH2980INS CIG020

BECMG 1314 20009KT 9999 SCT015 BKN035 OVC045 540109 620703
QNH2975INS CIG035

BECMG 1920 24007KT 9999 SCT020 BKN045 BKN150 T20/23 540009 620505
QNH2980INS CIG045

BECMG 0001 VRB05KT 9999 -SHRA SCT020 BKN040 OVC070 610505
QNH2983INS CIG040

BECMG 0405 VRB05KT 9999 NSW SCT020 BKN050 QNH2990INS CIG050=
```

There are several major differences between domestic FTs and TAFs and those issued by the military and foreign governments. Refer to Table 7-4. Wind speed of less than 6 knots might be forecast. Visibilities forecasts are in meters. In the example, the visibility through the forecast period is 9999, which means greater than 9,000 meters. Sky-condition forecasts do not use the summation principle and the ceiling is specified: BKN015 OVC080 BKN200 CIG015. The altimeter setting is forecast: QNH2980INS (altimeter setting 29.80" of mercury).

Maximum and minimum temperatures, icing, and turbulence are sometimes forecast. The temperature group follows sky conditions. For example, T12/13 (minimum temperature 12°C at 1300Z) or T20/23 (maximum temperature 20°C at 2300Z). Turbulence and icing forecasts may appear in a format illustrated in Table 7-4.

In the example, at 1200Z, 540109 appears. Five ("5"40109) represents the turbulence group. Four (5"4"0109) indicates the turbulence intensity, in this case, moderate turbulence in cloud, infrequent. Zero one zero (54"010"9) shows the base of the turbulence layer height in hundreds of feet, in this case, 1,000 feet. Nine (54010"9") represents the thickness of the turbulence layer in thousands of feet, in this case, 9,000 feet. Moderate turbulence is forecast between 1,000 and 10,000 feet.

In the example, at 1200Z, 610703 appears. Six ("6"10703) represents the icing group. One (6"1"0703) indicates the icing intensity, in this case, light icing. Zero seven zero (61"070"3) shows the base of the icing layer height in hundreds of feet, in this case, 7,000 feet MSL. Three (61070"3") represents the thickness of the icing layer in thousands of feet, in this case, 3,000 feet. Light icing is forecast between 7,000 and 10,000 feet.

Whether international or military, TAFs suffer from the same limitations as domestic FTs, or any forecast for that matter. It really can't be said that one or the other is necessarily more accurate; however, by comparing a domestic FT with a nearby military TAF, a pilot could get a second opinion.

USING THE TERMINAL FORECAST

Apparent inconsistencies arise because different forecasts serve different purposes. The following example compares FA and TWEB forecasts with a terminal forecast.

```
CHIC FA 141845
CLDS AND WX VALID UNTIL 150700...OTLK 150700-151300Z.
IN
AGL 45 SCT-BKN. CHC 3-5R-. 04Z CIGS 40 BKN VSBY 3-5F. OCNL CIGS 12
BKN VSBYS 3-5R-/RW. WDLY SCT TRW-. TOP CBS 400.

220 TWEB 141910 HUF-STL. ALL HGTS AGL EXCP TOPS. 40 SCT-BKN 100
SCT-BKN. AFT23Z...CIG 20-30 BKN-OVC...AREAS VSBY 4RW-F. HUF VCNTY
AFT 06Z...AREAS CIG 10 OVC VSBY 3TRW-F.
```

Below is the Terre Haute terminal forecast.

```
HUF FT 141717...45 SCT C100 BKN 200 OVC 2410 OCNL C45 BKN 100 OVC
CHC 5R-F.
```

02Z C30 BKN 70 OVC 2309 OCNL 10 SCT C25 OVC 4R-F.
06Z 10 SCT C25 OVC 4F 2908 OCNL C10 OVC 3TRW-F...

The following discussion is an analysis of the forecasts:

- FA AGL 45 SCT-BKN. CHC 3-5R-. 04Z...

- TWEB 40 SCT-BKN 100 SCT-BKN. AFT 23Z...

- FT 45 SCT C100 BKN 200 OVC 2410 OCNL C45 BKN 100 OVC CHC 5R-F. 02Z...

The TWEB and FT contain more detail with respect to the second cloud layer (100 SCT-BKN and C100 BKN OCNL 100 OVC respectively). The FA forecaster does not consider this layer to be extensive enough for inclusion in the FA for the entire state of Indiana. The FA and FT forecast reduced visibilities in light rain; the TWEB omits this phenomenon because the forecaster doesn't expect it to cover 10 percent of the route. The FT provides a surface wind forecast of 10 knots; ten knots is not within the scope of either the TWEB or FA. All three forecasts are consistent given the scope of each product.

The next time period products from the three samples:

- FA 04Z CIGS 40 BKN VSBY 3-5F. OCNL CIGS 12 BKN VSBYS 3-5R-/RW. WDLY SCT TRW-.

- TWEB AFT23Z...CIG 20-30 BKN-OVC...AREAS VSBY 4RW-F. HUF VCNTY AFT 06Z...AREAS CIG 10 OVC VSBY 3TRW-F.

- FT 02Z C30 BKN 70 OVC 2309 OCNL 10 SCT C25 OVC 4R-F . . 06Z 10 SCT C25 OVC 4F 2908 OCNL C10 OVC 3TRW-F.

All three forecast deteriorating conditions (0400Z, AFT 2300Z, and 0200Z). Ceilings are generally expected to be between 2,000 and 4,000 feet. FA and FT expect occasional ceiling between 1,200 and 2,500 feet; however, the TWEB and FT provide the specific time of 0600Z when conditions will further deteriorate to occasional ceilings 1,000 feet. Visibility forecasts are comparable: FA 3 to 5, TWEB more than 6 to areas of 4, and FT more than 6 to occasionally 4. All three forecasts indicate rain, fog, and thunderstorm activity.

From the forecasts, a pilot can conclude that the lower conditions and thunderstorm activity advertised in the FA will not occur in the Terre Haute area until around 0600Z. This is not inconsistent given that the FA must cover the whole state. From the FA, a pilot could only conclude that surface winds are expected to be less than 30 knots, and from the TWEB, less than 25 knots; however, from the FT, surface winds in the Terre Haute area are expected to be southwest to northwesterly at around 10 knots. Sky cover, ceiling, visibility and obstructions, weather, and wind forecasts are comparable. All three forecasts are perfectly consistent, given the scope and purpose of each product.

From these forecasts, there is no significant impact for either VFR or IFR operations in the Terre Haute area until 0600Z. From the FT, surface winds are expected to remain around 10 knots, which might be significant for a student pilot. And thunderstorms are not expected within 5 miles of the airport until between about 0600Z and 0800Z. This timeframe is deduced from the TWEB forecast AFT 0600Z and the FT 0600Z.

Let's say we're planning an IFR arrival to the Sullivan Co. Airport, about 25 miles southwest of Terre Haute. In accordance with FARs, we must determine if an alternate airport is required and select a suitable alternate if necessary. Our ETA is 2200Z. Because Terre Haute is farther than 5 miles, the HUF FT cannot be used to determine destination weather. The TWEB and FA, however, both cover the destination. From 1 hour before until 1 hour after the ETA, both forecasts predict ceilings of at least 2,000 feet and visibilities at least 3 miles; therefore, based upon the TWEB and FA, an alternate airport is not required.

What about an ETA of 0500Z? The FA forecasts OCNL CIGS 12 BKN; therefore, an alternate would be required. But take a look at the TWEB; for the ETA, the TWEB forecasts CIG 20-30 BKN-OVC...AREAS VSBY 4RW-F. If we derive our destination forecast from the TWEB, an alternate airport would not be required. Is the TWEB a legal destination forecast? Absolutely.

Finally, let's look at an 0800Z arrival. Now the forecast from both FA and TWEB call for a ceiling less than 2,000 feet. An alternate airport is required. Could Terre Haute be used as a legal alternate? Assuming standard alternate minimums for Terre Haute, the lowest forecast is contained in the remark OCNL C10 OVC 3TRW-F. Terre Haute would satisfy FAR requirements for an alternate airport.

The preceding comparison contradicts a misconception that in the absence of an FT there is no forecast on which to base destination, alternate, and fuel requirements. If not TWEBs, FAs are always available; the FAs might not be very detailed, requiring additional alternates and fuel, but the FAs are nonetheless available.

Forecasters consider four elements when writing FTs:

- Expected weather
- Local effects
- Climatology
- Amendment criteria

Attention to detail places a premium on the forecaster's time and judgment.

In addition to new FT amendment criteria, several other changes should improve the terminal forecast. The NWS recommends that forecasters restrict the use of conditional terms and keep "occasional" restricted to short periods. Forecasters are encouraged to amend when conditions do not materialize, even when amendment criteria have not yet been met. For readability, there is only one forecast period per line of text. The use of "chance" will not be used in the first 6 hours of the forecast. (The conditional term "slight chance" has been eliminated.) Maximum forecast visibility may be reduced to 5 miles.

A number of factors are to be expected that will cause inconsistencies in reported conditions. The forecaster might not believe observations are representative, or the forecaster might expect conditions to change rapidly. Even in the body of the forecast there is only a greater than 50 percent probability of occurrence during more than half the forecast period. Amendment criteria give the forecaster some latitude. The forecaster might be waiting until FT amendment criteria are reached—reported or expected. The time required to write and distribute an amendment is also a consideration; saturated computer and communications systems can hinder timely updates. An amendment might not be issued if it is less than 2 hours between the time an amendment is required and a new forecast becomes effective or the next portion of the forecast becomes effective. Time parameters on domestic FTs usually indicate phenomena will change within 2 hours of the specified time. And realize that forecasters have other duties that might hinder timely amendments.

FAs, TWEBs, and FTs are usually perfectly consistent given their individual purpose and criteria. Apparent inconsistencies arise because:

- Cloud heights in the FA are generally MSL, whereas in the FT cloud heights are always AGL.
- FTs can consider local effects not within the scope of other products.
- A pilot attempts to extrapolate an FT beyond 5 miles.

When inconsistencies develop, FSS specialists and forecasters coordinate to resolve differences. Pilots using DUATS will have to consult an FSS for resolution.

Domestic FTs do not directly forecast turbulence or icing; however, these phenomena are implied, which is the same as an SA. Strong surface winds, LLWS, and thunderstorms indicate turbulence. Surface temperatures close to freezing with cloud layers, freezing precipitation, and thunderstorms suggest icing.

GILBERT FT AND TAF

Recall that the beginning of this chapter presented an FT from Hurricane Gilbert's assault of the Texas Gulf Coast. The FT is translated after this re-presentation.

```
BRO FT 160808 20 SCT C60 OVC 4RW 0220G30 OCNL C20 BKN 2RW+ CHC
    C5 X 1/2TRW+ Q50. 00Z 3 SCT C12 OVC 2RW+ 0860Q75 OCNL C3 X 1/4
    T+RW+ G120. 02Z LIFR CIG TRW WND.
```

According to the FT, at 0800Z, expect 2,000 scattered, ceiling 6,000 overcast, visibility 4 in moderate rain showers, and wind 020° at 20 gusting to 30 is forecast. Remarks indicate occasional ceilings 2,000 feet broken, visibility 2 in heavy rain showers, and a chance of ceilings 500, sky obscured, visibility ½ mile in thunderstorms, and heavy rain showers with wind gusts to 50 knots.

VFR flight might be possible, but certainly not fun. An IFR pilot might get away with a nonprecision approach, but he or she better be prepared for a precision approach with a chance of 500 and ½. With either the "occasional" or "chance" forecast, an alternate is required; this airport would not qualify for filing as an alternate. Thunderstorm activity would have to be considered for both VFR and IFR operations.

Conditions are expected to deteriorate around 0000Z, so the remarks of the previous portion could be more prevalent during the end of that period. Hurricanes and other low-pressure systems demand caution. These systems often have repetitious bands of weather that cause conditions to deteriorate, improve, and deteriorate again; the eye of a hurricane is an ultimate example. (Refer to Fig. 10-18 in chapter 10 for a satellite view of Gilbert.)

At 0000Z conditions are expected to deteriorate to 300 scattered, ceiling 1,200 overcast, visibility 2 in heavy rain showers, and wind 080° at 60 with peak gusts in squalls to 75 knots.

Remarks indicate occasional ceilings 300 feet, sky obscured, visibility ¼ in severe thunderstorms, heavy rain showers, and wind gusts to 120 knots. Surely no one would argue with the briefer's probable statement that VFR flight is not recommended, and conditions will probably go below IFR landing minimums occasionally.

Here's how this FT might appear in TAF code:

KBRO FT 160545 0606 02020G30KT 4SM SHRA SCT020 OVC060 TEMPO 1023
    2SM +SHRA BKN020TCU PROB40 2023 02020G50KT 1/2SM +TSSHRASQ
    VV005CB BECMG 2300 08060G75KT 2SM +SHRASQ SCT003 OVC012TCU
    TEMPO 2308 08060G120KT 1/4SM +TSSHRAGRSQ VV003CB

The FT, the most detailed aviation forecast, is a valuable planning tool when used in conjunction with current reports and other forecasts, especially when limitations and amendment criteria are understood and considered.

A hurried pilot poked his head into the FSS and requested destination weather and terminal forecast. The briefer provided that specific information, which contained nothing significant. The pilot was halfway out the door when the briefer mentioned thunderstorms en route. Needless to say, the pilot was interested.

8
Winds and temperatures aloft forecasts (FD)

A NAVION PILOT CALLED FLIGHT SERVICE WITH A REQUEST FOR "WINDS aloft." The specialist asked, "Winds aloft for what altitudes?" The pilot rather indignantly replied, "Whatever's best for my direction of flight!" Perhaps the specialist's reply should have been, "It's 53,000 feet every time."

Winds and temperatures aloft forecasts (FD) for the contiguous United States, Alaska, and many oceanic areas are computer-generated at the National Meteorological Center outside Washington, D.C. FDs for the Hawaiian islands are produced by the Honolulu forecast office. The forecasts are based upon twice-daily radiosonde balloon observations (balloon with a radiosonde attached) that are normally released at 1100Z and 2300Z daily. A list of launch sites can be found in the *Airport/Facility Directory*. Because these sites and times are published, pilots cannot expect to receive launch warnings in the form of NOTAMs or broadcasts, except for unscheduled releases. Figure 8-1 shows the balloon preparation building and launch site at Norman, OK. FD forecast locations are contained in appendix B.

A computer program known as the Nested Grid Model (NGM) analyzes data to project weather system movements and produce a forecast. NGM replaced the Limited

Fig. 8-1. *A balloon with a radiosonde attached is launched twice daily from approximately 120 locations such as this one in the United States. Wind profilers that provide improved observations are scheduled to replace this system but not until the turn of the century.*

Fine Mesh (LFM) for FD-tabulated forecasts. NGM's smaller grid and better resolution, as compared to LFM, has improved forecast accuracy; however, the NGM can only consider synoptic (large-scale) weather systems. Large-scale terrain is considered to some extent, but local features are not considered; therefore, FDs tend to be less accurate in the western states, especially below 12,000 feet.

Approximately 750 upper-air stations worldwide take observations; about 120 stations are in the United States. Stations are generally located on land, which leaves great expanses of ocean without observations; satellite and aircraft reports help fill in the gaps. The computer must interpolate for locations without observations and sparse observational data hinders the accuracy of the forecast.

Winds and temperatures aloft forecasts can be obtained from a number of sources: flight service stations, transcribed weather broadcasts, pilots automatic telephone weather answering service, telephone information briefing service, direct user access terminals, and commercial vendors.

Emphasis must be placed on the term "forecast" because these are not "actual" winds and temperatures aloft observations. The subsection "Using the FD forecast" in this chapter will help you cope with the differences between forecast and actual.

TABULATED FORECASTS

FDs normally become available after their scheduled transmission times of 0440Z and 1640Z. They consist of three forecast periods (6, 12, and 24 hours) that are labeled FD1, FD2, and FD3 for levels through 39,000 feet, and FD8, FD9, and FD10 for 45,000 and 53,000 feet (Table 8-1).

Table 8-1.
Winds and temperatures aloft forecast schedule.

File type		Valid	For use
Forecasts based on 0000Z data, available at 0440Z.			
FD1	FD8	0600Z	0500–0900Z
FD2	FD9	1200Z	0900–1800Z
FD3	FD10	0000Z	1800–0500Z
Forecasts based on 1200Z data, available at 1640Z.			
FD1	FD8	1800Z	1700–2100Z
FD2	FD9	0000Z	2100–0600Z
FD3	FD10	1200Z	0600–1700Z

Note: FD1, FD2, FD3 for levels through 39,000 feet.
　　　 FD8, FD9, FD10 for levels 45,000 and 53,000 feet.

FAA automated flight service stations display levels 3,000 through 53,000 feet; the 45,000- and 53,000-foot levels are not available for all standard FD locations (appendix B). Other FSSs usually post levels through 39,000, and the two highest levels are normally available on request.

FD3s and FD10s are plagued with the same limitations as other forecasts and must be viewed with skepticism and used only for advanced preflight planning then updated with the latest forecasts immediately prior to departure. This requires the pilot to become familiar with the issuance times in Table 8-1. For example, the night before a flight, a pilot planning a 1900Z departure might obtain the FD3s that are based upon 0000Z data. The FD1s that are based upon 1200Z data become available at 1640Z. If the pilot fails to update the forecast, he or she could be wide open to a violation in the event of a problem.

Figure 8-2 contains FD1 and FD8 examples that are based upon 1200Z radiosonde data on the 30th day of the month (DATA BASED ON 301200Z). The "data based on" statement must always be checked. From time to time, old FDs fail to be purged and remain in the system; it's possible to receive data that is 24 hours old. The next line

WINDS AND TEMPERATURES ALOFT FORECASTS

DATA BASED ON 301200Z
VALID 301800Z   FOR USE 1700-2100Z.   TEMPS NEG ABV 24000

FT   3000   6000   9000   12000   18000   24000   30000   34000   39000

SFO 3513 3316+10 3220+06 3224+01 3138-11 3148-24 325539 315846 315954
RNO   0605   3308+02 3217-02 3134-16 3052-28 316542 316848 316853

DATA BASED ON 301200Z
VALID 301800Z   FOR USE 1700-2100Z.   TEMPS NEG ABV 24000

FT   45000   53000

SFO 314961 304263

Fig. 8-2. *Tabulated winds and temperatures aloft forecasts are available from just above the surface to 53,000 feet. These forecasts are based upon the twice-daily radiosonde observations.*

states VALID 301800Z FOR USE 1700–2100Z, which means that these FDs are for use between 1700Z and 2100Z. The computer does not forecast an average; the model predicts winds and temperatures for one specific time, in this case 1800Z (VALID 301800Z).

Realize that the reliance upon *expected* movement of synoptic systems might cause apparent errors in forecast preparation. With rapidly moving or intensifying systems, FDs can change significantly during the "for-use" period. This would be especially true for flights at the beginning or end of the period.

Forecast levels are true altitude through 12,000 feet MSL (true height above sea level). From 18,000 through 53,000 feet, levels are *pressure altitude* (height as indicated with an altimeter setting of 29.92"). There are no forecast levels between 12,000 MSL and 18,000 pressure altitude. Figure 8-2 shows FDs for SFO (San Francisco, CA) and RNO (Reno, NV).

Forecast levels within the area of frictional-effect of the wind and the earth's surface are omitted; therefore, forecast levels within approximately 1,500 feet of the surface do not appear. Temperatures are not forecast for the 3,000-foot level or levels within 2,500 feet of the surface. Refer to Fig. 8-2. RNO has an elevation of 4,412 feet. Because the 6,000-foot level is within 2,500 feet of the surface, no temperature is forecast.

Refer to the SFO 12,000-foot winds in Fig. 8-2. The first two digits of a wind group represent true direction (*from* which the wind is blowing) to the nearest 10° ("32"24+01); the third and fourth digits indicate speed in knots (32"24"+01); and the last two digits are temperature in degrees Celsius (3224"+01"). Temperatures are indicated as plus or minus (+ or –) through 24,000 feet. Because all temperatures above

24,000 are below 0°C, the minus sign is omitted starting at that level; therefore, 3224+01 is wind blowing from 320° true at 24 knots, and the temperature is 1°C.

Forecast speeds less than 5 knots are encoded 9900 and referred to as "light and variable." Briefers are periodically asked "What's the direction and speed of the light and variable winds?" One extreme case had a rather irate pilot demand to know what was "actually written on the paper." The specialist correctly replied "niner-niner-zero-zero!"

Pilots must *interpolate* (compute intermediate values) to determine forecast values between levels and reporting locations. For a route from Oakland to South Lake Tahoe at 13,500 feet, use the SFO and RNO FDs in Fig. 8-2. Pair up and average the 12,000-foot forecasts and 18,000-foot forecasts from each location.

At 12,000 feet, the forecasted directions are the same. A 7-knot speed difference results in an average of 21 knots: 7 ÷ 2 = 4 (averaged up from 3.5), and 17 + 4 = 21. The average temperature is 0°C.

At 18,000 feet above Oakland and South Lake Tahoe, the forecasted directions are the same. A difference in forecasted speed of 4 knots results in an average of 36 knots (4 ÷ 2 = 2, and 34 + 2 = 36). The difference in temperature is 5°C, resulting in an average of –14°C: –5 ÷ 2 = –3, and –11 + –3 = –14. Be careful with the algebraic sign; remember that you are adding two negatives in this case.

The final interpolated forecast results for the proposed route are:

- 12,000 feet—320°, 21 knots, and 0°C.
- 18,000 feet—310°, 36 knots, and –14°C.

The 13,500-foot level is ¼ of the way between 12,000 and 18,000 feet; therefore, divide the difference between levels by four and add the result to the 12,000-foot values. Wind direction is 320°, speed 25 knots (15 ÷ 4 = 4, and 21 + 4 = 25), and the air temperature is –04°C (–14 ÷ 4 = –4, and 0 + –4 = –4). Because direction is to the nearest 10°, speed is in knots, and temperature is in whole degrees Celsius, the result cannot have a value in smaller increments than the original data; values are rounded off. As already noted, be careful of the algebraic sign. FAA exams might require calculating wind direction to the nearest 5°; however, for practical purposes, this is not necessary.

Is a forecast of 701548 a misprint or garbled transmission? With forecast winds of 100 knots or more, 5 is added to the first digit of the wind-direction group. Decode by subtracting 5 from the first digit of the wind direction and adding 100 to speed. In this example, wind direction, speed, and temperature are:

Direction	Speed	Temperature
70	15	48°
– 5	+100	
200°	115 knots	–48°

No mathematical sign (+ or –) was specified, so the temperature must be negative. Maximum speeds for FD tabulated forecasts are 199 knots.

FORECAST CHARTS

Forecast winds and temperatures aloft are also available in graphic form, issued twice daily and valid at 1200Z and 0000Z. FD charts are excellent for determining forecast winds for long-distance flights. By visually depicting winds at various levels, favorable routes and altitudes can be determined. The eight panels contain forecast levels from 6,000 through 39,000 feet. Figure 8-3 illustrates the 30,000-foot pressure-level panel VALID 12Z 02 AUG 94. Because of valid times and computer models, some differences between the chart and tabular forecasts are to be expected.

Plotted data are standard. The inset in Fig. 8-3 shows the station model for DRT (Del Rio, TX). Forecasts are to the nearest 10° and 5 knots. Arrows with pennants and barbs are similar to those on other charts. The first digit of the wind direction is obtained from the general direction of the arrow, which is northeasterly at DRT. The number above the arrow (3) represents the second digit; therefore, the wind is blowing from 030°. Pennants (50 knots), barbs (10 knots) and half barbs (5 knots) denote speed. The wind speed at DRT decodes as 40 knots. The temperature appears just above the station: –33°C.

AMENDMENT CRITERIA

Although FDs are generated in Washington, regional NWS offices are responsible for amendments. FDs are amended when, in the forecaster's judgment, there is a change or an expected change in the wind or temperature that would significantly affect aircraft operations. Amendment procedures are complex.

Reviewing Table 8-2, it can be seen there must be a considerable difference between forecast and actual winds to require an amendment. In general, forecasts within 30° of direction, and plus or minus 20 knots are considered accurate. As with turbulence and icing, the only way to verify the forecast is through pilot reports.

Table 8-2. Winds and temperatures aloft amendment criteria.

With a forecast of:	Amend if:
Wind direction	
Wind speed 5 to less than 30 knots	Direction change of 45 degrees or more
Wind speed 30 knots or greater	Direction change of 30 degrees or more
Wind speed	
Wind speed of less than 70 knots	Speed changes by 20 knots or more
Wind speed between 70 and 100 knots	Speed changes by 30 knots or more
Wind speed between 100 and 135 knots	Speed changes by 40 knots or more
Wind speed greater than 135 knots	Speed changes by 50 knots or more
Temperature	
Amend if observed or forecast temperature changes by 5°C or more.	

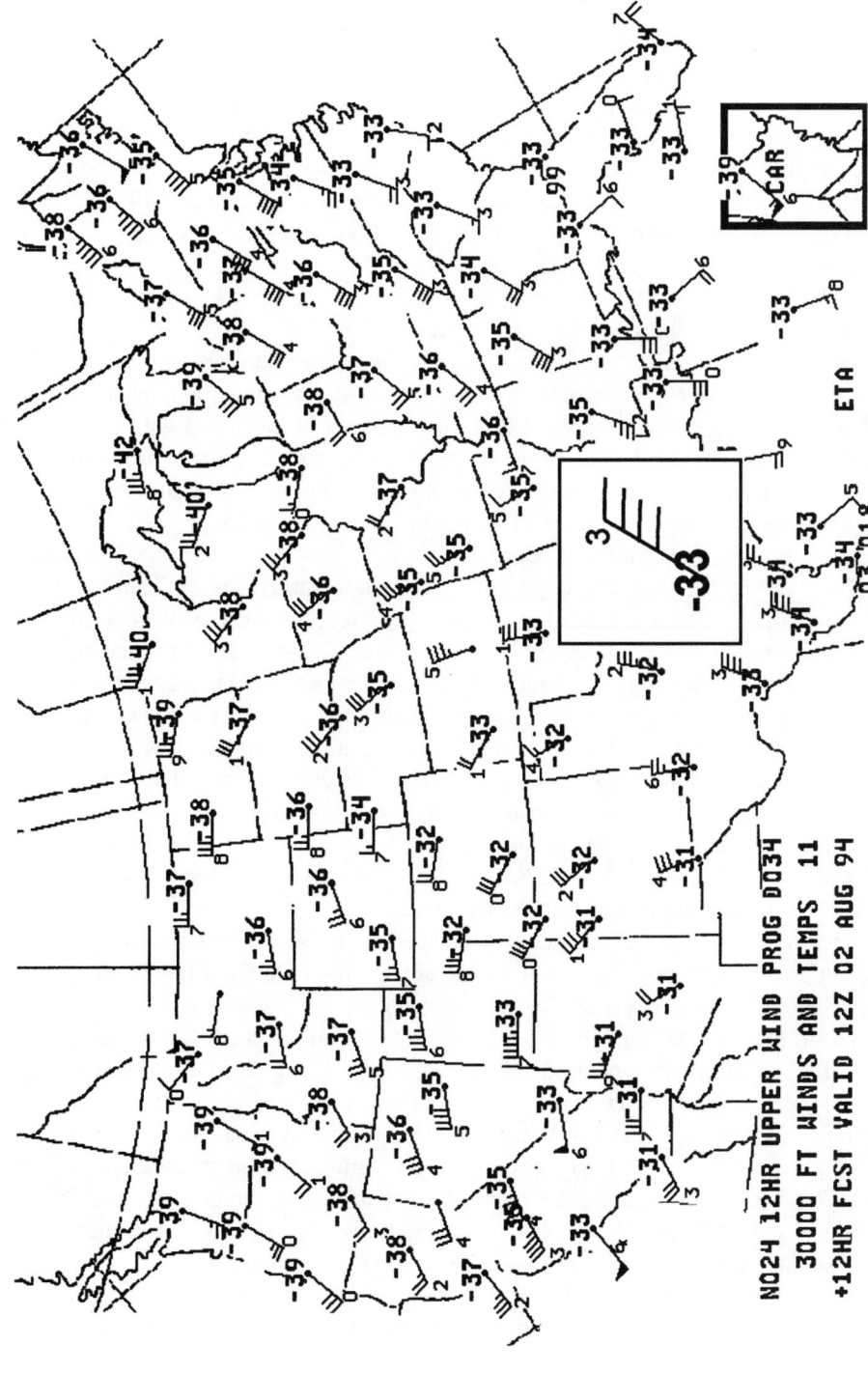

Fig. 8-3. *Winds aloft forecasts are available in chart form through 39,000 feet. Chart forecasts are based on a different computer model and have different valid times; therefore, differences between chart and tabulated values are to be expected.*

145

USING THE FD FORECAST

FDs provide the pilot with two valuable pieces of information: wind direction and speed, plus temperature. Both significantly affect aircraft operation and performance. Failure to properly consider and apply either can be potentially hazardous.

In spite of its limitations, the FD can never be ignored. Pilots are required by FARs to consider fuel requirements and are prohibited from beginning a flight either VFR or IFR unless (considering wind and forecast weather conditions) the aircraft will have enough fuel to fly to destination and an alternate, if required, and still have appropriate fuel reserves. FAR fuel-reserve minimums, which do not necessarily equate to "safe," in no way relieve the pilot from keeping careful track of ground speed and revising the flight plan accordingly.

When reserves are marginal, good operating practice dictates the careful tracking of position and ground speed. Marginal is not necessarily synonymous with legal; in sparsely populated areas, a fuel reserve of 30 minutes might be sufficient with clear weather reported and forecast. But a 30-minute reserve doesn't make any sense with marginal weather or thunderstorms and the nearest suitable alternate 35 minutes away. Chapter 7 discusses how legal alternates might not be satisfactory with a busted forecast; the same is true for legal fuel reserves.

The following situation illustrates how a series of small, seemingly insignificant factors at the time have the potential to lead to disaster. The flight from Van Nuys, CA, to Tonopah, NV, was based on 4 hours of fuel available and a 10-knot headwind; time en route was estimated at 3 hours and 15 minutes. My Cessna 150 was fueled Friday when I arrived at Van Nuys. During the preflight Sunday morning I noticed the fuel was not at the top of the fuller neck: factor no. 1. This was not unusual because the airplane was parked on a slight incline and some fuel tends to vent overboard. The departure required an IFR climb to on-top conditions, which added about 15 minutes to time en route: factor no. 2.

Over Trona, CA, about halfway, ground speed checks indicated winds were as forecast. Calculations indicated adequate fuel for Tonopah based upon the 4 hours of fuel and ignoring the extra time required for departure.

The Cessna 150 climbs like a wet mop, so I decided not to land at Trona: factor no. 3. The fuel gauges were bouncing on 0, and I still had 30 minutes to fly before landing at Tonopah. There were no suitable alternates: factor no. 4.

I made a straight-in approach with everything stowed ready to crash, but I landed safely in spite of some extremely poor planning. By the way, 22.6 gallons went into the 22.5-gallon-usable airplane. Never again.

A Grumman Tiger pilot—an instructor was with a student—was not so fortunate. On a flight from Salt Lake City to Tonopah, they crashed short of the airport, out of fuel. The pilot couldn't understand why, after calculating the airplane had 2 hours and 45 minutes of fuel, the engine quit after only 2 hours and 31 minutes. Needless to say, the FAA wanted to have a little chat with this flier.

The venturi effect at mountains and mountain passes accelerates winds over ridges and through passes. Stronger-than-forecast winds should be expected in these areas, especially within 5,000 feet of terrain.

Aircraft performance charts are based on the standard atmosphere, more precisely the *International Standard Atmosphere* (ISA). Standard atmosphere temperature and pressure at sea level are 15°C and 29.92" of mercury. The standard lapse rate, which is the decrease of temperature with height, in the troposphere is approximately 2°C per thousand feet. Temperature decreases to a value of –57°C at the tropopause altitude of approximately 36,000 feet; the tropopause is the boundary between the troposphere and the stratosphere. An isothermal lapse rate occurs in the stratosphere to about 66,000 feet (Fig. 8-4).

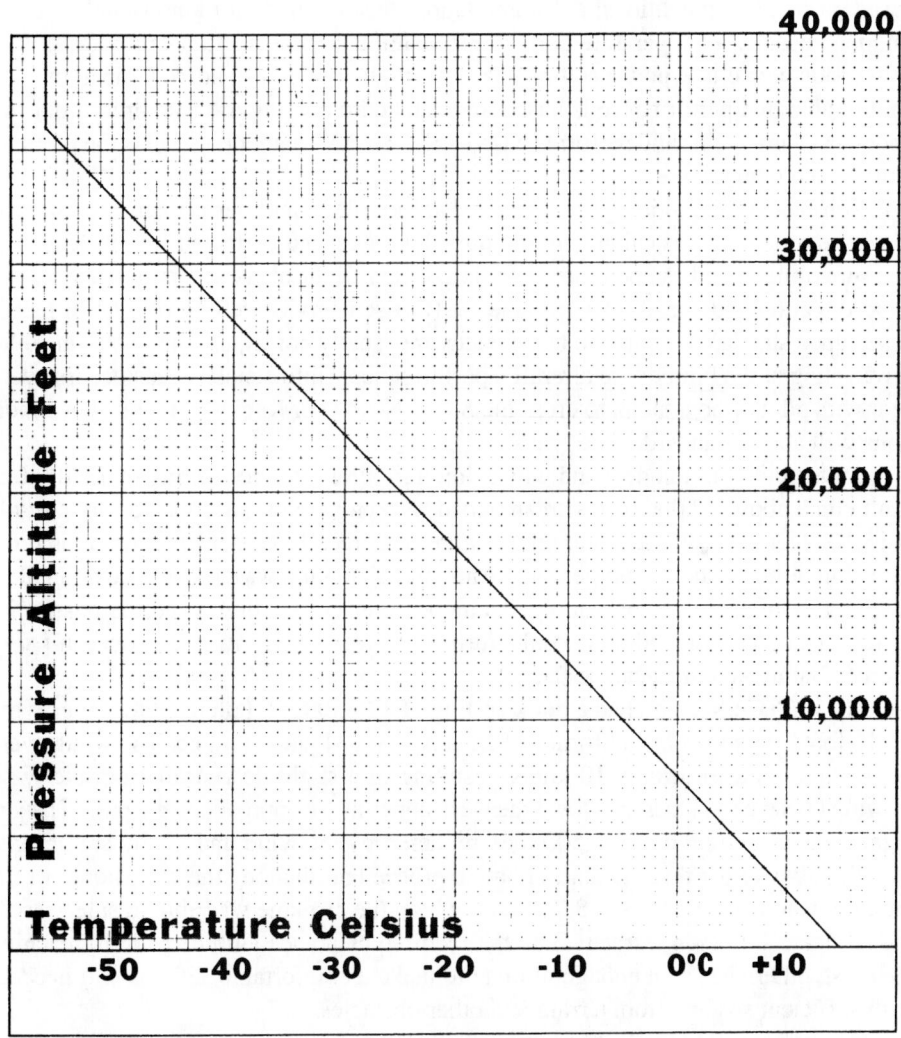

Fig. 8-4. *Because aircraft performance is based upon standard conditions and because standard conditions rarely exist in the real atmosphere, pilots need a means of calculating actual performance. This is most often accomplished by comparing standard conditions with actual conditions.*

Altitudes on the ISA chart are pressure altitudes. Pilots using the local altimeter setting fly indicated altitude through 17,500 feet; pressure altitude (29.92) is flown at 18,000 feet and above. Winds aloft forecasts are true altitude through 12,000 feet, so there will be a slight difference between the altitudes in the forecast (true altitude) and those flown by the pilot (indicated altitude). The difference is negligible unless atmospheric pressure is extremely high or low. A variance of 1" of mercury from standard would only result in a 1,000-foot altitude difference. Temperature is the biggest factor.

Standard conditions rarely occur in the real world, and performance charts are based on standard conditions; accommodation must be made for a nonstandard environment, usually a temperature correction. Manufacturers sometimes provide an ISA conversion with cruise power setting charts for high, low, and standard temperatures or simply note that performance is based upon standard conditions. The aircraft doesn't understand any of this and performs based upon the environment, which is the pressure altitude and temperature.

Nonstandard conditions affect true air speed (TAS) and performance, as well as power settings. (TAS is calibrated, or equivalent, airspeed corrected for air density— pressure altitude and temperature. TAS is used on flight plans and is independent of wind direction and speed.) An airplane's advertised service ceiling of 13,100 feet is based upon standard conditions. Differences are usually not significant unless the pilot is operating at the limit of the aircraft's performance. Unfortunately, this occurs every year with pilots who attempt to cross the Sierra Nevada or Rocky Mountains in conditions well above standard.

Some pilots can't understand why an aircraft with a service ceiling of 13,100 feet can't climb above 12,000 feet with an outside air temperature of 0°C. Density altitude is 13,100 feet at 12,000 feet and temperature 0°C; recall that density altitude is pressure altitude corrected for temperature. And this calculation was determined as an ideal case. A runout engine, poor leaning technique, over-gross weight, and the possibility of turbulence and downdrafts would further decrease performance. You can't fool Mother Nature; attempts can be fatal.

My 1966 Cessna 150 had a book ceiling of 12,650 feet. I planned to traverse the 9,941-foot Tioga Pass in California's Sierra Nevada Mountain range. The winds were out of the northeast at only 10 knots, resulting in a slight downdraft from the wind flowing up the east slopes and down the west slopes as we approached the pass. The airplane wouldn't climb out of 9,500 feet, hindered by the wind and a temperature that was slightly warmer than standard. I had to proceed north along the west slopes of the mountains to Ebbett's Pass at 8,732 feet, where the airplane was able to safely cross the mountains. Lesson learned: Gain the required clearance altitude prior to reaching the crest; otherwise, gain enough altitude to make a comfortable 180° turn, if needed, with sufficient spacing from terrain and other obstacles.

On another occasion, I had filed for an IFR flight from Lancaster Fox Field in the Mojave Desert to Ontario, CA. I requested 7,000 feet and planned to go through the San Fernando Valley because of lower minimum altitudes. The clearance came back, "cleared via the Cajon two arrival; climb and maintain 11,000." The surface temperature was 85°F and the Cessna 150 was not going to 11,000 that day. After negotiating with a perturbed ground controller, I received my requested routing. Pilots must know aircraft performance and not allow ATC or anyone else to push them into an untenable position that would compromise safety. I had been put in an *unobtainable* position.

All aircraft have limitations, including turbojet aircraft that often fly at the edge of their performance envelope where nonstandard conditions critically affect performance. The pilot's task is to determine aircraft performance based on forecast temperatures aloft. This requires the information in Fig. 8-4. From the discussion earlier in this chapter, the average temperature from Oakland to South Lake Tahoe at 13,500 feet was –04°C. Figure 8-4 indicates that the standard temperature for 13,500 feet is –10°C. The forecast temperature is 6° warmer, which is above standard. The air is less dense than standard, so aircraft performance will be less than performance charts advertise. Based on these conditions, density altitude is about 15,000 feet.

Certain performance charts require the pilot to determine temperature at altitude relative to ISA. For example, the forecast temperature over Reno at Flight Level 300 (30,000 feet) is –42°C. Figure 8-4 indicates that the standard temperature for that pressure altitude is –45°C. Certain flight computers can be used to determine ISA temperatures; the Jeppesen CR-3 has a true-altitude computation window. With the scales aligned (10 on the outer scale with 10 on the inner scale), standard temperature is read under pressure altitude. Underneath the pressure altitude of 30,000 feet, –45 appears; therefore, the forecast temperature is ISA +3°C.

FDs are a source of forecast freezing level. They should be in general agreement with the WA because they're based upon the same data. Differences result from FD freezing levels representing only one point in time, whereas WA forecasts take into account changes during the period; however, if a significant difference occurs, be alert for other possible forecast errors. FSS specialists coordinate with forecasters under such circumstances. Pilots using DUATS would normally have to consult an FSS for resolution. Figure 8-2 indicates the expected freezing level at 1800Z is approximately 12,500 feet over San Francisco, lowering to 11,000 feet in the Reno area.

FDs can be used to determine the approximate height of the tropopause. Winds are strongest just below the tropopause. Checking Fig. 8-2, San Francisco winds at 39,000 feet are 59 knots, and at 45,000 feet, the winds have decreased to 49 knots. Also note that speed continues to decrease at 53,000 feet, and temperature remains almost constant; therefore, the tropopause is between the 39,000- and 45,000-foot pressure levels.

FD forecast limitations and amendment criteria must be understood for effective flight planning. Product preparation plus the advantages of—and to some extent, the limitations of—computer models have already been discussed. Surface heating as well as terrain affect winds aloft. Today's technology cannot account for the effects of land/sea and mountain/valley winds or frictional effects between wind and the surface. Other forecast problems include the extent/availability of data and timing.

FDs are based upon the expected movement of weather systems, so errors result when weather systems move faster or slower than forecast. This is one reason why flight watch specialists are required to continually solicit reports of winds and temperatures aloft. Specialists are specifically trained to recognize these situations and provide updated information. A pilot's careful tracking of position and ground speed will verify the accuracy of the FDs.

Inertial navigation, LORAN, global positioning system, and other electronic navigation systems can provide immediate wind readouts. Anyone remember how to calculate winds with the E6B flight computer? Electronic flight computers make the calculation easy. This is a reminder for pilots to become actively involved in the system with PIREPs. Observed winds aloft, whether they confirm or contradict the forecast, should be routinely passed to flight watch. Only two upper-air observations are made each day, so PIREPs are the only other direct source of observed winds, and the only way to verify the forecast.

The situation might change in the future with wind profilers that will automatically and almost continuously provide high-resolution upper-air wind measurements. Naturally, the closer to observation time, the more accurate the forecast. Accuracy normally deteriorates with time; the FD2s and especially the FD3s are vulnerable to the clock's effects. The evening forecast becomes available after 0440Z; it doesn't do much good to request winds any earlier for the following day. Flights departing after 1640Z, must consult the new FDs. Weather patterns can change significantly in 12 hours.

The pilot has no option but to use FDs for flight planning according to regulations. Local or short flights might mean nothing more than an eyeball interpolation. Exams and flight tests require computer calculations. Flights toward the limit of aircraft range will require a careful interpolation and calculation, plus a close watch on progressive checkpoints en route.

With the general criticism of winds aloft forecasts, it's amazing how many pilots call and must absolutely have winds two or even three days in the future. Then there's the pilot who contacts flight watch and can't understand why the winds are 20° and 5 knots off the forecast. Oh well, you can't please everyone. Hopefully this chapter provided some insight into FD limitations, the causes of inaccuracies, and perceived errors of this valuable but often-maligned product.

9
Radar and convective analyses

"A DISASTROUS THUNDERSTORM ACCIDENT CLOSE TO BOWLING GREEN, Kentucky, in 1943 that involved an American (Airlines) DC-3 started a chain of events that eventually led to the first systematic research into thunderstorm behavior. The plane crashed onto the ground either near or under a severe thunderstorm. Buell (C.E. Buell, chief meteorologist, American Airlines 1939–1946) initiated a letter to the Civil Aeronautics Board pointing out the appalling dearth of understanding of what actually occurs inside a thunderstorm, as evidenced by the accident investigation. He recommended a massive research effort be organized to probe into thunderstorms and document their internal structure." ("Airline Weather Services 1931–1981," by Peter E. Kraght.)

Early airborne radars that evolved from research in the late 1940s and early 1950s had many technical problems, but by the beginning of the 1960s, most airliners were equipped with acceptable airborne weather radar. Ground-based radar was also developed.

Access to radar information has increased over the years. NWS radars in the East and FAA radars west of the Rockies supplemented by a few NWS sites are used to compile a national radar-summary chart. The radar network in the West was thought unjustified because severe weather is relatively rare in that area. Radar weather reports (RAREPs) and convective analysis charts are routinely transmitted on NWS and FAA circuits and are available through many private services.

Many twin and high-performance single-engine aircraft are being equipped with airborne weather radar or lightning detection systems. There have been several attempts to place ground-based radar displays in aircraft.

By the middle to late 1990s, the next-generation weather radar (NEXRAD) will become operational. NEXRAD will be a Doppler-radar system with coverage from coast to coast. Each system, product, or service has its own particular application and limitations that must be thoroughly understood for safe and efficient flight.

RADAR

Radar displays an image dependent upon reflected energy or back scatter. Intensity depends on several factors and among them are particle or droplet size, shape, composition, and quantity. Particles must be at least precipitation-size for detection on most of today's weather radars; therefore, a precipitation-free area does not translate into a cloud-free sky. Droplets reflect more energy than snow. There is almost no relation between the intensity of snow and back scatter; therefore, an intensity is never assigned to a radar report of snow. On the other hand, hail coated with water produces the best back scatter and the notation "hail shaft" might appear on reports and charts.

Three radars are used in the aviation weather system:

- FAA ATC en route
- National Weather Service
- Airborne

Each system has a specific purpose and its own application and limitations.

ATC radar is specifically designed to detect aircraft; a narrow fan-shaped beam reaches from near the surface to high altitudes. ATC radars have a wavelength of 23 centimeters (cm), which is ideal for detecting aircraft but which reduces the intensity of detected precipitation; additional features reduce the radar's effectiveness to see weather.

To efficiently detect aircraft and eliminate distracting targets, ATC radars use circular polarization (CP), moving target indicator (MTI), and sensitivity time control (STC) features.

CP results in a low sensitivity to light and moderate precipitation. MTI only displays moving targets; unless droplets have a rapid horizontal movement they remain undetected; even rapidly moving precipitation will not be observed when advancing perpendicular (tangentially) to the radar beam. STC further eliminates light and decreases the intensities of displayed precipitation. Naturally, controllers engage these features during poor weather to accomplish their primary task of aircraft separation, especially at approach facilities.

NWS radars, on the other hand, with wavelengths of 5 or 10 cm and a narrow linearly polarized beam are ideal for detecting precipitation-size particles. Figure 9-1 shows a typical weather radar installation. This site houses a WSR-88 NEXRAD unit at Norman, OK. Sensitivity time control on NWS radars compensates for range attenuation, which is loss of power density due to distance from echoes. STC-displayed intensity remains independent of range; therefore, targets at different ranges with the same intensity appear the same to the radar specialist. NWS radars can detect targets up to 250 nautical miles (nm); however, due to range and beam resolution, which is the ability of the radar to distinguish individual targets at different ranges and azimuth, an effective range of 125 nm is used.

Figure 9-2 shows a WSR-57 display console. The large center scope is the plan position indicator (PPI). The PPI displays distance (range) and direction (azimuth) of the target in relation to the antenna. The smaller scope on the right is the range height indicator (RHI). The RHI determines approximate tops and occasionally determines bases of precipitation. Tops are approximate due to nonstandard refraction, bending of the radar beam, by the earth's atmosphere. Tops can be in error as much as several thousand feet.

The amplitude modulation scope on the left of Fig. 9-2 is used to distinguish between liquid precipitation, snow, and hail. Melting level (MLT LVL), which translates into freezing level for aviation purposes, might appear in radar reports. This scope is also used to determine nonprecipitation echoes, such as aircraft and ground clutter, or *anomalous propagation* (AP), which is ground targets that are not normally present in the usual ground clutter pattern.

NEXRAD will be a quantum leap in providing early warning of severe weather. NEXRAD will be the standard for the next 20 to 25 years with a wavelength of 10 cm. The WSR-88 network should consist of up to 195 units: 113 NWS sites in the contiguous United States, with additional sites in Alaska, Hawaii, the Caribbean, and western Europe, and 22 Department of Defense units. The NEXRAD network will fill radar gaps in the West.

NEXRAD is a Doppler radar that detects the relative velocity of precipitation within a storm. Additionally, NEXRAD has the capability of detecting cloud particles, as well as precipitation. It will increase the accuracy of severe thunderstorm and tornado warnings and has the capability of detecting wind shear. The radar specialist using a WSR-88 will have radar data acquisition, radar product generation, and display units as shown in Fig. 9-3.

Notice the trend in technology from the raw-data cathode-ray tube to the computer-generated display era. NEXRAD will provide hazardous and routine weather radar data above 6,000 feet east of the Rockies and above 10,000 feet in the western United States, although the details of how this information will be relayed to the FSS specialist and pilot have not yet been established. It is also expected to be available in the aircraft directly through Mode S transponders with automatic data-link capability.

Fig. 9-1. *National Weather Service weather radars are designed to detect and display precipitation-size particles.*

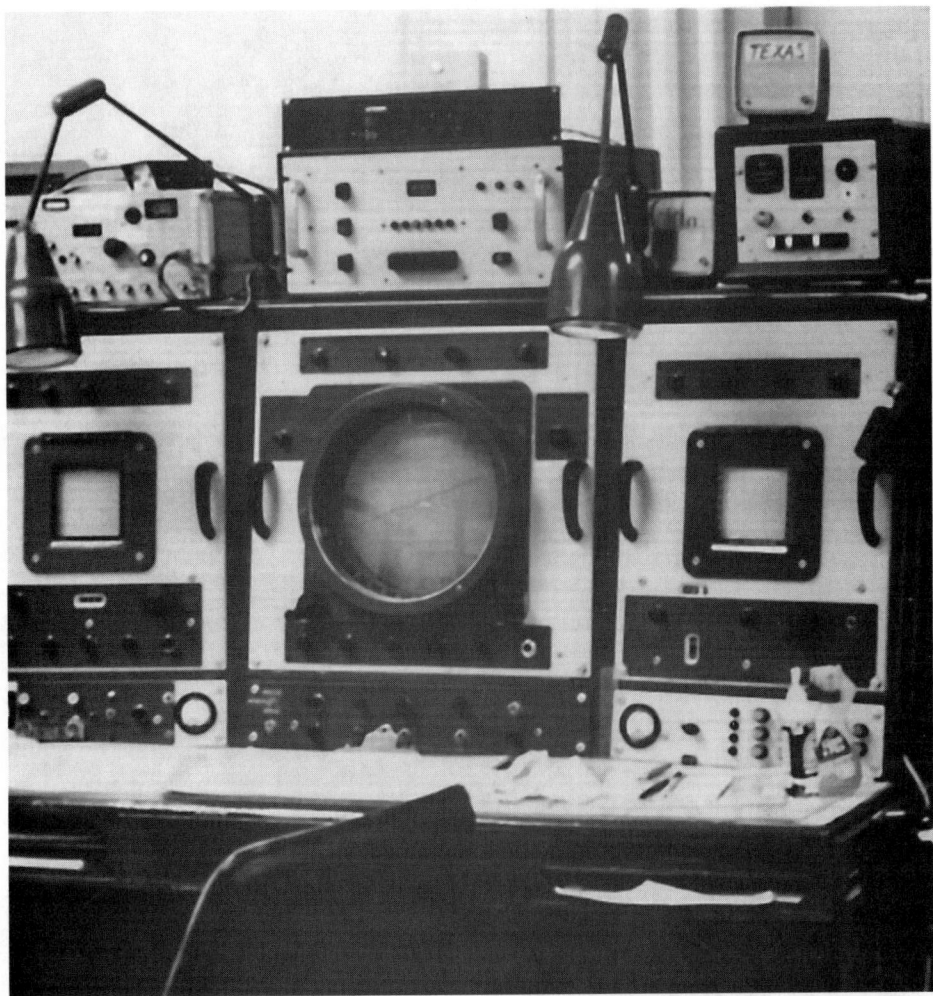

Fig. 9-2. *This is a National Weather Service WSR-57 display console. The PPI center screen displays echo azimuth and distance; the RHI scope on the right displays echo height; and the amplitude modulation scope on the left allows the radar specialist to determine echo type.*

Airborne weather radars are low-power, generally with a wavelength of 3 centimeters. Precipitation attenuation, which is directly related to wavelength and power, can be a significant factor. Precipitation attenuation results from radar energy being absorbed and scattered by close targets, and the display becomes unreliable in close proximity to heavy rain or hail. Intensity might be greater than displayed, and distant targets might be obscured. An accumulation of ice on the radome causes additional distortion.

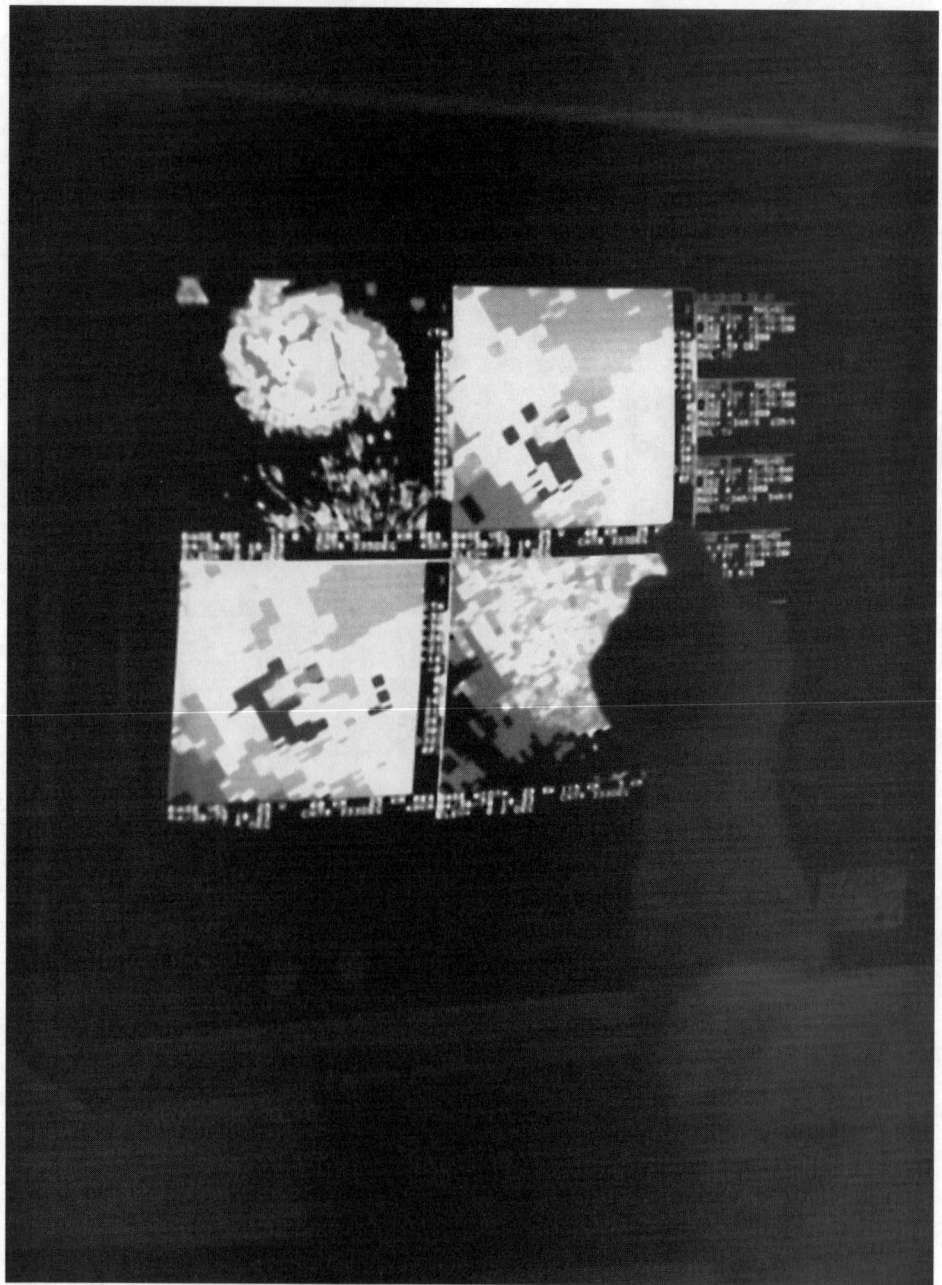

Fig. 9-3. *The NEXRAD WSR-88 radar provides the radar specialist with a computer display of radar echoes. This Doppler radar allows the observer to see relative movement within the storm. The WSR-88 system has the potential to greatly improve storm prediction and warning.*

Figure 9-4 illustrates the effects of precipitation attenuation, showing how a heavy-precipitation pattern with a very strong gradient might appear on an NWS 10-cm radar, compared with the same weather system as seen on 5- and 3-cm units. A pilot seeing the pattern on a 3-cm set might elect to penetrate the weather at what appears to be the weakest point only to find the most severe part of the storm or find additional severe weather where the radar showed clear.

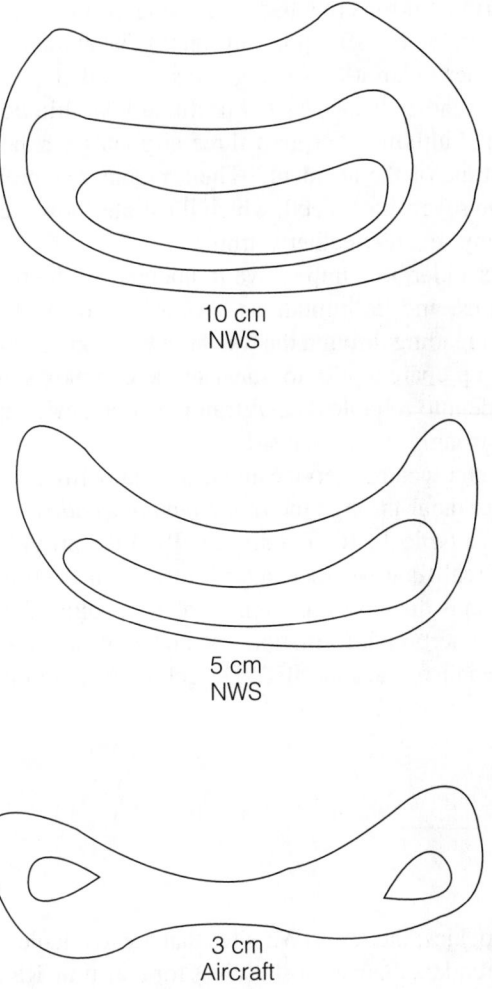

10 cm
NWS

5 cm
NWS

3 cm
Aircraft

Fig. 9-4. *Precipitation attenuation, caused by close targets absorbing and scattering the radar's energy, can be a serious problem with low-power, short-wave-length sets. Attenuation is not significant with high-power NWS radars.*

According to the National Transportation Safety Board, precipitation attenuation was a contributing factor in the crashes of a Southern Airways DC-9 in 1977 and an Air Wisconsin Metroliner in 1980. Precipitation attenuation is not significant with NWS 10-cm high-power units; however, it can be a serious problem with units of 5 centimeters or less, especially in heavy rain. The NTSB stated and recommended that "in the terminal area, comparison of ground returns to weather echoes is a useful technique to identify when attenuation is occurring. Tilt the antenna down, and observe ground returns around the radar echo. With very heavy intervening rain, ground returns behind the echo will not be present. This area lacking ground returns is referred to as a *shadow* and might indicate a larger area of precipitation than is shown on the indicator. Areas of shadowing should be avoided."

In August 1985, a Delta Air Lines L-1011 crashed at the Dallas-Fort Worth Airport. The NTSB was unable to determine if the crew had been using airborne weather radar at the time of the crash; however, the NTSB report did state: "The evidence concerning the use of the airborne weather radar at close range was contradictory. Testimony was offered that the airborne weather radar was not useful at low altitudes and in close proximity to a weather cell," although "at least three airplanes scanned the storm at very close range near the time of the accident." The accident was probably caused by a microburst from a single severe storm cell, which illustrates how weather can develop rapidly, often without any severe-weather warning.

When using an airborne weather radar, it is imperative to understand the particular unit, its operational characteristics, and its limitations. According to the March-April 1987 *FAA Aviation News*, "Just reading through the brochure that comes with the equipment is certainly not enough to prepare a pilot to translate the complex symbology presented on the [airborne] scope into reliable data. A training course with appropriate instructors and simulators is strongly recommended."

Flight watch control stations, center weather service units, and many flight service stations have access to direct weather radar through the *radar remote weather display system* (RRWDS). Figure 9-5 shows a typical RRWDS display. RRWDS provides the FSS or flight watch specialist with real-time weather-radar information from one of several different sites; the information is displayed on high-resolution color monitors. Specialists are specially trained to interpret information for preflight and in-flight briefings. FSS and flight watch specialists translate RRWDS echo coverage into the following categories:

- Widely scattered (less than $\frac{1}{10}$)
- Scattered ($\frac{1}{10}$ to $\frac{5}{10}$)
- Broken ($\frac{6}{10}$ to $\frac{9}{10}$)
- Solid (more than $\frac{9}{10}$)

Limitations exist. West of the Rockies, the few NWS sites that are available suffer from extreme ground clutter, which renders them almost useless for real-time RRWDS use; facilities using RRWDS with FAA radars suffer from the limitations of ATC units. Implementation of the NEXRAD network will help eliminate many of these limitations. Observations are made 35 minutes past each hour and charts are normally available 15–20 minutes later; however, real-time information, which is available in the East, is not yet available in the West.

Fig. 9-5. *The radar remote weather display system (RRWDS) provides FSS and Flight Watch specialists with real-time weather radar.*

Private vendors also have access to radar data. Information from radars is coded and transmitted. The limitations of RRWDS apply to this system. When using an RRWDS, it's important to know how the unit displays information such as intensity, whether it's displaying an ATC or NWS radar, and if it is indeed a real-time observation or a freeze or memory display.

The violent nature of thunderstorms causes gust fronts, strong updrafts and downdrafts, and wind shear in clear air adjacent to the storm out to 20 miles with severe storms and squall lines. Precipitation, which is detected by radar, generally occurs in the downdraft, while updrafts remain relatively precipitation free. Clear air or lack of radar echoes does not guarantee a smooth flight in the vicinity of thunderstorms.

LIGHTNING DETECTION EQUIPMENT

Lightning detection equipment, trade-named Stormscope, was invented in the 1970s by Paul A. Ryan as a low-cost alternative to radar. The Stormscope senses and displays electrical discharges in approximate range and azimuth to the aircraft. Stormscope also has limitations. One misconception proclaims that in the absence of dots or lighted bands, there are no thunderstorms; however, NASA's tests of the Stormscope differed. Precipitation intensity levels of 3 and occasionally 4 (VIP levels) would be indicated on radar without activating the lightning detection system. A clear display only indicates the absence of electrical discharges. This does not necessarily mean convective activity and associated thunderstorm hazards are not present. Even tornadic storms have been found that produced very little lightning. The lack of electrical activity, as with the absence of a precipitation display on radar, does not necessarily translate into a smooth ride.

Many authorities agree that a combination of radar and Stormscope is the best thunderstorm detection system. It cannot be overemphasized that these are avoidance devices, not penetration devices. Thunderstorms imply severe or greater turbulence, and neither radar nor Stormscope directly detect turbulence.

RADAR WEATHER REPORTS (RAREPs)

The National Weather Service routinely takes radar observations from NWS and ATC radars at 35 minutes past the hour. These observations are coded and transmitted over the FAA's weather distribution system. Locations are contained in appendix B of this book and the inside back cover of the *Airport/Facility Directory*. (Locations might change with the implementation of NEXRAD.) All locations are not available on a continuous basis. Local-warning radars normally only report when convective activity occurs in the area. The radar report, or SD for *storm detection*, contains the following information:

- Location of the radar
- Time of observation
- Configuration of echoes
- Coverage of echoes (subjective by the radar specialist)
- Type of precipitation
- Intensity of precipitation
- Intensity trend
- Location of echoes
- Movement of echoes
- Height of echoes (not available from ATC radars)
- Remarks

The following illustrates a coded SD report. All reports begin with the radar site. This report for OKC (Oklahoma City) was observed at 2235Z.

OKC 2235 LN 8TRWX/+ 350/100 230/112 25W L2930 C2535 MT 500 AT 320/67
TOP 480 AT 235/100 AREA 3TRW/NC 080/75 110/100 140/100 170/75 100/50
A2525 C2330 MT 340 AT 155/75
IK32 JK42 KK5 LK4 MP21 NJ11 NQ1 OJ4 OQ1 PJ5 PO21 QI13=

Echo configuration and coverage

Echo configuration falls into three categories: cell, area, and line. A single isolated area of precipitation, clearly distinguishable from surrounding echoes, constitutes a cell. The following illustrates how cells are indicated on a RAREP.

GGG 1057 SPL CELL TRWXXA/NC 323/95 D30 C2730 MT 550 HOOK 321/91...

The Longview, TX (GGG) 1057Z special (SPL) reports a cell with a 30 nm diameter (D30) exhibiting a hook echo. *Cell* diameter refers to precipitation, not necessarily the diameter of the cloud, which could be considerably larger. A hook echo is the signature of a mesolow, often associated with severe thunderstorms that produce strong gusts, hail, and tornadoes. An *area* consists of a group of echoes of similar type that appear to be associated. A *line* (LN) defines an area of precipitation more or less in a line, straight, curved, or irregular, at least 30 miles long, five times as long as it is wide, with at least 30 percent coverage. Echo coverage is estimated by the observer and reported in tenths. In the Oke City example, the line has ⁸⁄₁₀ coverage (LN "8"TRWX/+) and the area coverage is ³⁄₁₀ (AREA "3"TRW/NC).

Precipitation type, intensity, and intensity trend

The radar specialist determines precipitation type from scope presentations and other sources (SAs and satellite pictures). Standard aviation weather symbols are used. In the Oke City example, the line and the area contain thunderstorms and rain showers (TRW). Following precipitation type, one of the six standard video integrator and processor (VIP) levels describe intensity. VIP-level definitions and descriptions are in Table 9-1. The line is intense (TRW"X") and the area moderate (TRW-no symbol). Intensity trend follows the solidus (/). Refer to Table 9-2 for intensity trend symbols. The OKC example indicates that the trend of the line is increasing (+). The + indicates an increase of one or more VIP levels during the past hour. The intensity of the area has not changed (NC) during the preceding hour.

Echo location and movement

The RAREP defines the location of precipitation by points, azimuth (true), and distance (nm) from the reporting station (Fig. 9-6). The line extends from a point 350° at 100 nm (350/100) to a point 230° at 112 nm (230/112), and is 25 nm wide (25W). The points (080/75 110/100 140/100 170/75 100/50) encompass the area. Line or area echo movement indicates the long-term progress of the system; cell movement indicates short-term motion of cells within the line or area. The line in Fig. 9-6 is moving from 290° at 30 knots (L2930); cell movement within the line is from 250° at 35 knots (C2535).

Table 9-1. Radar precipitation intensity (VIP) levels.

VIP* level	RAREP symbol	Echo intensity	Precipitation intensity	Rainfall rate stratiform**	Rainfall rate convective**	Associated weather
1	—	WEAK	LIGHT	less than 0.1	less than 0.2	LGT-MDT TURBC PSBL LTNG
2	NO SYMBOL	MODERATE	MODERATE	0.1–0.5	0.2–1.1	LGT-MGT TURBC PSBL LTNG
3	+	STRONG	HEAVY	0.5–1.0	1.1–2.2	SVR TURBC & LTNG
4	++	VERY STRONG	VERY HEAVY	1.0–2.0	2.2–4.5	SVR TURBC & LTNG
5	X	INTENSE	INTENSE	2.0–5.0	4.5–7.1	SVR TURBC LTNG WIND GUSTS HAIL
6	XX	EXTREME	EXTREME	more than 5.0	more than 7.1	SVR TURBC LTNG EXTENSIVE WIND GUSTS LG HAIL

*VIP (Video Integrator and Processor) A system used to determine the intensity of precipitation.
**Precipitation in inches per hour.

Table 9-2. RAREP/radar summary chart plotted data.

_____ Intensity trend _____

Symbol	Trend	Symbol	Trend
–	Decreasing	NC	No change
+	Increasing	NEW	New echo

_____ Operational status of radar _____

Symbol	Meaning	Symbol	Meaning
PPINE	Equipment normal—no echoes observed	PPIOM	Out of service for maintenance
PPINA	Observation not available	ROBEPS	Radar operation below standards
ARNO	Azimuth/range indicator inoperative	RHINO	Range/height indicator inoperative
NE	No echoes observed	NA	Observation unavailable
OM	Out for maintenance	NS	Nonsignificant echoes

_____ Remarks _____

Symbol	Meaning	Symbol	Meaning
LEWP	Line echo wave pattern	BWER	Bounded weak echo pattern
WER	Weak echo region	MLT LVL	Melting level
MA	Echoes mostly aloft	PA	Echoes partly aloft

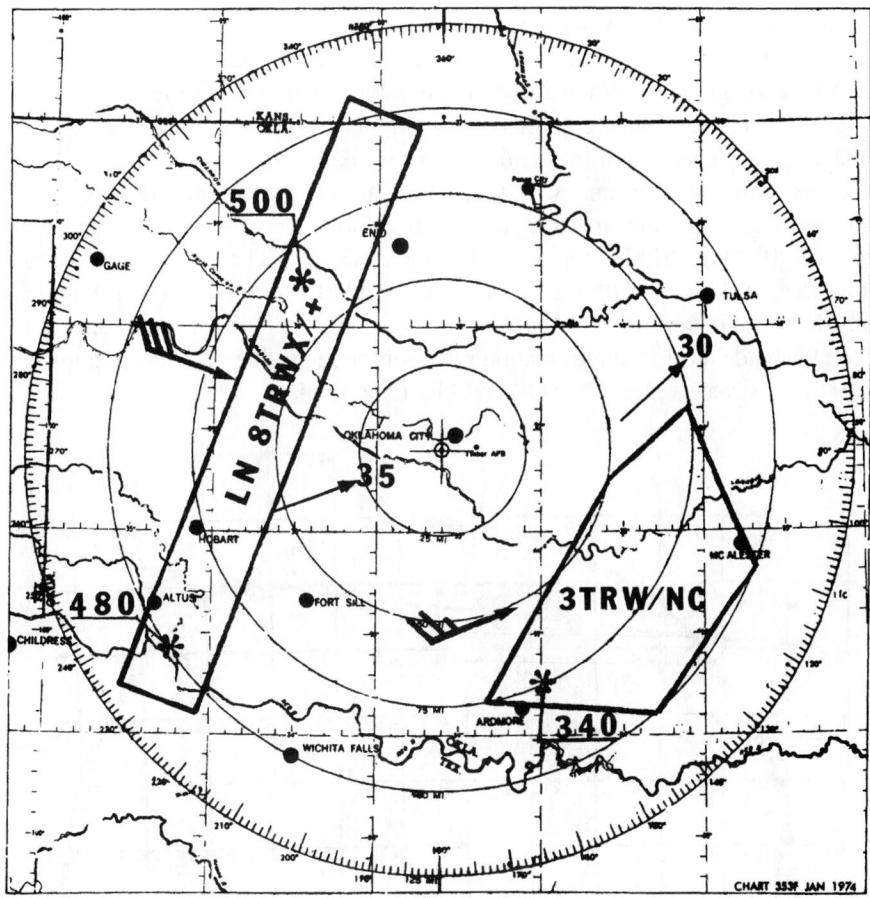

Fig. 9-6. *As with most coded data, plotting the RAREP helps visualize conditions.*

Echo height

Maximum heights are reported in relation to azimuth and distance from the reporting station, with approximate elevation in thousands of feet MSL (MT 500 AT 320/67). Tops within a stable air mass are usually uniform, indicated by the letter U: MT U120 (uniform tops to 12,000 feet MSL). It's important to remember that these are precipitation tops, not cloud tops. Precipitation tops will be close to cloud tops within building thunderstorms; however, precipitation in dissipating cells will normally be several thousand feet below cloud tops.

Remarks

Remarks that are added by the radar specialist elaborate on or explain the report in plain language or standard contractions. Table 9-2 contains some of the remarks used on the report.

RAREP digital data appears at the bottom of the report. A grid, as shown in Fig. 9-7, centers on the reporting station. Each block, 22 nm on a side, is assigned the maximum VIP level observed. When 20 percent of a block contains VIP level 1, that level is assigned; therefore, from digital data alone, all that can be concluded from VIP level 1 is that at least 20 percent of that grid contains weak echoes.

Letters represent coordinates; numbers indicate the maximum VIP level for that and succeeding coordinates to the right. The first block (Fig. 9-7) containing precipitation is IK ("IK"32). Grid IK contains VIP level 3 (IK"3"2). The next coordinate to the right (IL) has a VIP level 2 (IK3"2"). Occasionally a VIP 8 or 9 appears; an 8 indicates an echo believed to be severe, and a 9 is a nonsevere echo. An 8 or 9 will only appear in a block outside the 125-nm operation range of the radar that contains precipitation. Figure 9-7 illustrates a plot of the OKC RAREP digital data.

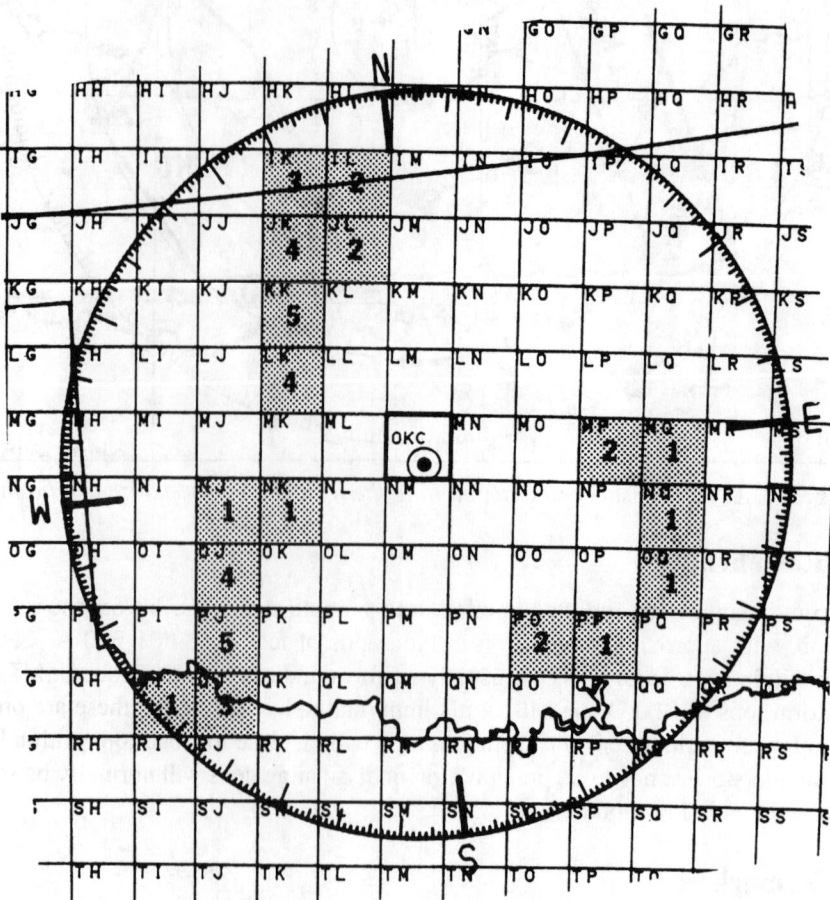

Fig. 9-7. *Plotting RAREP digital data provides additional insight into the strength and location of convective activity. This plot shows a weak area in the center of the line west of Oklahoma City.*

This second example of a RAREP is from the Sacramento, CA (SAC), NWS radar site. Sacramento radar was not showing any significant weather prior to the following observation.

SAC 0125 CELL TRW++/+ 349/80 D9 C2920 MT 250 AREA 1R-S/NC 10/115
46/30 80W C2920 MT 130 AT 33/79
IN41 KM101=

Look what popped up, a level 4 cell (very strong). Thunderstorms were not forecast. This single cell with a diameter of 9 miles and approximate precipitation tops to 25,000 feet "didn't read the forecast." Figure 9-8, a single thunderstorm cell, illustrates what a pilot flying in this area could expect to see. The cell should be easily circumnavigable and only presents a hazard if it is in the vicinity of the departure or destination airport. A pilot should not fly into, close to, or underneath this cell.

SAC 0225 CELL RW/- 346/81 D6 C2820 MT 170 AREA 1R-S/NC 6/95/87/15 80W
C2820 MT 120 AT 22/89
IN21 LO1=

Fig. 9-8. *Isolated and widely scattered thunderstorms are normally easily circumnavigable. But, don't get too close; severe turbulence and hail can reach many miles into the clear air surrounding the cell.*

One hour later (0225), the cell has deteriorated to moderate rain showers and continues to decrease in intensity with maximum precipitation tops down to 17,000 feet MSL. Unforecasted convective activity can obviously develop and dissipate at an alarming rate.

RAREPs will be computer generated with the NEXRAD implementation. NEXRAD RAREPs are identified by the "AUTO" between the narrative and digital sections.

RADAR SUMMARY CHART

The radar summary chart graphically displays a computer-generated summation of RAREP digital data. The date and time of the observation appear on the chart; time is important because the transmission system might make the report several hours old. Figure 9-9 illustrates a May 9, 1986, summary based upon 0935Z data. Similar to the RAREP, the chart contains information on precipitation type, intensity and trend, configuration, coverage, tops and bases, and movement.

Line and area movement, echo movement, and tops are depicted using the RAREP symbology. Arrows with flags, barbs, and half barbs indicate line or area movement (flag is 50 knots, barb is 10 knots, and a half barb is 5 knots). Figure 9-9 shows that the line in Texas is moving from the northwest at 50 knots; the area in Louisiana is moving from the west at 25.

An arrow with the speed printed at the arrowhead represents echo or cell movement. Echoes within the line (Fig. 9-9) are moving from the southwest at 25 knots and within the area from the northwest at 15 knots; the contraction LM indicates little movement (see the echoes in Idaho). Maximum tops in Fig. 9-9 within the line vary between 37,000 and 53,000 feet MSL (370, 440, and 530). When bases can be determined, the height MSL will appear below the line. West of the Rockies, echo heights are usually missing because ATC radars are used; heights are included with NWS radar data.

Echo configuration is graphically depicted. Echoes reported as a line are drawn and labeled solid (SLD) when at least 9/10 coverage exists, for instance, in southwestern Texas (Fig. 9-9). The computer plots lines of equal value to indicate echo coverage and intensity; however, unlike the RAREP, the chart only contains three levels. The first contour includes VIP levels 1 and 2 (echoes in Utah), the second contour is levels 3 and 4 (echoes in Iowa), and the third contour is levels 5 and 6 (echoes in Louisiana and around the line in Texas).

The advisory circular *Aviation Weather Services* states: "When determining intensity levels from the radar summary chart, it is recommended that the maximum possible intensity be used." In the western United States, nonsignificant weather echoes (VIP 1) might be indicated by the notation NS and contouring will not appear. This is interpreted as light precipitation occurring without significant convective activity.

Precipitation type and intensity trend use the same symbology as the RAREP, with one exception. The + indicates an increase in intensity of at least one VIP level since the last observation, or the + indicates a new echo. The radar summary chart can contain remarks that have special significance (Table 9-2. See also the glossary).

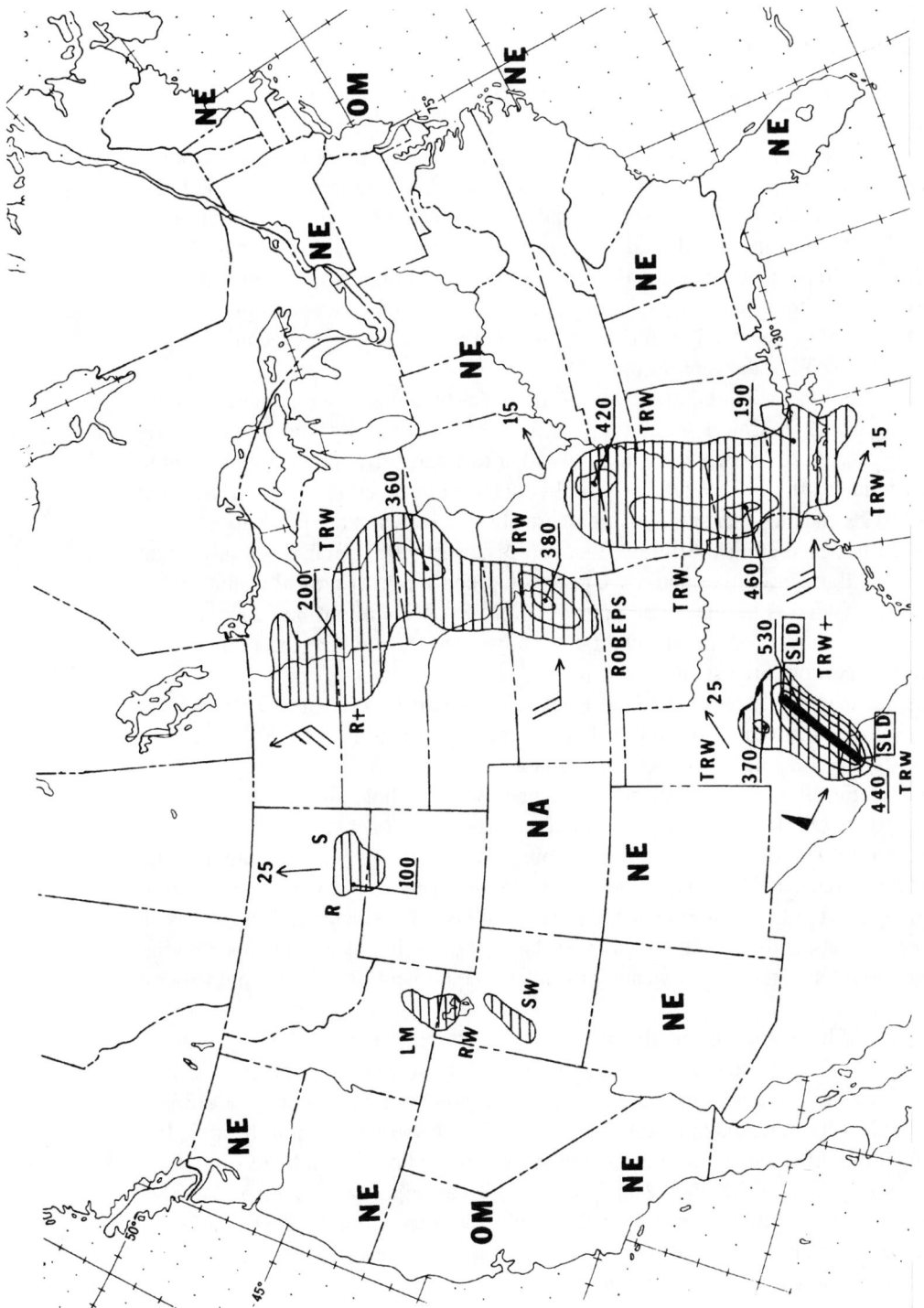

Fig. 9-9. *The radar summary chart is a computer analysis of RAREP digital data. It is always old and should be used for pre-planning purposes only; always update the radar summary with current observations.*

USING RAREPs, RRWDS, AND
THE RADAR SUMMARY CHART

Pilots can expect to find holes in what the RAREP or radar summary chart portray as an area of solid echoes. This apparent inconsistency is due to several factors. Targets farther from the antenna might be smaller than depicted due to range and beam resolution. NWS weather radars, at a range of 200 miles, cannot distinguish between individual echoes less than 7 miles apart. A safe flight between severe thunderstorms requires 40 miles, so this provides adequate resolution to detect a safe corridor. Recall that as little as 20 percent coverage of VIP 1 requires the entire grid to be encoded, so holes also occur with isolated and scattered precipitation. On the radar summary chart, large areas might be enclosed by relatively isolated echoes. This is especially true in the West, where ATC radars are used.

Assumptions can never be made with RAREPs or the radar summary chart. Refer to Figs. 9-6 and 9-7. Can a pilot fly west from Oklahoma City and avoid severe weather by at least 20 miles? Not without weather avoidance equipment, contact with a facility with real-time RRWDS, or visual contact with the convective activity because RAREPs and the radar summary chart are observations that report what occurred in the past, not predicting the future as a forecast map would; it has already been demonstrated that convective activity can develop and move at an astonishing rate. Time of observation is an important consideration; RAREPs might be as much as 2 hours old and the radar summary chart from 2 to 4 hours old. On the other hand, RRWDS observations are usually real time.

Most radar information west of the Rockies comes from ATC sites. SDs as such are not available from the ATC radars, but a summary of activity is distributed. Regardless, some SDs from NWS sites are available.

Refer to Fig. 9-9. A pilot can never assume there are holes in the solid line in southwestern Texas. The chart can only be interpreted as $\frac{9}{10}$ or more coverage within the line. A pilot could expect to see something similar to Fig. 9-10. By reviewing RAREP digital data, a pilot can quickly assess echo intensity and coverage. For example, let's say the digital data for the line read MJ24542 NI23532 NH23632 PI3532. Notice that each consecutive group contains levels 5 or 6. The line is continuous with $\frac{10}{10}$ coverage. No pilot, regardless of weather avoidance equipment, should attempt to penetrate this activity.

Nor should he or she assume that it is OK to fly through the weather in southern Kansas by avoiding thunderstorms without weather avoidance equipment or visual contact with convective activity. If the RAREP for southern Kansas were JN13531 KM11311 LM1111, it would indicate weak activity at the southern end of the area. Interpreting the chart alone, a pilot must conclude that the echoes in Utah are solid with moderate precipitation (VIP level 2); however, the areas might only contain scattered or widely scattered light precipitation. This could be verified from the RAREP.

The radar summary chart provides general areas and movement of precipitation for planning purposes only. The summary chart must be updated by hourly RAREPs or

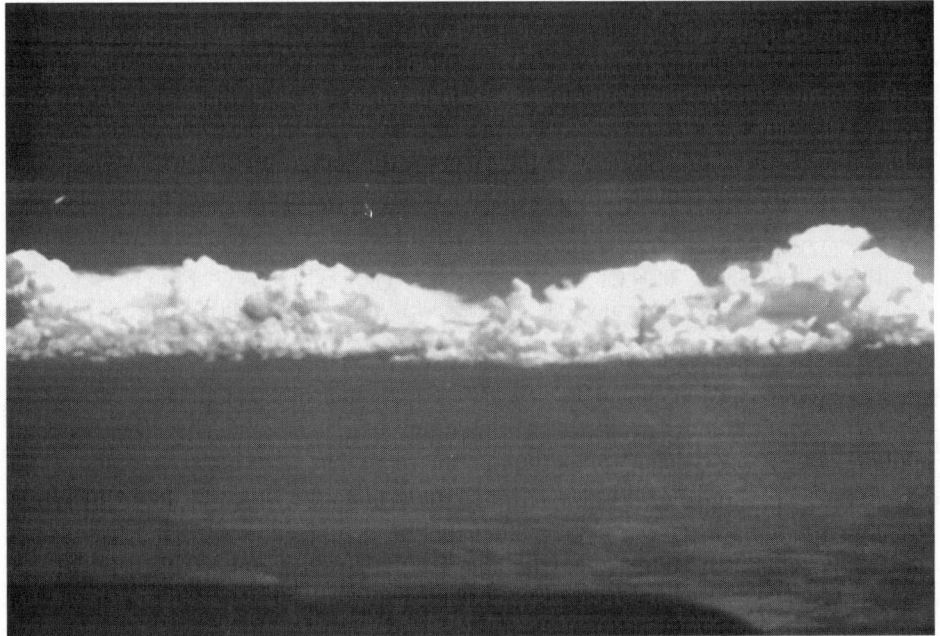

Fig. 9-10. *A pilot should never attempt to negotiate a line of convective activity without airborne storm-avoidance equipment or direct access to real-time ground-based weather radar information.*

RRWDS. Chart notations, such as OM (out of service for maintenance) or NA (not available), must be considered. The chart must always be used in conjunction with other charts, reports, and forecasts. When airborne, in-flight observations (visual or electronic) and radar information from an FSS or flight watch must be used.

Precipitation should not be cause to cancel a flight; however, the following must be considered. What is the coverage and intensity of precipitation? What is the weather expected to do (improve or deteriorate)? What is the pilot's experience level, the capability of his or her aircraft, and the time of day? It can be difficult to see clouds at night; although lightning flashes are visible, any distance can be difficult to judge. How familiar is the pilot with the terrain and weather patterns over the intended route? Is an alternate available if the planned flight cannot be completed? Is the pilot mentally prepared to divert if it becomes necessary? What about the pilot's physical condition? Tired and anxious to get home is a potentially fatal combination.

CONVECTIVE OUTLOOK NARRATIVE AND SEVERE WEATHER OUTLOOK CHART

The convective outlook (AC) is prepared by the National Severe Storm Forecast Center at 0700Z, 1500Z, and 1930Z daily. The 0700Z issuance covers a 24-hour period from 1200Z to 1200Z, and the 1500Z and 1930Z ACs update the original issuance. The AC describes the potential for thunderstorm activity and describes areas where thunderstorms might approach severe limits. Risk categories are defined: slight risk is 2–5 percent coverage, moderate risk is 6–10 percent coverage, and high risk is more than 10 percent coverage.

```
MKC AC 030700
CONVECTIVE OUTLOOK . . . REF AFOS NMCGPH940
VALID 031200 - 041200Z
THERE IS A SLGT RISK OF SVR TSTMS TO THE RGT OF A LN FM ERI
35 SE FDY 10 ENE LAF UIN OJC P28 25 WSW GAG 45 S DHT 30 SSE RTN 10
SW 4FC CPR GCC 30 E 81V 20 NNE SNY 20 WSW MCK EAR FSD
45 WSW AXN 20 SE GFK 35 W D45.
GEN TSTMS ARE FCST TO THE RGT OF A LN FM LRD 20 W CLL 10 S PGO 20
W MKO FSI MAF 45 W MRF . . . CONT . . . 50 SW TUS 55 NE PHX 10 E BCE 25
ENE BAM MHS 35 S EUG 25 ENE BLI.
TAIL END OF VIGOROUS SHRTWV MOVING ACRS E CNTRL CANADA IS
PROGD TO SWEEP SEWD ACRS UPR MS VLY . . . INTO THE GRTLKS REGION
BY THE END OF THE FCST PERIOD.  ASSOCD WITH THIS SYS . . . PRE-
FRONTAL SFC TROF CURRENTLY OVER THE NRN PLAINS IS FCST TO EX-
TEND FM THE WRN GRTLKS INTO THE CNTRL PLAINS BY LATE AFTN . . .
WITH COLD FRONT SURGING INTO THE NRN PLAINS/UPR MS VLY.  WHILE
MID/UPR FLOW STRENGTHENS ACRS THIS REGION . . . 20 TO 30 KT LLJ DE-
VELOPING FM THE MID MS VLY INTO ONTARIO EXPCD TO ADVECT UPR
60S/LWR 70S SFC DWPTS NEWD FM THE E CNTRL PLAINS.  COMBINATION
OF SFC HEATING/INCREASE IN LOW LEVEL MOISTURE/MID LEVEL COLD AD-
VECTION SHUD RESULT IN MDT INSTABILITY (SFC BASED LIS OF –4 TO –8)
ACRS PORTIONS OF THE UPR MS VLY/WRN GRTLKS BY LATE AFTN.  GIVEN
INSTBY . . . CONVERGENCE ALONG BOTH SFC TROF AND COLD FRONT EX-
PCD TO RESULT IN STG/SVR TSTM DEVELOPMENT . . . AIDED BY INCREAS-
INGLY FVRBL VERTICAL SHEAR.  AS SFC WAVE DEVELOPS ALONG COLD
FRONT OVER THE GRTLKS REGION WITH APPROACH OF UPR TRO F. . . AC-
TIVITY MAY CONT SVR INTO THE OVERNIGHT HOURS.
ELSEWHERE . . . SCT STG/ISOLD SVR TSTMS PSBL ALONG PRE-FRONTAL
SFC TROF INTO THE CNTRL PLAINS . . . WHERE AMS IS EXPCD TO REMAIN
MDTLY UNSTABLE UNDERNEATH DIVERGENT UPR FLOW.  ISOLD SVR TSTMS
PSBL ACRS THE FRONT RANGE OF THE ROCKIES/ADJACENT CNTRL HIGH
PLAINS . . . ASSOCD WITH CONTG UPSLOPE FLOW COMPONENT.
. . KERR . . 08/03/94
```

The AC was issued on the third day of the month at 0700Z and is valid from 031200Z until 031200Z. There is a slight risk of severe thunderstorms to the right of a line described by the location identifiers. General thunderstorms are forecast to the right of a line specified by the location identifiers. (*General thunderstorms* are those not expected to reach severe limits.) The tail end of a vigorous shortwave trough is moving across east central Canada that is forecast to sweep southeastward across the upper Mississippi Valley into the Great Lakes region by the end of the forecast period. Associated with this system is a prefrontal surface trough currently over the northern plains that is forecast to extend from the western Great Lakes into the central plains by late afternoon with a cold front surging across this region.

A 20- to 30-knot low-level jet is developing from the middle Mississippi Valley into Ontario, which is expected to advect upper 60s and lower 70s dew points from the east central plains. A combination of surface heating and an increase in low-level moisture along with middle-level cold-air advection should result in moderate instability (surface-based lifted indices of –4 to –8) across portions of the upper Mississippi Valley and western Great Lakes by late afternoon.

Given the instability and convergence along both the surface trough and the cold front, conditions are expected to result in strong to severe thunderstorm development, aided by increasingly favorable vertical shear. As a surface wave develops along the cold front over the Great Lakes region with the approach of an upper trough, activity may continue to be severe into the overnight hours.

Elsewhere, scattered strong to isolated severe thunderstorms are possible along the prefrontal surface trough into the central plains where the air mass is expected to remain moderately unstable underneath a divergent upper flow. Isolated severe thunderstorms are possible across the Front Range of the Rockies adjacent to the central High Plains. The storms are associated with a continuing upslope flow component.

This narrative falls into the same category as the outlook portion of convective SIGMETs, providing synoptic details in meteorological terms directed toward forecasters more than pilots. It would be helpful to have surface and upper-level analysis charts to help visualize the discussion.

Severe weather outlook chart

The severe weather outlook chart provides a preliminary 48-hour thunderstorm probability potential. The chart is prepared manually. The left panel gives a 24-hour outlook for general and severe thunderstorms. Figure 9-11 is valid for the same time frame as the AC in the previous discussion. As seen in Fig. 9-11, there is a slight risk of severe thunderstorms in the upper Midwest. The right panel indicates only severe thunderstorm probability for the next 24 hours.

This chart indicates areas where conditions are right for the development of convective activity sometime during the period. It is basically a pictorial display of the AC; however, the chart is not amended or updated.

A potential for thunderstorms exists within depicted areas. This does not necessarily mean thunderstorms will develop. The chart, like the AC, is strictly for advanced planning to alert forecasters, pilots, briefers, and the public to the possibility of future storm development. Appropriate FAs, FTs, WSTs, and severe weather watches must be consulted just prior to and during flight for details on convective activity.

COMPOSITE MOISTURE STABILITY CHART

The composite moisture stability chart is available twice a day and is computer-generated based upon radiosonde data. The chart consists of four panels:

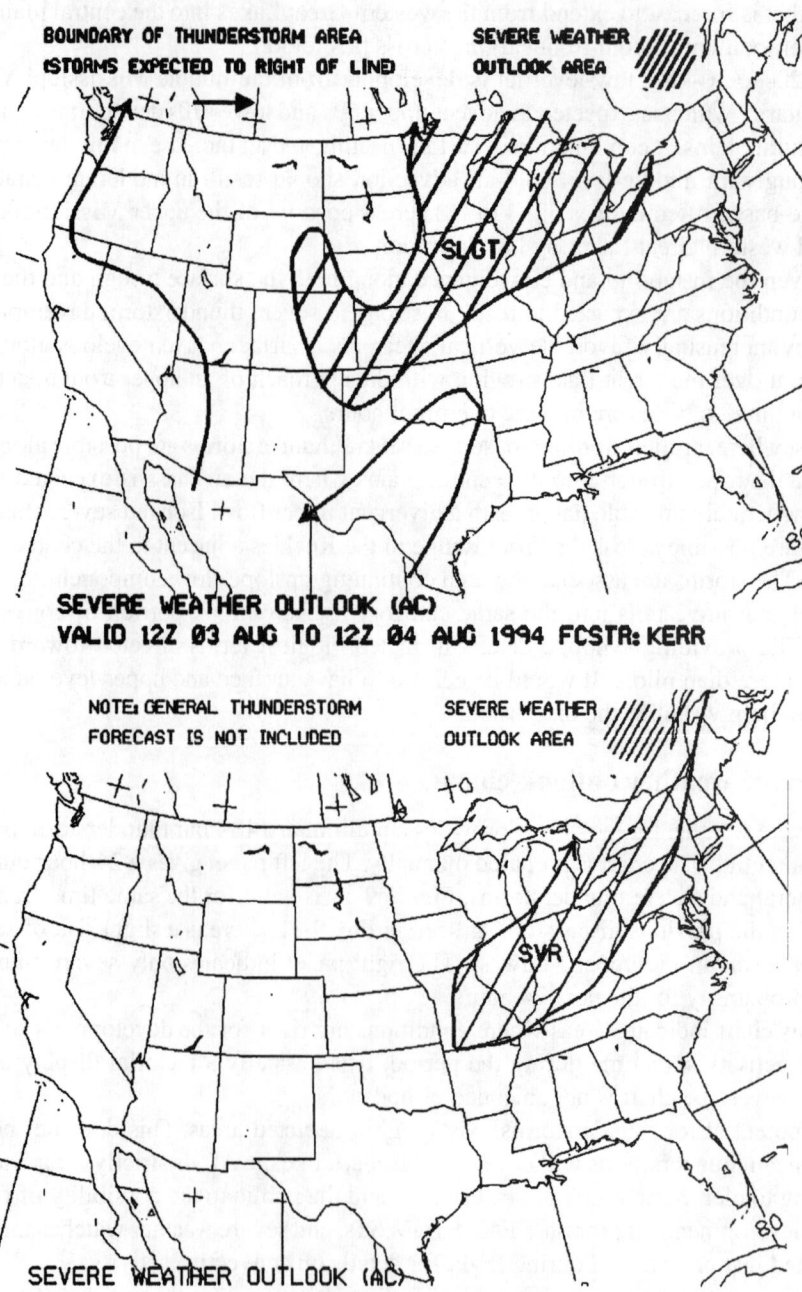

Fig. 9-11. *The severe weather outlook chart provides a preliminary look at general and severe thunderstorm potential for 24 hours and severe thunderstorm potential for 48 hours.*

- Lifted index analysis
- Precipitable water
- Freezing level
- Average relative humidity

Notice in Fig. 9-12 that the analysis is based upon the 0000Z observation on Wednesday, August 3, 1994. Due to computation and transmission times, this chart is about 4½ hours old by the time it becomes available.

The lifted index provides an indication of atmospheric moisture and stability, essentially the potential for thunderstorms at the time of observation. The lifted index compares the temperature that a parcel of air near the surface would have if lifted to the 500-mb level and cooled adiabatically, with the observed temperature at 500 mbs. The index indicates stability at the 500-mb level. The index can range from +20 to –20, but generally remains between +10 and –10. (The figures are strictly an index, not a representation of temperature.) A positive index indicates a stable condition; high positive values indicate very stable air. A zero index indicates neutral stability. Values from 0 to –4 indicate areas of potential convection. Large negative values from –5 to –8 represent very unstable air, which could result in severe thunderstorms if convection develops.

The "K index" evaluates moisture and temperature. The higher the K index, the greater the potential for an unstable lapse rate and the availability of moisture. The K index must be used with caution; it is not a true stability index. Large K indexes indicate favorable conditions for air mass thunderstorms, during the thunderstorm season. K values can change significantly over short periods due to temperature and moisture advection.

Refer to the lifted index panel in the upper left portion of Fig. 9-12. The lifted index appears above the K index in the plotted data. *Isopleths*, which are lines equal in number or quantity, of stability are plotted beginning at zero then for every four units (plus and minus). Negative lifted indexes and large K values exist over Florida and the upper Mississippi Valley; therefore, the chart would indicate a potential for thunderstorms in those areas. Conversely, positive lifted indexes and small K values were observed over the southwestern United States, indicating little moisture and a stable lapse rate.

The precipitable water panel in the upper right portion of Fig. 9-12 analyzes water vapor content from the surface to 500 mbs. Darkened station circles indicate large amounts of available water. Isopleths of precipitable water are drawn at one-quarter intervals. The panel is more useful for meteorologists concerned with flash floods; however, a pilot can get an excellent indication of changes in moisture content. For example, considerable moisture exists over the southeastern United States.

The freezing level panel in the lower left portion of Fig. 9-12 plots the lowest observed freezing level. Multiple entries indicate inversions and above-freezing temperatures aloft. Figure 9-12 is a graphical plot of the RADAT discussed in chapter 1; implications and analysis are the same. Remember that for flight planning, this *observed* data must be used with *forecast* information contained in the FA and other forecast products.

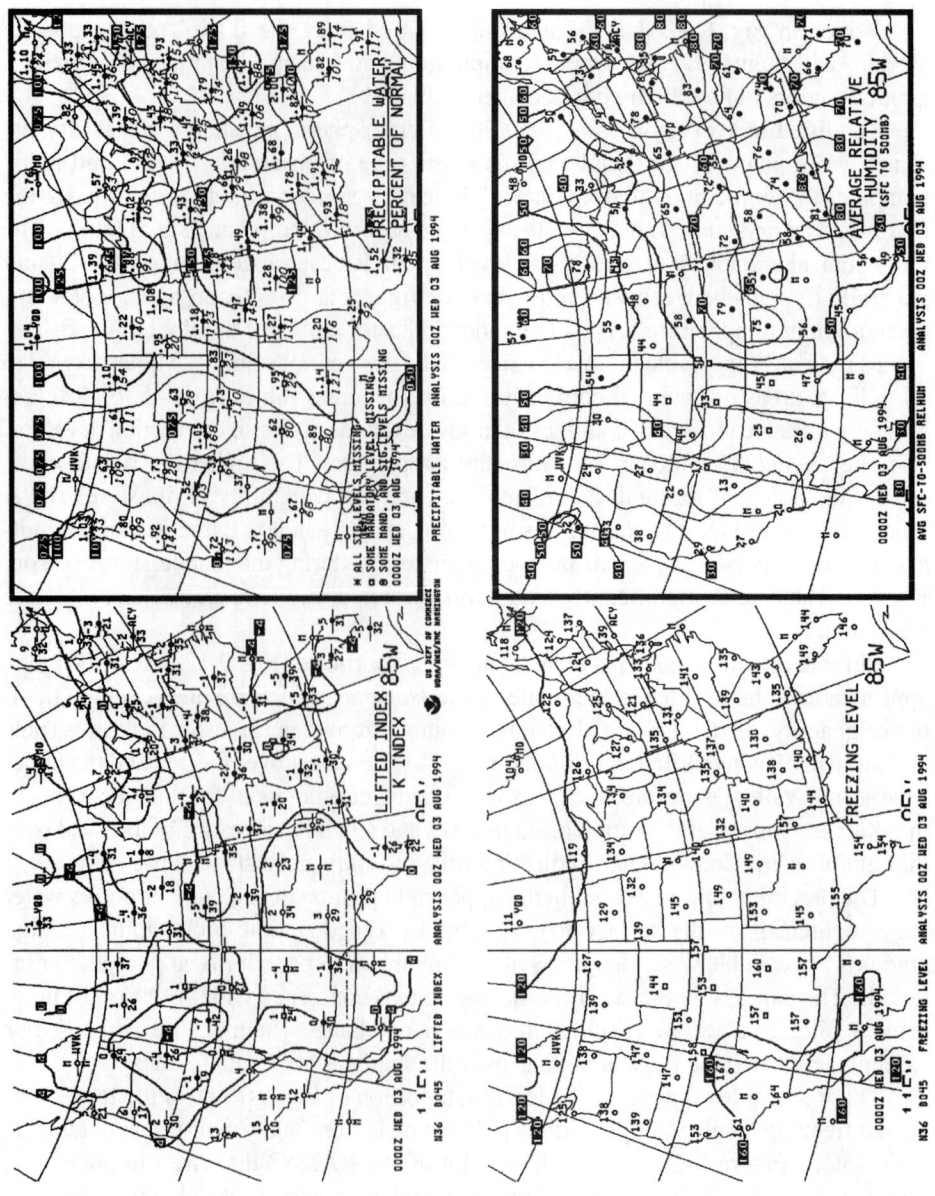

Fig. 9-12. *The composite moisture stability chart provides a pictorial analysis of stability, moisture, and freezing level based upon the twice-daily radiosonde observations. The chart is always old by the time it becomes available and must never be substituted for current observations and forecasts.*

The average relative humidity panel in the lower right portion of Fig. 9-12 analyzes the average relative humidity from the surface to 500 mbs. Darkened station plots indicate high humidity. *Isohumes*, which are lines of equal relative humidity, are drawn at 10 percent intervals. The chart indicates large-scale moisture content in the lower part of the troposphere. Clouds and precipitation are indicated in the southeastern United States with a lifting mechanism available and 70-percent relative humidity.

Figure 9-12 indicates several likely areas for significant convective activity. Florida and the upper Mississippi Valley are prime candidates, due to abundant moisture and instability.

Freezing levels across the country vary from 12,000 feet in the north to 16,000 feet in the south; therefore, potential icing exists in clouds and precipitation to 21,000 feet in the southern states. Icing intensities and the heights at which icing would be encountered increase with convective activity. This is a typical summer case where pilots must be alert for icing into the lower flight levels.

The lifted index analysis is only one element used to develop the AC severe weather outlook chart and other forecasts. Instability below 500 mbs might not be indicated. The chart is several hours old when received and the lifted index does not consider a lifting mechanism nor can it consider modifications by the development, dissipation, and movement of systems. AC 00-45 *Aviation Weather Services* states: "It is essential to note that an unstable index does not automatically mean thunderstorms."

Certain pilots have attached great significance to the AC and lifted index analysis. They infer some additional insight into thunderstorm activity and severity from these products. The lifted index is an observation and only one element used to prepare the AC and other products. The convective outlook narrative and severe weather outlook chart are just that, outlooks. They merely provide a statement of potential and must never be used in place of weather advisories or area and terminal forecasts.

MICROBURSTS AND LOW-LEVEL WIND SHEAR

As aircraft instruments and navigational system capabilities improved, pilots began "taking on" more and more weather "challenges." Hazards of fog and low clouds were solved with the instrument landing system and increased fuel reserves. Icing was overcome in the 1930s with properly equipped airplanes. Turbulence and thunderstorms were mastered for the most part with higher cruise altitudes by higher-performance airplanes and the development of airborne radar. Major air-carrier accidents in the 1990s most often fall into the categories of mechanical failure, pilot error, and wind shear.

The hazards of wind shear can be even more hazardous to general aviation. *Wind shear*, which is the rapid change in wind direction or speed, has always been around. Convective activity produces *severe shear*, which is a rapid change in wind direction or velocity that causes an airplane's airspeed to change more than 15 knots or the vertical speed to change more than 500 feet per minute.

The *microburst* (fully defined in the next paragraph) produces the most severe wind shear threat. Rain-cooled air within a thunderstorm produces a concentrated rain or virga shaft that is less than one-half mile in diameter, which forms a downdraft. The downdraft, or *downburst*, has a very sharp edge and forms a ring vortex upon contact with the ground. The downburst spreads out, causing gust fronts that are particularly hazardous to aircraft during takeoff, approach, and landing. Upon reaching the ground, the burst continues as an expanding outflow.

A microburst consists of a small-scale and severe downburst that is less than 2½ miles across. This flow can be 180° from the prevailing wind with an average peak intensity of about 45 knots. Microburst winds intensify for about 5 minutes after ground contact and typically dissipate about 10 to 20 minutes later. Microburst wind speed differences can seem almost astronomical; indicated differences close to 200 knots were observed on August 1, 1983, at Andrews Air Force Base near Washington. Some microburst events are beyond the capability of any aircraft and pilot to recover. Microbursts can occur anytime in any season, although most normally occur during midafternoon, midsummer events.

The FAA in cooperation with aviation specialists has developed AC 00-54 *Pilot Wind Shear Guide*. Although primarily for the airlines, much of the information can be applied to general aviation. Avoidance is the best defense against a microburst encounter. When the possibility of microbursts exists, the pilot must continually check all clues; therefore, he or she must learn to recognize situations favorable to this phenomenon.

Wind shear recognition

The following discussion is based on the FAA's *Pilot Wind Shear Guide* and the Department of Commerce's *Microbursts: A Handbook for Visual Identification*. The latter publication is for sale by the Superintendent of Documents; it contains an in-depth, technical explanation of the phenomenon along with numerous color photographs depicting microburst activity. It should be part of every pilot's library.

Microbursts can develop whenever convective activity such as thunderstorms, rain showers, or virga occur associated with both heavy and light precipitation. Approximately 5 percent of all thunderstorms produce microbursts, and more than one microburst can occur with the same weather system; therefore, pilots must be alert for additional microbursts if one has already been encountered or reported and must be prepared for turbulence and shear as subsequent microbursts interact. Microbursts are characterized by:

- Precipitation or dust curls carried up toward the cloud base.
- Horizontal bulging in a rain shaft forming a foot-shaped prominence near the surface.
- An increase in wind speed as the microburst expands over the ground.
- Abrupt wind gusts.

Microbursts can occur in extremely dry or wet environments. The lack of low clouds does not guarantee the absence of shear. Microbursts can develop beneath clouds with bases as high as 15,000 feet. As virga or light rain falls, intense cooling causes the cold air to plunge, resulting in a dry microburst. (Evaporative-cooling turbulence that is associated with this phenomenon has already been discussed.) Anvils of large dry-line thunderstorms can produce high-level virga and result in dry microbursts. High-based thunderstorms with heavy rain should be of particular concern; they produced intense wind shear in the 1985 Delta accident at DFW.

Embedded microbursts are produced by heavy rain from low-based clouds in a wet environment. A wet microburst might first appear as a darkened mass of rain within a light-intensity rain shaft. A characteristic upward curl appears as the microburst moves out along the surface.

The potential for wind shear and microbursts exists whenever convective activity occurs. Pilots should review forecasts (FAs, FTs, WSTs, and AWWs) for thunderstorms, which imply low-level wind shear, or specific mention of LLWS. Check surface reports and PIREPs for wind shear clues: thunderstorms, rain showers, gusty winds, or blowing dust. Dry microbursts are more difficult to recognize. Check surface reports for convective activity (RWU, CBs, or VIRGA) and low relative humidity (30°F to 50°F temperature/dew point spread).

The low-level wind shear alert system (LLWAS) has been installed at 110 airports in the United States. The system detects differences between wind speed around the airport and a reference center-field station. Differences trigger an alert. Sensors are not necessarily associated with specific runways; therefore, descriptions of remote sites are based on the eight points of the compass: "Center-field wind three one zero at one five. North boundary wind zero niner zero at three five."

The *Airport/Facility Directory* advertises the availability of LLWAS under Weather Data Sources. The lack of an LLWAS alert does not necessarily indicate the absence of wind shear. LLWAS has limitations. Magnitude of the shear might be underestimated. Surface obstructions can disrupt or limit the airflow near the sensor. Due to the placement of sensors, microburst development might go undetected, especially in the early stages. Sensors are located at the surface; therefore, microburst development that has not yet reached the surface will be undetected and because coverage only exists near the runways, microbursts on approach will not be observed. Even with these limitations, LLWAS can provide useful information about winds near the airport.

Development continues on an automated terminal Doppler weather radar (TDWR) system that is based upon the same principle as NEXRAD. TDWR will provide wind shear and microburst warnings to controllers and should become operational during the last half of the 1990s.

Airborne radar returns of heavy precipitation indicate the possibility of microbursts. Potentially hazardous dry microbursts might only produce weak radar returns. Strong wind shear might occur as far as 15 miles from storm echoes. Radar echoes can be misleading by themselves, and it might require a Doppler radar to spot the danger of a dry microburst. The southwest edge of an intense storm can appear weak visually and on radar; however, this area is known to spawn tornadoes and severe wind shear. Convective weather approaching an airport, the downwind side, tends to be more hazardous than activity moving away.

No quantitative means exist for determining the presence or intensity of microburst wind shear. Pilots must exercise extreme caution when determining a course of action. Microburst wind shear probability guidelines have been developed by the FAA and apply to operations within 3 miles of an airport, along the intended flight path, and below 1,000 feet AGL. Probabilities are cumulative; therefore, when more than one point exists, probability increases.

The following indicate a high probability of wind shear with the presence of convective weather near the intended flight path:

- Localized strong winds reported, or observed blowing dust, rings of dust, or tornadolike features
- Visual or radar indications of heavy precipitation
- PIREPs of airspeed changes 15 knots or greater
- LLWS alert or wind velocity change of 20 knots or greater

A pilot must give utmost attention to these observations. A decision to avoid, divert, or delay is wise.

The following indicate a medium probability of wind shear with the presence of convective activity near the intended flight path.

- Rain showers, lightning, virga, or moderate or greater turbulence reported or indicated on radar
- A temperature/dew point spread of 30°F to 50°F
- PIREPs of airspeed changes less than 15 knots
- LLWS alert or wind velocity change less than 20 knots

A pilot should consider avoiding these conditions. Precautions are indicated.

The FAA states: "Pilots are . . . urged to exercise caution when determining a course of action." Probability guidelines ". . . should not replace sound judgment in making avoidance decisions." In aviation weather, there are no guarantees. The lack of high or medium probability indicators in no way promises the absence of wind shear when convective weather is present or forecast. Avoidance is the best precaution.

Review the following Colorado Springs (COS) SA and UUA for wind shear probability indicators.

COS SA 2150 M90 BKN 250 OVC 45TRW– 74/42/3612G24/019/T OVHD MOVG E OCNL LTGCG SW AND E

COS UUA /OV COS/TM 2156/ . . . TP PA60/RM AIRSPEED +– 40 KTS, THOUGHT I WAS IN THE TWILIGHT ZONE

A thunderstorm with rain showers, a 32° temperature/dew point spread, strong and gusting surface winds, and lightning are being reported. Add the PIREP, and there are two high and two moderate probability indicators. This is an example of the dry environment. A pilot must be watchful for visual microburst and LLWS clues.

This SA from chapter 1 illustrates a wet microburst environment.

PHL RS 2250 W6 X 1TRW+F . . . 3618G24/ . . . FQT LTGICCG

A thunderstorm with heavy precipitation, strong, gusty winds, and lightning is being reported. There is certainly a high probability of wind shear and microbursts.

Takeoff, approach, and landing precautions

Select the longest suitable runway for takeoff. Determine at what point the takeoff can be aborted with enough runway remaining to safely stop the aircraft. Certain manufacturers provide tables for takeoff calculations; otherwise, the pilot will have to base this distance on landing-roll tables and experience. Use the recommended flap setting, if available, for gusty wind or turbulent conditions. Use maximum-rated takeoff power to reduce the takeoff roll and overrun exposure.

Consider increased airspeed at rotation to perhaps improve the ability of the airplane to negotiate wind shear or turbulence after liftoff. Do not use a speed-reference flight director. Be alert for airspeed fluctuations, which might be the first signs of wind shear. If shear is encountered with sufficient runway remaining, abort the takeoff; however, this decision can only be made by the pilot based upon training and experience. After takeoff, use maximum-rated power and rate of climb to achieve a safe altitude, at least 1,000 feet AGL.

Select the longest suitable runway to land. Consider a recommended approach configuration with a faster-than-normal approach speed for a widened margin of safety. Turbulent air penetration or maneuvering speed should be considered. Establish a stabilized approach at least 1,000 feet AGL with power and trim, plus other elements of the landing configuration, set to follow the glideslope without additional changes. Any deviation from the glideslope or an airspeed change will indicate shear. The autopilot, except for autoflight systems, should be disengaged. The pilot should be closely monitoring vertical speed, altimeter, and glideslope displacement. Ground speed and airspeed comparisons can provide additional information for wind shear recognition. Finally, remember that the faster approach speed will require a longer-than-normal landing distance.

Wind-shear recovery technique

A wind-shear recovery technique has not yet been developed for small aircraft. The following wind-shear recovery technique developed for air carriers has been adapted from AC 00-54; however, it is applicable to most wind shear encounters in practically any aircraft.

Wind shear recognition is crucial to making a timely recovery decision. Encounters occur infrequently with only a few seconds to initiate a successful recovery. The objective is to keep the airplane flying as long as possible in hope of exiting the shear.

The first priority must be to maintain airplane control. The following guidelines were developed for the airlines; exact criteria cannot be established. Whenever these parameters are exceeded, recovery and/or abandoning the takeoff or approach should be strongly considered.

It must be emphasized that it is the responsibility of the pilot to assess the situation and use sound judgment in determining the safest course of action. It might be necessary to initiate recovery before any of these parameters are reached:

- ±15 knots indicated airspeed
- ±500 feet per minute vertical speed
- ±1 dot glideslope displacement
- Unusual throttle position for a significant period of time

If any condition is encountered, aggressively apply maximum rated power. Avoid engine overboost unless required to avoid ground contact. While on approach, do not attempt to land. Establish maximum rate of climb airspeed. As with any turbulent condition, pitch up in a smooth, steady manner.

If ground contact is imminent, pitch up to best-angle-of-climb airspeed being careful not to stall the airplane because controlled contact with the ground is preferable to an uncontrolled encounter.

When airplane safety has been ensured, adjust power to maintain specified limits. When the airplane is climbing and ground contact is no longer an immediate concern, cautiously reduce pitch to the desired airspeed.

The key to the thunderstorm and LLWS hazard is avoidance. A superior pilot uses superior knowledge to avoid having to use superior skill. At the first sign of severe shear, reject the takeoff or abandon the approach. It is easier to explain an aborted takeoff or missed approach to passengers rather than explain an accident to the FAA and insurance company.

Avoidance is the operative word with thunderstorms, microbursts, and wind shear. A pilot's proper application of many resources—training, experience, visual references, cockpit instruments, weather reports and weather forecasts—make avoidance possible.

10
Air analysis charts

CONSIDERABLE MISUNDERSTANDING ARISES BECAUSE MANY AVIATION weather texts fail to adequately describe and explain nonfrontal weather-producing systems. A pilot presented with such a situation will commonly ask the briefer, "Where's the front?" Only to be told, "There is no front." Weather occurs at all altitudes within the troposphere; the surface analysis chart often cannot solely explain the weather, even weather occurring at or near the surface.

Surface and upper-air analysis charts graphically display a three-dimensional view of the atmosphere based on selected locations at the time of observation. These charts provide a primary source for locating areas of moisture and vertical motion. Products normally available to aviation are:

- Surface analysis
- Weather depiction
- 850 mb
- 700 mb
- 500 mb
- 300 mb
- 200 mb constant-pressure charts

Each provides details on phenomena occurring at that level. The most complete description of the atmosphere can only be obtained from a combined analysis, which should include the radar summary chart.

SURFACE ANALYSIS CHART

The surface analysis chart provides a first look at weather systems. Sea-level pressure is the key element. Observed station pressure, converted to sea level, allows analysis from a common reference. The sea-level conversion introduces errors, especially in mountainous areas. The data is computer-analyzed for a first guess at isobars, the lines of equal sea-level pressure. NWS meteorologists (Fig. 10-1) manually correct the chart before transmission and annotate the position of fronts based on pressure patterns, wind shifts, the previous chart, and satellite imagery. The chart also contains wind flow, temperature, and moisture patterns, providing a primary source for the synopsis.

Fig. 10-1. *Meteorologists at the National Meteorological Center manually correct computer generated isobars and annotate fronts on the surface analysis chart before transmission.*

Often the exact location and sometimes even the presence of fronts are a matter of judgment; a front can be a zone several hundred miles across. Additionally, fronts do not necessarily reach the surface; they might be found within layers aloft. This is especially true in the western United States and in the Appalachians where mountain ranges break up fronts; therefore, there might be differences between the charted position of fronts and their location as described in the FA or TWEBs. In such cases, it would be advisable to compare chart, FA, and TWEB analysis, plus the time of each product.

The surface analysis chart is prepared and transmitted every 3 hours and is available at flight service stations and through most commercial vendors with graphics capability. The amount of detail available will depend on the vendor. Observed data must be plotted and analyzed, so the chart is always old, sometimes several hours, by the time it becomes available. The chart should always be updated with current reports.

An overall perspective of the history of system movements can be obtained by reviewing previous charts. Care must be used analyzing the apparent movement of low-pressure centers, fronts, and troughs, especially across the western United States because their movement is subjectively analyzed by the meteorologist.

Surface analysis station model

Most pilots have forgotten how to read station models. Pilots might need to brush up on this skill with the advent of DUAT and commercial weather-information vendors. Flight watch specialists, trained in their interpretation, use surface analysis chart station models when SAs are not available. Pilots don't need to decode the entire model, just the details significant to aviation.

Figure 10-2 provides an explanation of the station model. Information on the right (in small letters) is primarily for the meteorologists. Cloud types are above and below the model. Information that is most significant to aviation is on the left in capital letters; this includes temperature, dew point, present weather, total sky cover, and wind direction and speed.

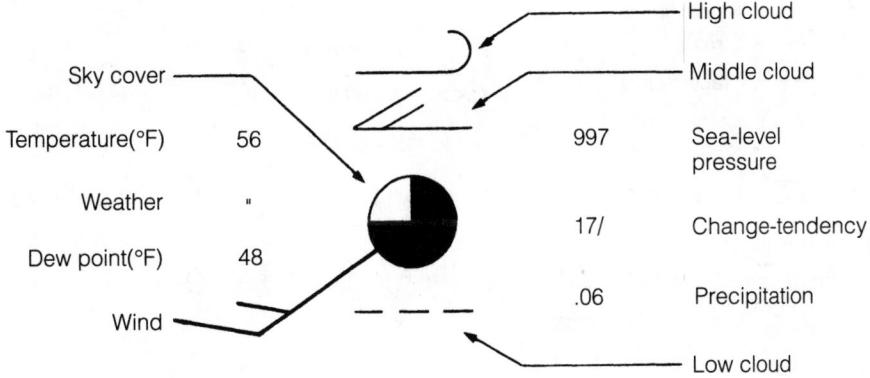

Fig. 10-2. *Surface analysis chart station models contain much valuable information. Pilots should be able to decode details significant to aviation.*

From Table 10-1, total sky cover contains more detail than sequence reports. There is a general correlation, but heights are not provided, and it is the summation of all cloud layers. Automated sites are depicted with a square. A diagonal line in the upper left corner of the square indicates "clear below 12,000 feet." Table 10-1 also decodes cloud-type symbols used with station models. Considerable information can be deduced from reported cloud types. A pilot should have a general understanding of the codes. Unfortunately, cloud-type codes will disappear with many automated observations.

Meteorologists divide clouds into four main groups:

- Low clouds (bases near the surface to about 6,500 feet)
- Middle clouds (bases from 6,500 feet to 20,000 feet)
- High clouds (based at or above 20,000 feet)
- Clouds with vertical development (based near the surface, tops of cirrus, Cu and Cb)

Table 10-1 Station model symbols.

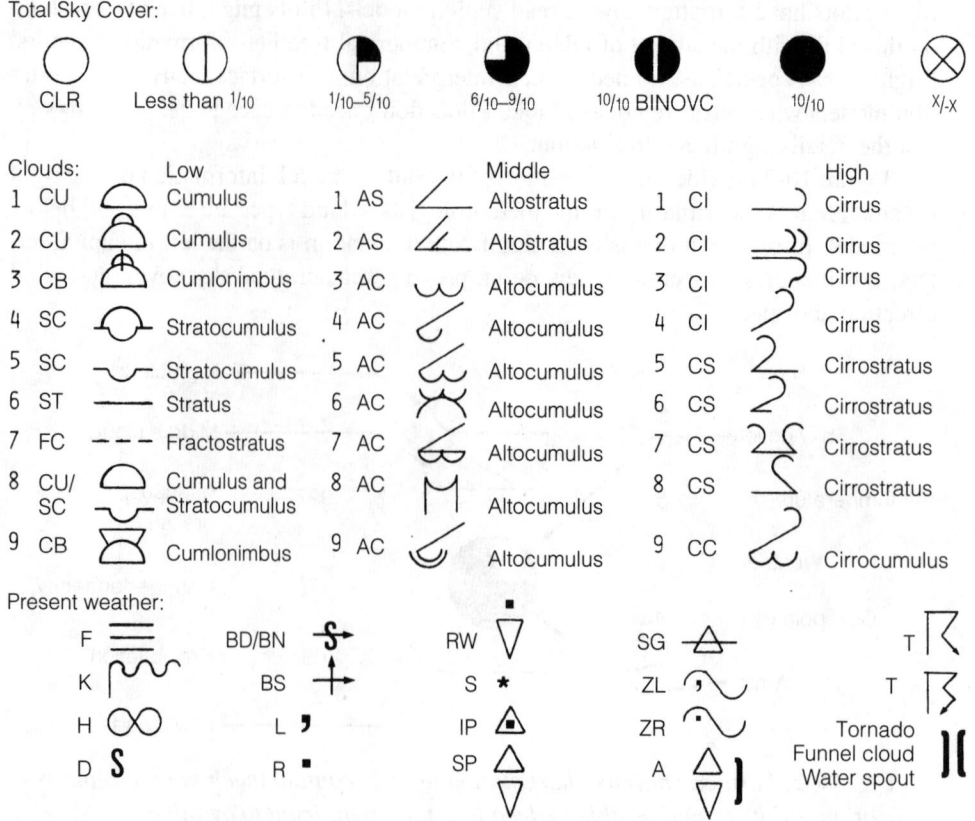

184

Notice that in Table 10-1, Cu and Cb are listed with low clouds associated with their bases.

The following discussion, based on the cloud types in Table 10-1, might refer to a code number (1 through 9) to the left of the contraction and symbol. The code starts with low clouds with vertical development. Cumulus (1) describes fair weather Cu, seemingly flattened, with little vertical development: no significant weather, turbulent below the bases, and smooth on top.

Cumulus (2) contain considerable vertical development, generally towering. This type precedes the development of cumulonimbus and thunderstorms. SAs often contain the remark TCU (towering cumulus). This refers to growing cumulus that resemble a cauliflower but with tops that have not yet reached the cirrus level.

Cumulonimbus (3) exhibit great vertical development with tops at least partially composed of ice crystals. Tops no longer contain the well-defined cauliflower shape. Cumulonimbus (9) have clearly fibrous (cirroform) tops and are often anvil shaped. Regardless of vertical development, a cloud is classified as cumulonimbus only when all or part of the top is transformed, or in the process of transformation, into a cirrus mass. Any cumulonimbus cloud should be considered a thunderstorm with all the ominous implications.

Stratocumulus (4) and (5), and cumulus and stratocumulus (8) represent a moist layer with some convection. Stratocumulus (4) forms from the spreading of Cu, which indicates decreasing convection. Stratus indicates low-level stability. Fractostratus and fractocumulus (scud) are normally associated with bad weather. Scud are shreds of small detached clouds moving rapidly below a solid deck of higher clouds.

Middle clouds fall into two general types: altostratus and altocumulus. Altostratus indicate a stable atmosphere at middle levels. Altostratus (1) is thin, semitransparent, while altostratus (2) is thick enough to hide the sun or moon. Rain or snow, even heavy snow, can fall from altostratus. When the cloud layer thickens and lowers, it becomes nimbostratus (Ns).

Altocumulus indicate vertical motion at mid levels. Altocumulus (3) is thin, mostly semitransparent. Altocumulus numbers (5, 6, and 7) describe cloud thickness, development, and altocumulus associated with other cloud forms.

High clouds are known as cirrus, cirrostratus, and cirrocumulus. Cirrus (1) consists of filaments, commonly known as mares' tails. Cirrus (2) and (3) are often associated with cumulonimbus clouds. Cirrus (4) usually indicates a thickening layer, which might be associated with the approach of a front. Cirrostratus (5 through 8) describe sheets or layers of cirrus. Sun or moon halos often appear in these layers. Cirrocumulus indicate vertical motion at high levels and might indicate high-altitude turbulence.

Cirrus has no significance to low-level flights; however, cirrus is often associated with the jet stream and high-altitude turbulence. Cirrus clouds that form as transverse lines or cloud trails perpendicular to the jet stream indicate moderate or greater turbulence. These clouds might be reported as cirrocumulus. Cirrus streaks, parallel to the jet, are long and narrow streaks of cirrus frequently seen with jet streams. Jet-stream cirrus and cloud trails are easily identified in satellite imagery.

Selected present-weather symbols and their SA-contraction equivalents are contained in Table 10-1. Up to four drizzle, rain, or snow symbols can indicate intensity and trend; the station model in Fig. 10-2 shows two drizzle symbols. Symbols can be combined to indicate more than one phenomenon occurring. Precipitation representations will appear above the thunderstorm symbol to indicate type and intensity of precipitation accompanying the "thunder bumper." An inverted, elongated triangle indicates showers. The type of showers (rain, snow, etc.) appears above the triangle. Unknown light precipitation reported by automated sites is indicated by "?-." Volcanic ash is a mountain with an ash plume. Interpretation of present-weather symbols is the same as SA symbols.

Temperature and dew point in degrees Fahrenheit appear to the left of the station model; they will be in degrees Celsius with METAR. Wind is indicated as the direction (true) *from which* the wind is blowing, with barbs, half barbs, and flags representing speed in knots.

Information to the right of the station model is less significant to the pilot. Sea-level pressure in millibars is decoded the same as SAs; Fig. 10-2 shows 997, which means that the sea-level pressure is 999.7 millibars. (Refer to chapter 1 for additional explanation.) The pressure change during the past 3 hours in 10ths of millibars and the pressure's tendency (increasing, decreasing, or steady) appear below sea-level pressure. Any precipitation during the past 6 hours appears in the lower right; the measurement is to the nearest hundredth of an inch.

Let's decode, translate, and interpret the station model in Fig. 10-2; refer to Table 10-1. Total sky cover is broken (%0 to %0). Cloud bases are not provided, but they can be inferred from cloud type. The low clouds consist of fractostratus, or scud. Thick altostratus and cirrus are also being reported, so the scud can't be too extensive. The higher cloud types might indicate an approaching front. Wind is out of the southwest at 15 knots. Visibility, which does not directly appear, is probably good in spite of a relatively close temperature/dew point spread (8°F). Moisture is being added by drizzle, however. If the wind remains constant, additional moisture could increase the amount of fractostratus, but if the wind subsides, fog and reduced visibility might result. Except for low-level mechanical turbulence due to wind, the atmosphere appears stable from the cloud types reported; therefore, mostly smooth flying conditions and light icing in clouds and precipitation above the freezing level are indicated.

Table 10-2 contains frontal symbols, provides a description of the front, and defines hues used for color displays. *Frontogenesis*, the initial formation of a front or frontal zone, is depicted by a broken line with the appropriate front-type symbol. *Frontolysis*, the dissipation of a front or frontal zone, is represented by a broken line with the appropriate front-type symbol on every other line. As previously mentioned, frontal zones can occur in layers aloft. Hollow bumps and triangles indicate a frontal boundary aloft.

186

Table 10-2. Surface analysis chart symbols.

Symbol	Description	Color*
H	High pressure center	Blue
L	Low pressure center	Red
	Cold front	Blue
	Warm front	Red
	Stationary front	Red/Blue
	Occluded front	Purple
	Cold frontogenesis**	
	Cold frontolysis**	
	Squall line	Purple
	Trough (Trof)	Brown
	Ridge	Yellow

* Suggested colors for color presentations. Bumps and triangles are normally omitted with a color display.

** Symbol, with appropriate bumps and triangles, also used to depict warm, stationary and occluded frontogenesis, and frontolysis.

Note: Open bumps and triangles indicate location of front aloft.

187

Surface chart analysis

Refer to Fig. 10-3, which is the 0900Z, August 6, 1994, chart. Pressure patterns indicate areas at the surface that are under the influence of high or low pressure, troughs, or ridges. The terms high and low are relative. A *high* is defined as an area completely surrounded by lower pressure. Conversely, a *low* is an area surrounded by higher pressure. Lines known as isobars connect areas of equal sea-level pressure.

Beginning with 1,000 mbs, lines are drawn at four-millibar intervals (00, 04, 08, etc.). A weak pressure gradient may be drawn at two-mb intervals using dashed isobars. Pressures at each pressure center are indicated by a four-digit number. For example, the high center over Nebraska has a central pressure of 1,018 mbs, the low over southern California, 1,007 mbs.

A *trough* consists of an elongated area of low pressure. Figure 10-3 shows a trough extending south into Wyoming from a low center in eastern Montana. A *ridge*, the opposite of a trough, is an elongated area of high pressure and is almost always associated with anticyclonic (clockwise in the Northern Hemisphere) wind flow. Figure 10-3 shows that a ridge extends from the high center over the upper Mississippi Valley into eastern Canada.

Wind blows across isobars due to friction between the wind and surface causing convergence and divergence. *Convergence* is an upward vertical motion that destabilizes the atmosphere, increasing relative humidity that can lead to the formation of clouds and precipitation. *Divergence* is a downward vertical motion that stabilizes the atmosphere, decreasing relative humidity and reducing the likelihood of clouds and precip. Convergence of itself does not necessarily produce poor weather, nor does divergence produce good weather. Other factors, such as moisture, must be considered. Surface convergence and divergence only affect the atmosphere below about 10,000 feet, but they are a major factor in the "weather machine" of vertical motion.

Convergence occurs along curved isobars surrounding a low or trough. Maximum convergence takes place at the low center or along the trough line. Figure 10-3 shows surface winds blowing into the low in western Canada.

Troughs are not fronts, although fronts normally lie in troughs. A *front* is the boundary between air masses of different temperatures, whereas a trough is simply a line of low pressure. Both phenomena produce upward vertical motion.

Note the anticyclonic, outward flow from both high centers in Fig. 10-3. Maximum divergence takes place at the high center. The symbol for a ridge is a continuous zigzag line, which is normally not depicted on the surface analysis. A high or ridge implies surface divergence.

There is a misconception that high pressure always means good flying weather. Although good weather often occurs, there are exceptions. Strong pressure gradients at the edge of high cells can cause vigorous winds and severe turbulence. Near the center of a high, or with weak gradients, moisture and pollutants can be trapped at lower levels causing reduced visibilities and even producing zero-zero conditions in fog for days or even weeks.

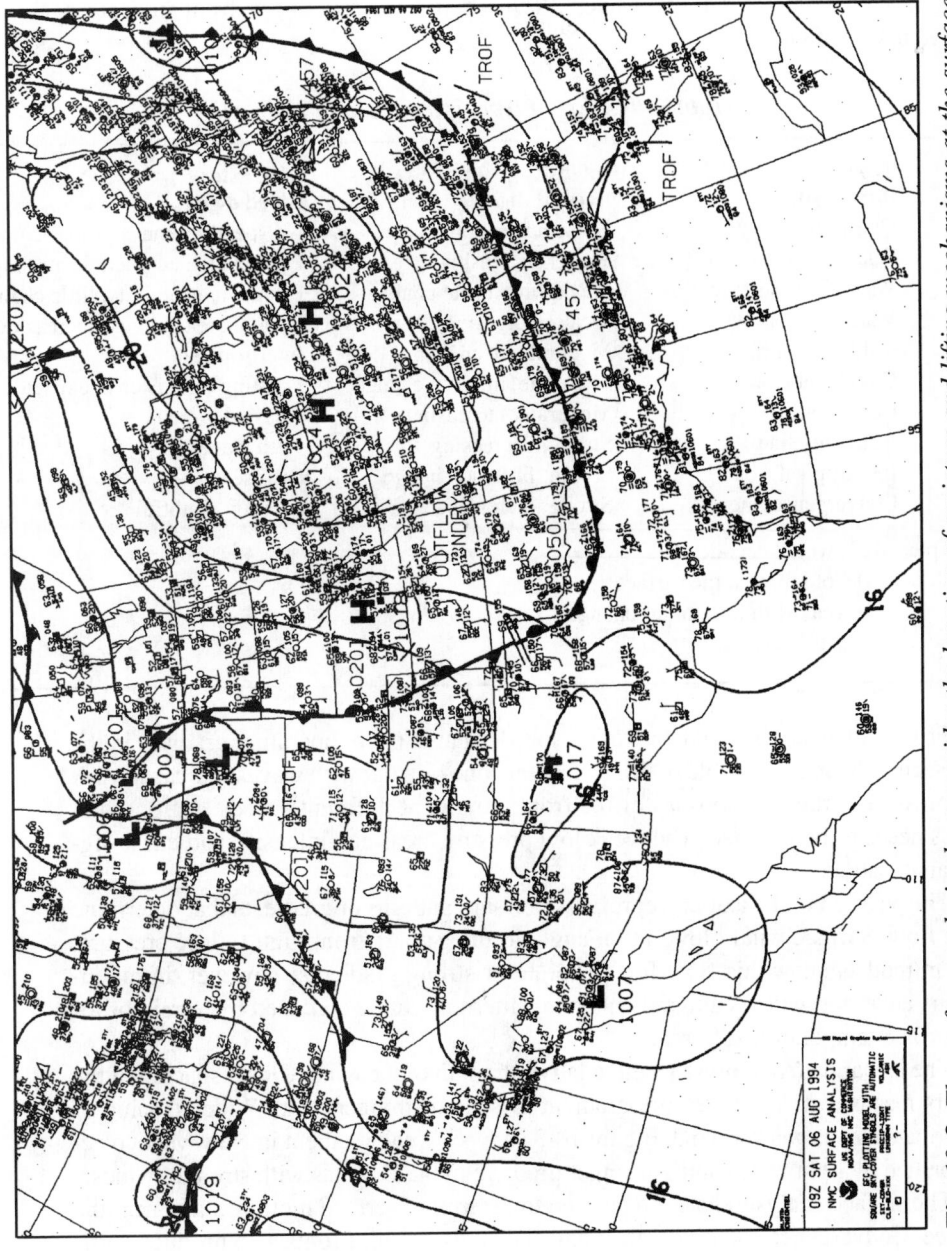

Fig. 10-3. *The surface analysis chart provides the location of moisture and lifting mechanisms at the surface and in the lower atmosphere.*

189

A three-digit code entered along the front specifies type, intensity, and character (Table 10-3). For example, the type, intensity, and character of the front in Alabama is 457. The type is 4 (cold front at the surface), intensity 5 (moderate with little or no change), and character 7 (with waves). Two short lines crossing a front indicate a change in classification.

Table 10-3. Front type, intensity, and character.

Code	Type (first digit)	Intensity (second digit)	Character (third digit)
0	Quasi-stationary surface	No specification	No specification
1	Quasi-stationary aloft	Weak, decreasing	Frontal area activity, decreasing
2	Warm front surface	Weak, little or no change	Frontal area activity, little change
3	Warm front aloft	Weak, increasing	Frontal area activity, increasing
4	Cold front surface	Moderate, decreasing	Intertropical
5	Cold front aloft	Moderate, little/no change	Forming or existence expected
6	Occlusion	Moderate, increasing	Quasi-stationary
7	Instability line	Strong, decreasing	With waves
8	Intertropical front	Strong, little/no change	Diffuse
9	Convergence line	Strong, increasing	Position doubtful

Example: 427] would decode:
(4) Cold front at the surface;
(2) weak with little or no change;
(7) with waves.

Frontal intensity is based on frontal speed. That is the temperature gradient in the cold sector, the region of colder air at a frontal zone. A front with waves indicates weak low pressure centers or portions of the front moving at different speeds. A front with waves needs to be watched. The weak low pressure areas can intensify and cause significant weather.

The position of the isobars represents pressure patterns or gradients that determine wind flow. Surface wind blows at an angle to the isobars from high to low pressure. Station models show this flow for moderate or strong gradients; however, in mountainous areas and with weak gradients, the pattern might be confused by terrain or local surface-temperature differences.

The isobar pattern represents the relative strength of the wind. Closer spacing of the isobars means a stronger pressure-gradient force and stronger wind. Wider spacing of the isobars means weaker wind. Figure 10-3 shows a weak gradient in New Mexico accompanied by light winds and a stronger gradient in the Dakotas with stronger winds.

The surface analysis chart can be used to determine vertical motion at and near the surface. Convergence and divergence have been discussed. Fronts also produce vertical motion. Tables 10-2 and 10-3 describe how this phenomenon is depicted on the surface analysis chart.

A front is generally the zone between two air masses of different density. Temperature is the most important density factor; therefore, fronts almost invariably separate air masses of different temperatures. Other factors can distinguish a front, such as a pressure trough, change in wind direction, moisture differences, and cloud and precipitation forms.

Norwegian meteorologist Vilhelm Bjerknes and his son Jakob developed the polar front theory at the beginning of the twentieth century. World War I had begun and it was popular to use the language of the conflict; thus, weather was described using words like fronts, advances, and retreats. The weather did resemble a war between air masses.

The Earth's atmosphere is a giant heat exchanger, moving cold air down from the poles and warm air up from the tropics. Typically in the Northern Hemisphere, the cold air pushes down from the northwest, lifts, and replaces the warm air. The boundary where this action takes place is known as a *cold front*; however, to accomplish this, at some point, warm tropical air must replace the colder air. This typically takes place ahead of the cold front as the warm air, moving from the south, rises above and replaces the colder, retreating air; this boundary is known as a *warm front*. Cold fronts move faster than warm fronts; sometimes the cold front will overtake the warm front, and an occlusion occurs, and the front is known as an *occluded front*. When frontal speed decreases to 5 knots or less, it is labeled *stationary*. This action is more or less continuous around the world at middle latitudes. Fronts produce vertical motion from the surface to about the middle troposphere.

The intensity and movement of fronts are affected by many factors, such as temperature, moisture, stability, terrain, and upper-level systems. Another factor is slope; the average cold-front slope is 1 mile vertically for every 100 miles horizontally. The closer that a front is to a jet stream, the slope steepens and typically the front becomes proportionally stronger.

Fronts run the spectrum from a complete lack of weather to benign clouds that are ideal for a novice instrument pilot to lines of severe thunderstorms that no pilot or aircraft can negotiate. Each front—and for that matter, any weather system—must be evaluated separately, then a flight decision can be made based on the latest weather reports and forecast plus the pilot's and airplane's capabilities and limitations.

The chart might depict a dry line, appearing as a broken line and labeled "DRY LINE." The *dry line*, or temperature-dew point front, marks the boundary between moist, warm air from the Gulf of Mexico and dry, hot air from the southwestern United States. A dry line usually develops in the New Mexico, Texas, and Oklahoma region during the summer months. Thunderstorms and tornadoes can be triggered by the temperature-dew point front.

Figure 10-3 shows an outflow boundary in Nebraska. An *outflow boundary* is a surface boundary left by the horizontal spreading of thunderstorm-cooled air. The boundary is often the lifting mechanism needed to generate new thunderstorms.

Upslope and downslope flow causes vertical motion. This can be determined from the chart when we are familiar with the terrain. Upslope produces the same characteristics as convergence and downslope divergence. Strong downslope flows are sometimes given local names, such as the *Chinook* (Native American name for "snow eater") that develops along the eastern slopes of the Rockies and the *Santa Ana* of Southern California. The Chinook flows have been known to raise temperatures as much as 20°F in 15 minutes, and melt and evaporate a foot of snow in a few hours. The strong winds can exceed 85 knots, causing extreme damage.

Forecast synopses often include "upslope" and might contain "Chinook" or "Santa Ana" when these conditions occur. Figure 10-3 shows upslope flow over Texas from the Gulf of Mexico.

Onshore and offshore flows of moderate or greater intensity can be determined from the chart. An onshore flow can translate into advection fog, upslope, or convection with the development of thunderstorms (depending on conditions); an offshore flow means clear skies. An onshore flow can be seen in Fig. 10-3 along the Texas Gulf Coast.

Temperature and moisture patterns are determined by analyzing station model temperatures and dew points. Refer to Fig. 10-3. Considerable moisture at the surface exists along the Gulf Coast with light surface winds that have caused fog to develop. Because 0900Z is 4 a.m. central daylight time, what kind of weather can be expected during the morning? What kind of weather can be expected in the vicinity of the outflow boundary in Nebraska? The answers are in the next subsection regarding the weather depiction chart.

WEATHER DEPICTION CHART

The weather depiction chart is computer-generated, computer-analyzed, and transmitted every 3 hours as a record of observed surface data. Frontal positions are obtained from the previous surface analysis. The information is hours old by the time the chart becomes available; data should always be updated with current reports. The chart is analyzed into three categories the same as the area forecast outlook: IFR, MVFR, and VFR.

Because the chart is computer-analyzed, it cannot consider terrain, nor is it intended to represent conditions between reporting locations. Gross errors between depicted categories and actual weather can occur. Compare Fig. 10-4 and Fig. 10-5; both represent approximately the same period in time. Figure 10-4 indicates extensive IFR and MVFR conditions in central California and western Oregon; compare that depiction with the visual satellite image in Fig. 10-5. The fog that has created the conditions has almost dissipated in central California, and the fog in Oregon is restricted to the Willamette Valley. The rather significant difference between the coverage in Fig. 10-4 and Fig. 10-5 is due to the limitations of the weather depiction chart. The weather depiction chart is not a substitute for current observations because conditions can improve or deteriorate.

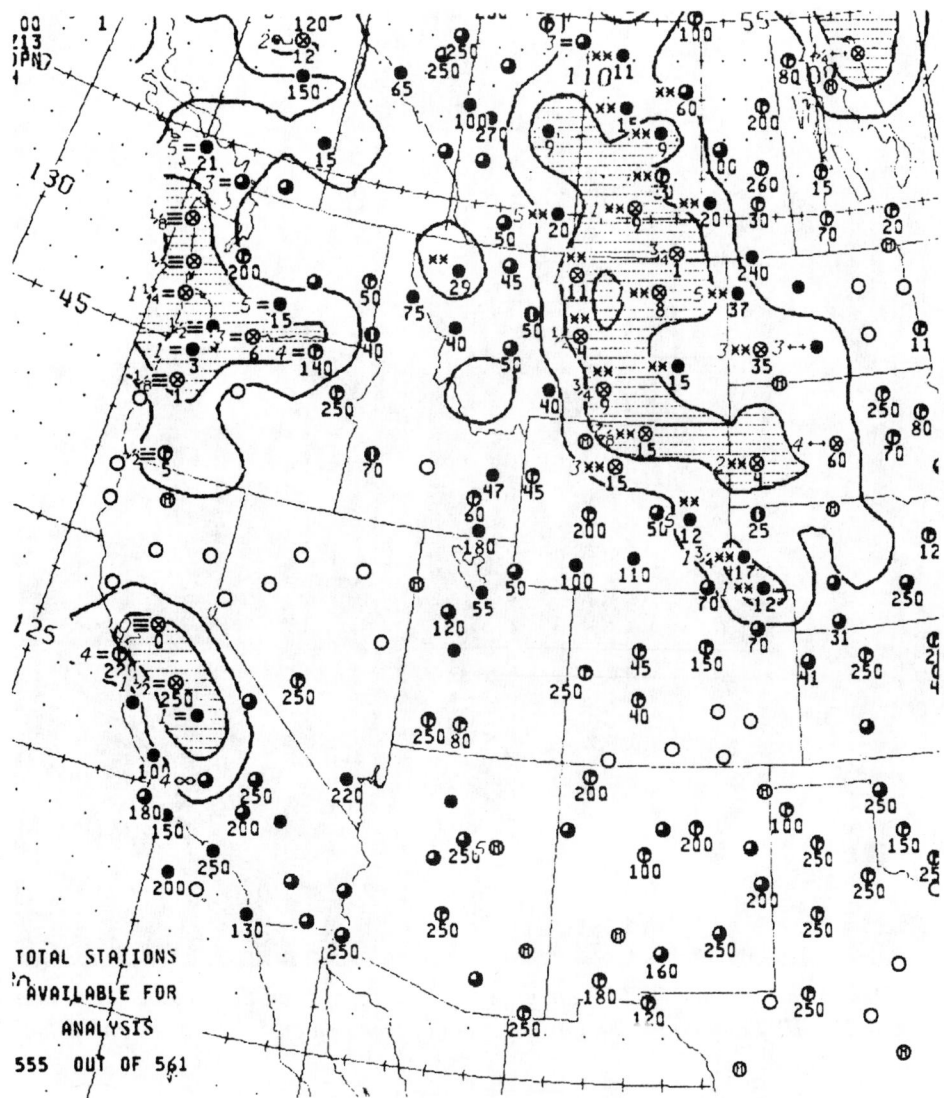

Fig. 10-4. *The computer program that generates the weather depiction chart cannot consider terrain and is not intended to represent conditions between reporting locations.*

Weather depiction station model

Cloud bases in hundreds of feet AGL appear beneath the station circle of a station model on the weather depiction chart. When total sky cover has few or scattered clouds (less than ⅒), the base of the lowest layer appears. Visibilities of 6 miles or less and present weather are entered to the left of the station circle. Sky cover and present weather symbols are the same as used on the surface analysis, as illustrated

Fig. 10-5. *Compare this satellite image with the excerpt from the weather depiction chart in Fig. 10-4 and notice the gross errors between depicted categories and actual weather.*

in Table 10-1, with one exception. Automated observations eliminate much useful information; automated sites are indicated by a bracket "]" to the left of the station circle (Fig. 10-6).

The station model in extreme southern Texas (Fig. 10-6) shows the sky condition broken (station circle) and the cloud base at 6,500 feet AGL (65, underneath station circle). Visibility is reduced in fog to 3 miles (3, left of fog symbol). Because the number of stations analyzed exceeds the number plotted, contoured areas might appear without station models; a station has been analyzed but not plotted for the area of MVFR in central Oklahoma.

194

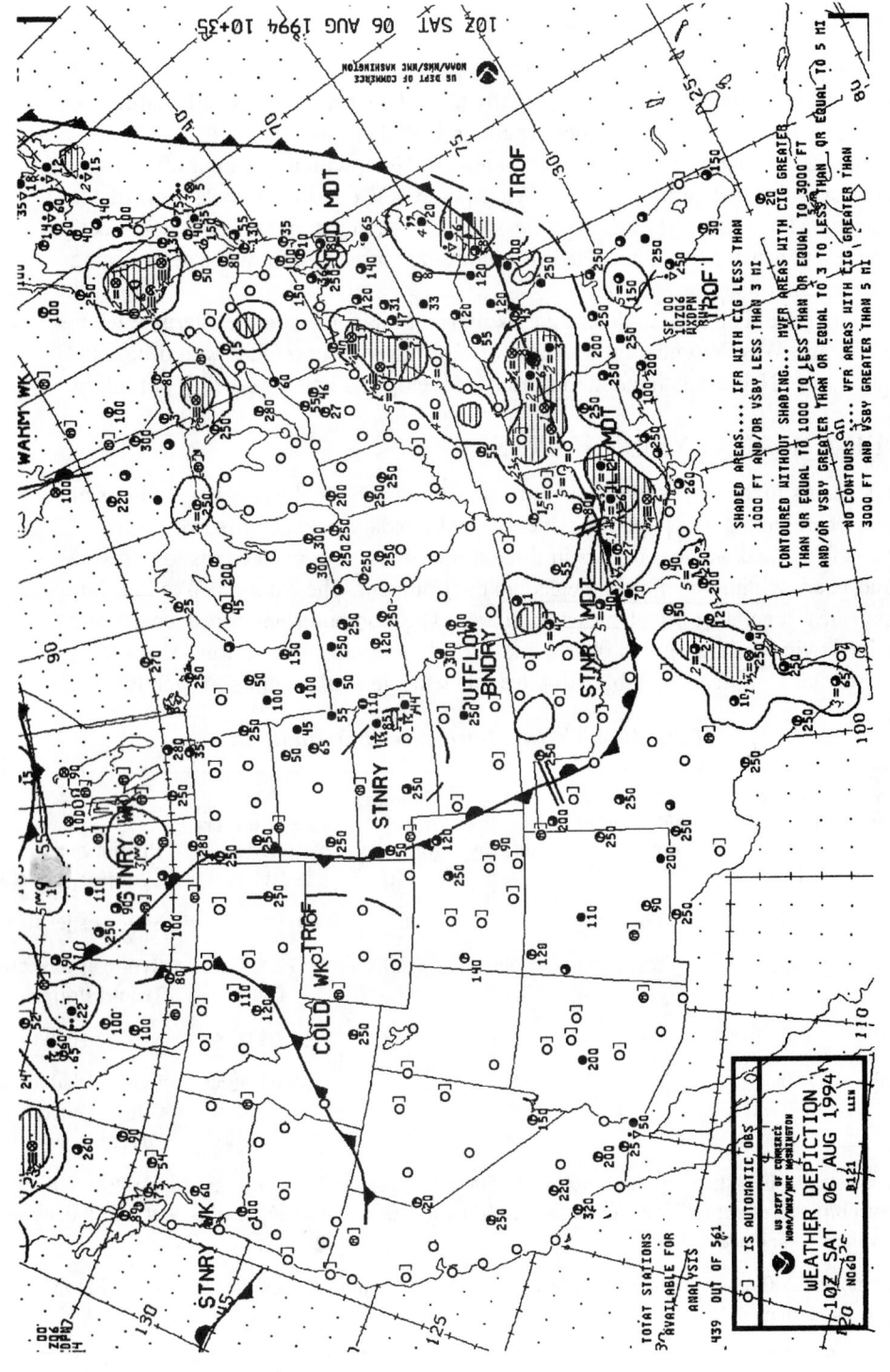

Fig. 10-6. *The computer-generated weather depiction chart analyzes surface weather into three categories: IFR, MVFR, and VFR.*

Weather depiction chart analysis

The weather depiction chart provides a big, simplified picture of surface conditions. It alerts pilots and briefers to areas of potentially hazardous low ceilings and visibilities. The chart is a good place to begin looking for an IFR alternate.

Figure 10-6 contains 1000Z data for August 6, 1994, which is 1 hour after a surface analysis on that date in Fig. 10-3. Widespread IFR conditions exist west of the Appalachian Mountains and in the Southeast. The computer program does not necessarily represent actual conditions. Satellite pictures are often helpful when determining the actual extent of stratus and fog.

The chart could be used to determine likely locations for a suitable alternate when flying IFR into West Virginia; an alternate toward the west into Ohio would be indicated. In eastern Nebraska, thunderstorms have developed along the outflow boundary.

UPPER-AIR ANALYSIS CHARTS

Weather occurs in the troposphere and stratosphere, which are the two lower layers of the atmosphere, and the tropopause, which is the boundary between the two layers.

Pilots fly in and weather occurs in three dimensions, so a need exists to describe the atmosphere within this multidimensional environment. The National Weather Service prepares several constant-pressure charts. The computer-generated charts are transmitted twice daily based on 0000Z and 1200Z upper-air observations. Each level has a particular significance; Table 10-4 describes the general features of each level.

Table 10-4. Constant-pressure chart analysis.

Pressure altitude (feet)	Pressure level (mb)	Temp/dew point spread	Isotachs	Contour interval (meters)	Height meters plotted/decode	Primary uses
39,000	200	Yes	Yes	120	192/11,920	Synopsis jet stream
30,000	300	Yes	Yes	120	911/9,110	Synopsis jet stream
18,000	500	Yes	No	60	572/5,720	Synopsis advection Troughs/ridges
10,000	700	Yes	No	30	928/2,928	Synopsis advection
5,000	850	Yes	No	30	585/1,585	Synopsis advection Convergence Divergence

To decode station height: prefix 850 mb level with a "1"; prefix a "2" or a "3" to the 700 mb height, whichever brings it closer to 3,000 meters; add a "0" to 500 mb and 300 mb heights; and, for 200 mb level prefix with a "1" and add a "0."

For 300 mb and 200 mb the temperature/dew point spread is omitted when the air is too cold to measure dew point (less than –41°C).

Each chart represents a constant-pressure level, so the chart is analyzed for altitude or height in meters above sea level. Lines, known as contours, connect areas of equal height. Contours are analyzed in the same way as isobars, the closer the spacing of the contours, the stronger the wind; however, wind blows *parallel* to the contours, due to the lack of friction; only pressure gradient and Coriolis forces are present.

Constant-pressure analysis station model

Constant-pressure analysis charts depict radiosonde data. Figure 10-7 illustrates standard station plots. Wind direction and speed use standard symbology, except for the contraction LV, which indicates "light and variable."

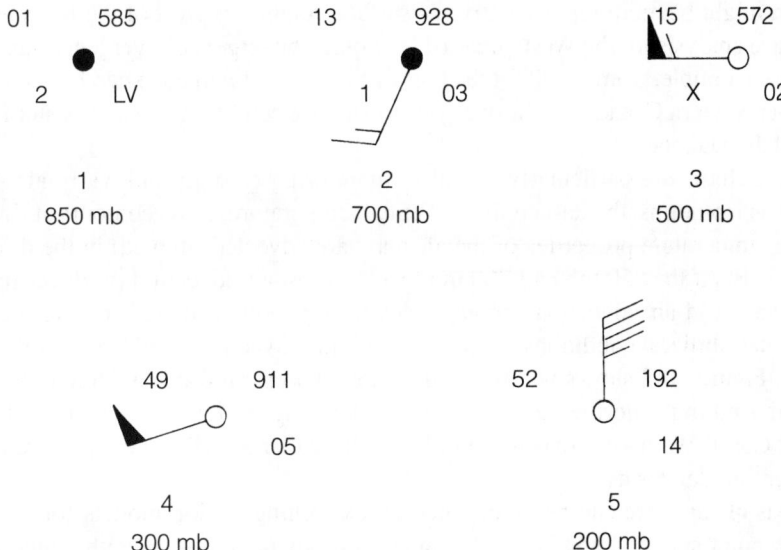

Fig. 10-7. *Constant-pressure-chart station models depict radiosonde data. The models provide height, wind, temperature, and moisture for the constant-pressure surface.*

Temperature is plotted in degrees Celsius; however, unlike surface station plots, the temperature/dew point spread, or depression, appears instead of the dew point temperature. For example, in Fig. 10-7, the 850-mb station model has a temperature of −1°C and a dew point depression of 2°C; therefore, the dew point temperature is −3°C. Darkened station circles, plotted with a temperature/dew point spread of 5°C or less, indicate a moist atmosphere. This alerts pilots and forecasters to potential clouds, precipitation, and icing depending on temperature. An X indicates a temperature/dew point spread greater than 29°C (Fig. 10-7, the 500 mb plot). With temperatures less than −41°C, air is too dry to measure a dew point; therefore, the dew point depression is omitted in the 300 mb and 200 mb plots in Fig. 10-7.

The upper right corner of the plot contains the height of the constant-pressure level. Table 10-4 decodes these values. The number in the lower right corner represents height change in tens of meters during the past 12 hours. For example, in Fig. 10-7, the height of the 700 mb surface has lowered 30 meters (–03). In general, lowering heights indicate deteriorating weather, and rising heights indicate improving weather. The greater the fall or rise, the more rapid the change, and systems tend to move in the direction of greatest height change.

850 mb and 700 mb constant pressure charts

Table 10-4 shows that the 850 mb and 700 mb charts represent the lower portion of the troposphere, providing a synopsis for approximately 5,000 and 10,000 feet. The 850 mb chart might be more representative of surface conditions west of the Rockies than the surface analysis. In the West, areas of frictional convergence/divergence can be located. For example, from the 850 mb chart in Fig. 10-8, divergence can be seen in the ridge over western Canada. Additionally, a downslope condition exists in eastern Montana and the Dakotas.

These charts are particularly useful in monitoring cold-air and warm-air advection. When isotherms, the dashed lines of equal temperature, cross contours at right angles, the temperature properties of the air mass are advected (moved) in the direction of the winds. At the 850 mb and 700 mb levels, warm-air advection produces upward motion and cold-air advection produces downward motion; therefore, warm-air advection destabilizes conditions, whereas cold-air advection tends to stabilize the weather. Figure 10-8 shows warm-air advection in western Canada from a tongue of warm air. And in the northeastern United States, a tongue of cold air advances toward the East Coast. Warm-air advection might be all that's needed as a lifting mechanism to trigger thunderstorms.

Areas of moisture can be determined by examining station models for temperature/dew point spread. Icing is implied in areas of visible moisture with temperatures between 0°C and –10°C. The 700 mb chart in Fig. 10-9 shows temperatures within this range at this level over the entire United States. The air is dry along the West Coast, implying little potential for icing; however, over Kansas and Oklahoma there is considerable moisture present for an increased icing potential.

These charts can be used to determine the potential for turbulence and mountain wave activity. The 700 mb chart is usually the reference level for mountain waves. Winds in excess of 40 knots imply moderate or greater mechanical turbulence. When winds of these speeds blow perpendicular to a mountain range, accompanied by cold-air advection (a stabilizing condition), a strong potential exists for mountain waves and associated turbulence.

Air-mass thunderstorms tend to move with the 700 mb winds. And in the middle latitudes, a 700 mb temperature of 14°C or warmer tends to inhibit convection at this level. If convection occurs below, clouds tend to stop rising and spread out at about the 700 mb height.

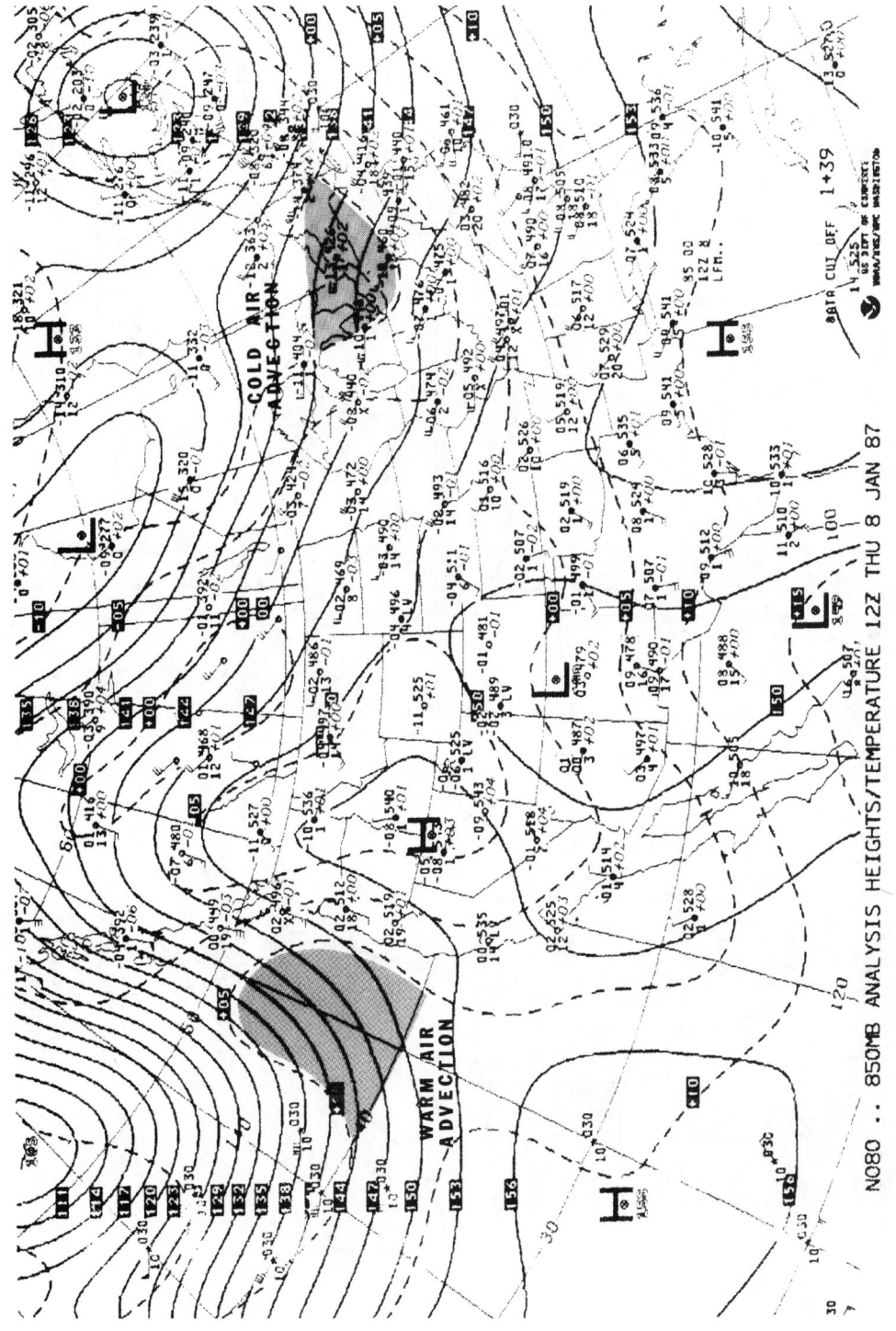

NO80 .. 850MB ANALYSIS HEIGHTS/TEMPERATURE 12Z THU 8 JAN 87

Fig. 10-8. *The 850-mb chart might be more representative of surface conditions west of the Rockies than the surface analysis.*

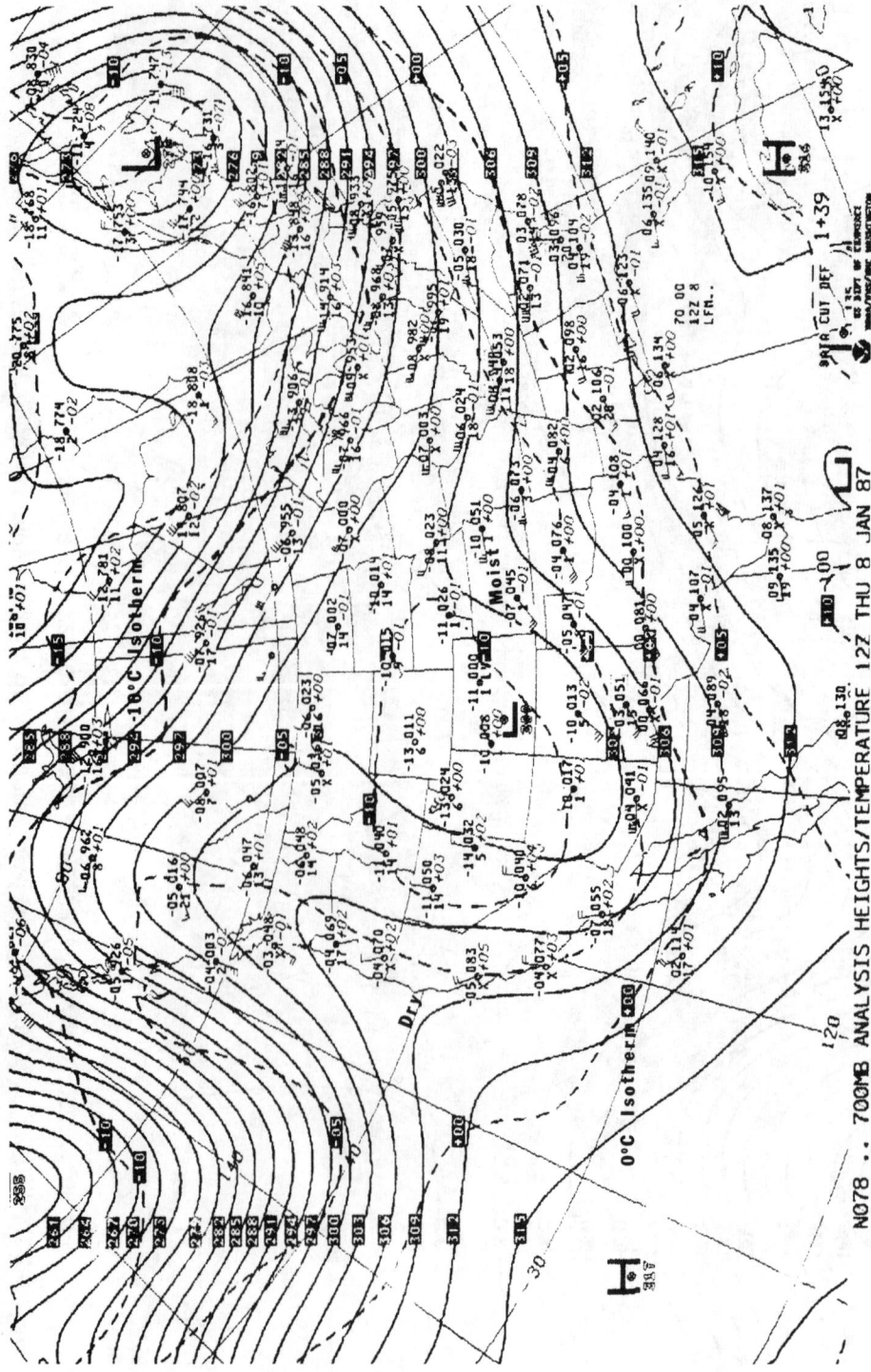

N078 .. 700MB ANALYSIS HEIGHTS/TEMPERATURE 12Z THU 8 JAN 87

Fig. 10-9. *The 700-mb chart is usually the reference level for mountain waves. The 700-mb and 850-mb charts describe the atmosphere in the lower troposphere.*

500 mb constant pressure chart

Probably the most important and useful chart—maybe even more important to meteorologists than the surface analysis—is the 500 mb chart, which describes the atmosphere in the middle troposphere, an altitude of approximately 18,000 feet. This chart provides important pressure, wind flow, temperature, and moisture patterns and can be used to determine areas of vertical motion at this level.

Troughs and ridges are easily seen in Fig. 10-10. Unlike the surface analysis chart where upward vertical motion takes place along a trough line, upward vertical motion at the 500 mb level takes place between the trough and ridge line (off the West Coast and to a lesser degree ahead of the trough moving through Arizona). Troughs transport cold air down from the north and warm air up from the south. Warm air rides northward on the east side of the trough (trough-to-ridge flow), so the air is lifted as it moves northward, producing upward vertical motion.

When moisture is present, such as the area outlined in gray over western Canada, clouds and precipitation can develop in the middle troposphere. Conversely, in the ridge-to-trough flow over California, cold air sinks southward, producing downward vertical motion with clear and dry conditions in the middle troposphere. Clouds and precipitation frequently accompany upper-level lows and troughs, even without surface frontal systems or storm systems.

Three to seven major waves in each hemisphere encircle the globe. These global, or *long-wave*, ridges and troughs extend for thousands of miles. Long waves move generally eastward at up to 15 knots, but can remain stationary for days or even retreat.

Short-wave troughs, embedded in the overall flow, tend to pass through the long-wave pattern at speeds of 20 to 40 knots. Most surface lows and frontal systems are associated with upper-level short-wave troughs. Figure 10-10 illustrates two short-wave troughs in southern Canada. Although usually best seen on the 700 mb chart, significant waves usually extend to the 500 mb level. Upper troughs are a key to the evolution of weather systems.

When a short-wave trough moves through a long-wave trough, upward vertical motion is amplified, even upward vertical motion as it moves through a ridge. Short waves can be strong vertical-motion producers. Information on short waves often appears in FA and TWEB synopses.

Cold-air advection destabilizes at the 500 mb level and warm-air advection stabilizes the atmosphere. This is opposite to the effects of cold- and warm-air advection near the surface. Rising air will be warmer than surrounding air, so cold-air advection enhances thunderstorms by promoting vertical development, sometimes referred to as a "cold low aloft." A cold low aloft tends to be slow and move erratically. Warm-air advection at this level strengthens high-pressure ridges and diminishes low-pressure troughs. In Fig. 10-10, cold-air advection is occurring in the trough-to-ridge flow off the West Coast.

Moisture at the 500 mb level can be determined by darkened station models. In Fig. 10-10, the gray areas of western Canada and over Colorado and Kansas show ar-

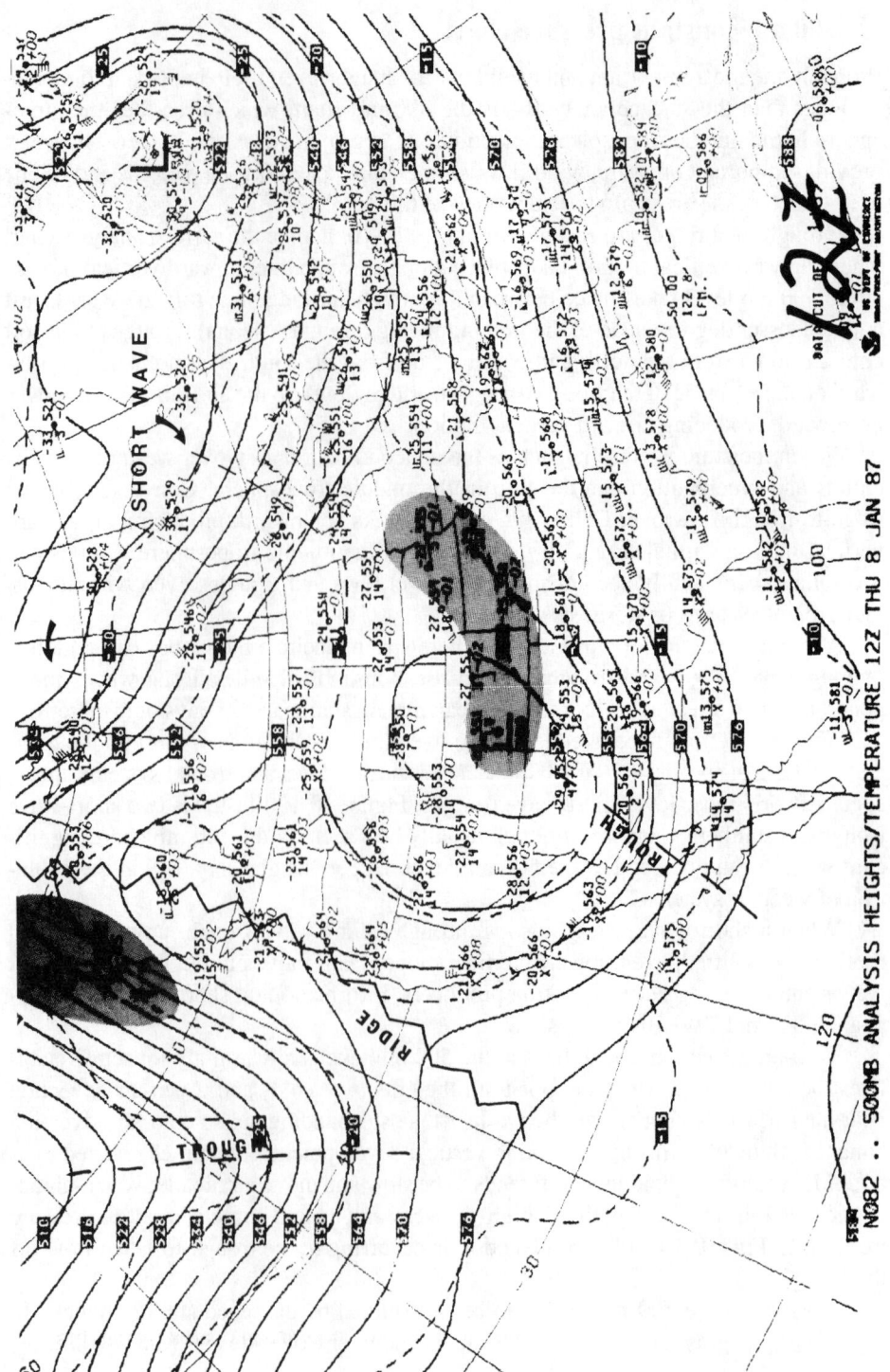

Fig. 10-10. *The 500-mb chart describes the atmosphere in the middle troposphere and might be more important to meteorologists than the surface analysis.*

eas where the temperature/dew point spread is 5° or less. This chart is a good indicator of high-level icing in summer months with storms that are either well-developed or contain tropical moisture.

Surface weather systems tend to follow the 500 mb flow, and organized thunderstorms tend to move in the direction of the 500 mb winds. The 12-hour height change provides a general trend for system movement. Rising heights indicate a building ridge or weakening trough, and lowering heights indicate a deepening trough or weakening ridge.

At the 500 mb level, the shape of the contours rather than wind speed determines the potential for turbulence. Wind shear turbulence at 500 mb occurs as an airplane flies through an area of changing wind direction or speed; therefore, the greater the curvature (directional change), the greater the potential for turbulence with proportionally greater intensity. The horizontal distance where this change occurs is crucial (Fig. 10-10). The greater the curvature of contours, the greater the probability of turbulence; therefore, more turbulence potential exists in the trough over the western Pacific than over the southwestern United States. The probability of turbulence also exists in the sharp ridge over western Canada. Developing low-pressure troughs moving from the northwest are particularly dangerous.

Areas of potential turbulence occur in merging flows or the neck of a cutoff low. In Fig. 10-10, a merging flow exists east of the trough in the southwestern United States and northern Mexico. Turbulence could also be expected in the neck of the cutoff low over eastern Montana and the Dakotas.

FSS specialists have been criticized for not appreciating the importance of or providing information from the 500 mb chart and even discarding the product. I know of no FSS with access to charts that doesn't post the 500 mb product; however, like pilots, some specialists are better schooled than others. Translation and interpretation of the chart over the phone is difficult. The best answer is direct access, which should be available through DUATS; unfortunately there will be a fee for graphics.

300 mb and 200 mb constant pressure charts

The 300 mb and 200 mb charts provide details of pressure, wind flow, and temperature patterns at the top of the troposphere and occasionally into the lower stratosphere. The charts indicate the strength of features in the lower atmosphere. Strong storm systems on the surface are reflected in the 300 mb and 200 mb patterns, whereas weaker systems lose their identity at these levels. The 500 mb low over the Rockies in Fig. 10-10 is reflected in Fig. 10-11 at the 300 mb level, but has lost its identity in Fig. 10-12 at the 200 mb level. Rather than become a closed low (surrounded by a closed contour), it has weakened into a trough.

At middle latitudes, such as the United States, the jet stream can usually be found on the 300 mb and 200 mb charts. Wind speeds and curvature of contours provide a clue to clear air turbulence. Because wind speed and direction is of primary importance, areas with observed wind speeds of 70 to 110 knots are indicated by hatching. A

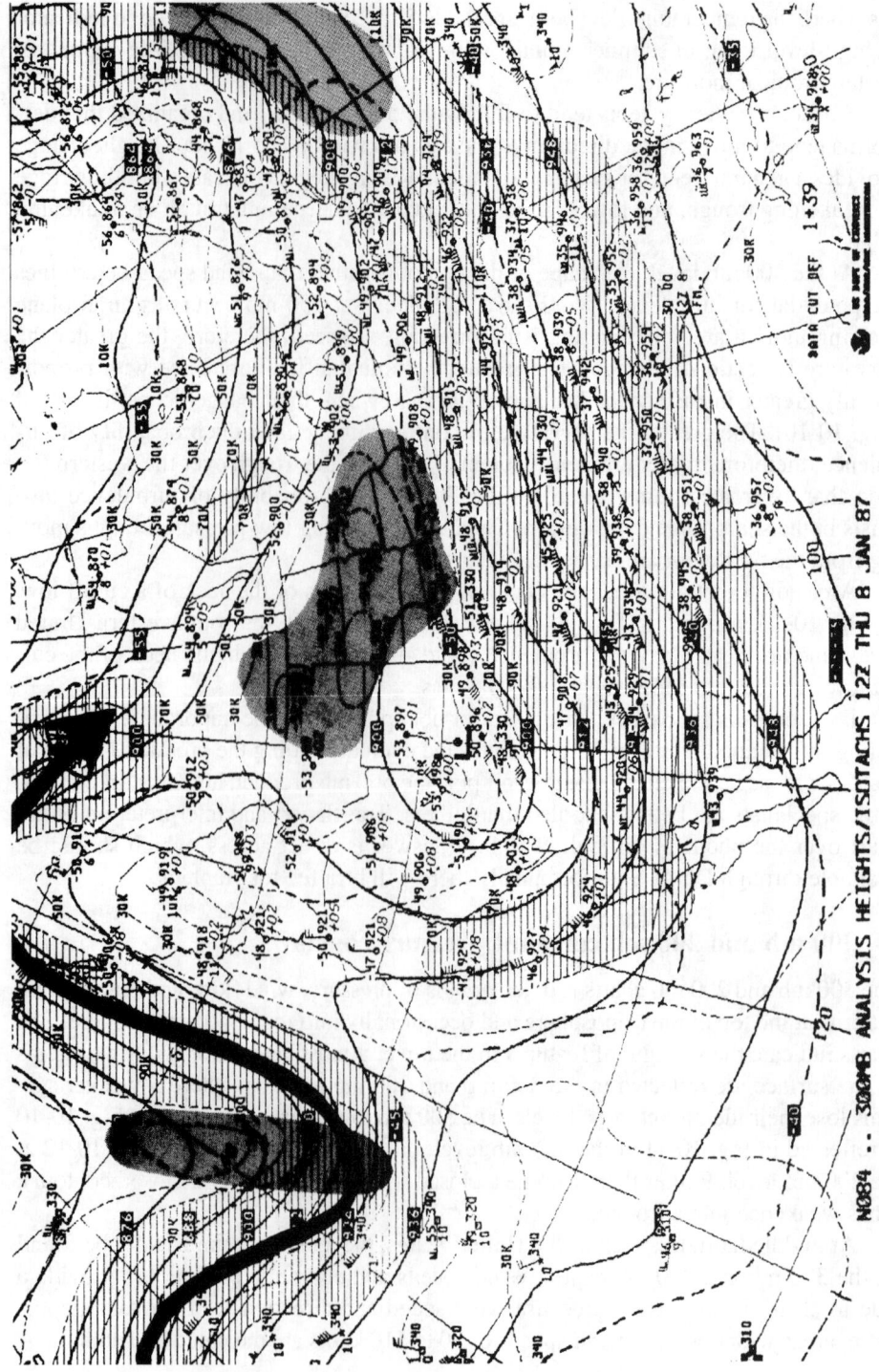

Fig. 10-11. The 300-mb and 200-mb charts describe the atmosphere at the top of the troposphere and occasionally the lower stratosphere.

clear area within the hatching identifies speeds of 110 to 150 knots. If speeds exceed 150 knots, a second hatched area appears. Areas of potential turbulence occur in:

- Sharp troughs (in Fig. 10-11, off the United States-Canadian West Coast and over New England).
- The neck of cutoff lows (in Fig. 10-11, over the north-central United States).
- A divergent flow (in Fig. 10-12, over southern Canada).

Turbulence in these areas can exist despite relatively slow wind speeds.

Jet stream

The jet stream was virtually unknown until World War II when pilots flying at high altitudes reported turbulence and tremendously strong winds. These winds blew from west to east near the top of the troposphere. Not until 1946 was the jet stream fully recognized as a meteorological phenomenon.

Sharp horizontal temperature differences cause the jet stream; across strong temperature gradients, temperature changes rapidly with height. In such zones, the slope of constant-pressure surfaces increases with height. The 500 mb slope is greater than that at 700 mb, and the 300 mb slope is greater than that at 500 mb. The slope of the pressure surface determines approximate wind speed. When pressure surface slope increases with height, wind speed increases with height. This is the general case in the troposphere. Winds are light or calm in areas of little or no horizontal temperature difference, and in some cases, winds can decrease with height in the troposphere. This tends to occur within large high-pressure areas. Because fronts lie in zones of temperature contrast, the jet is closely linked, or associated, with frontal boundaries. When wind speed becomes strong enough, the flow is termed a jet stream.

A jet stream is a narrow, shallow, meandering area of strong winds embedded in breaks in the tropopause. Two such breaks occur in the Northern Hemisphere; the polar jet located around 30°–60°N at an approximate height of 30,000 feet, associated with the polar front, and the subtropical jet around 20°–30°N at approximately 39,000 feet. To be classified a jet stream, winds must be 50 knots or greater, although winds generally range between 100 and 150 knots. Winds can reach 200 knots along the East Coasts of North America and Asia in winter when temperature contrasts are greatest.

A "jet" is most frequently found in segments 1,000 to 3,000 miles long, 100 to 400 miles wide, and 3,000 to 7,000 feet deep. The strength of the jet stream increases in winter in middle and high latitudes when temperature contrasts are greatest, and the jet shifts south with the seasonal migration of the polar front. The troposphere varies in depth from an average 55,000 feet at the equator to 28,000 feet at the poles: deeper in summer than in winter. The 300 mb and 200 mb charts are ideal for locating the jet. Jet cores are shown by thick black lines in Fig. 10-11 and Fig. 10-12.

As a general rule, locations north of a jet stream associated with a surface front are likely to be cold and stormy; locations south of this boundary tend to be warm and dry.

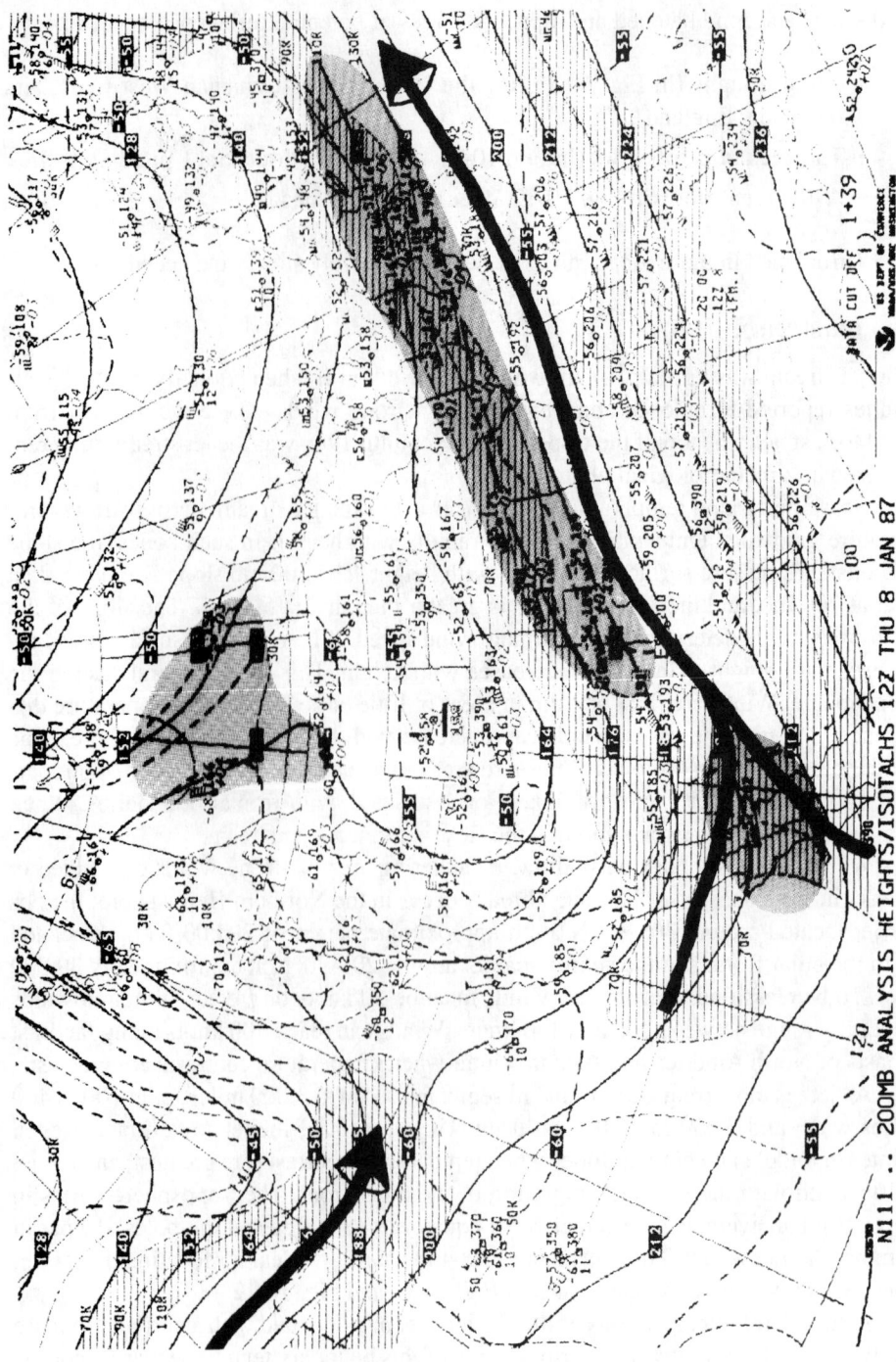

Fig. 10-12. *The 200-mb chart often depicts the location and speed of the jet stream.*

A jet embedded in a long wave can remain relatively stationary for weeks; this usually brings long periods of bad weather to the north of its location and good weather to the south. The movement of surface high- and low-pressure areas and fronts is related to the movement of the jet stream. Low-pressure areas tend to move with the jet stream flow. As the wave with the jet passes, a ridge builds aloft usually bringing high pressure and good weather; however, as high pressure builds, surface pressure gradients are often steep, causing strong, sometimes destructive surface winds.

The presence of jet streams has a significant impact on the aircraft operations. The jet stream can cause a significant headwind component for westbound flights, increasing fuel consumption and requiring additional landings.

Another factor associated with the jet is wind shear turbulence. With an average depth of 3,000 to 7,000 feet, a change in altitude of a few thousand feet will often take the aircraft out of the worst turbulence and strongest winds. The gray shading in Fig. 10-11 and Fig. 10-12 illustrates areas of potential turbulence. Maximum jet-stream turbulence tends to occur above the jet core and just below the core on the north side, as in Fig. 10-12 over the Central and East United States. Additional areas of probable turbulence occur where the polar and subtropical jets merge or diverge, such as western Mexico in Fig. 10-12.

OBSERVED WINDS ALOFT CHART

The observed winds aloft chart, which is transmitted twice daily, plots radiosonde data. This four-panel chart provides observed winds at four levels:

- Second standard level
- 14,000 feet (600 mb)
- 24,000 feet (400 mb)
- 34,000 feet (250 mb)

The second standard level (lower left panel of Fig. 10-13) occurs between 1,000 and 2,000 feet AGL. The chart provides observed winds above the surface, but within the frictional layer. This chart supplements the constant pressure charts by providing observed wind and temperatures between constant pressure levels; therefore, observed winds are available for the following heights:

- 5,000*
- 10,000*
- 14,000
- 18,000*
- 24,000
- 30,000*
- 34,000
- 39,000*
  * Obtained from constant pressure charts.

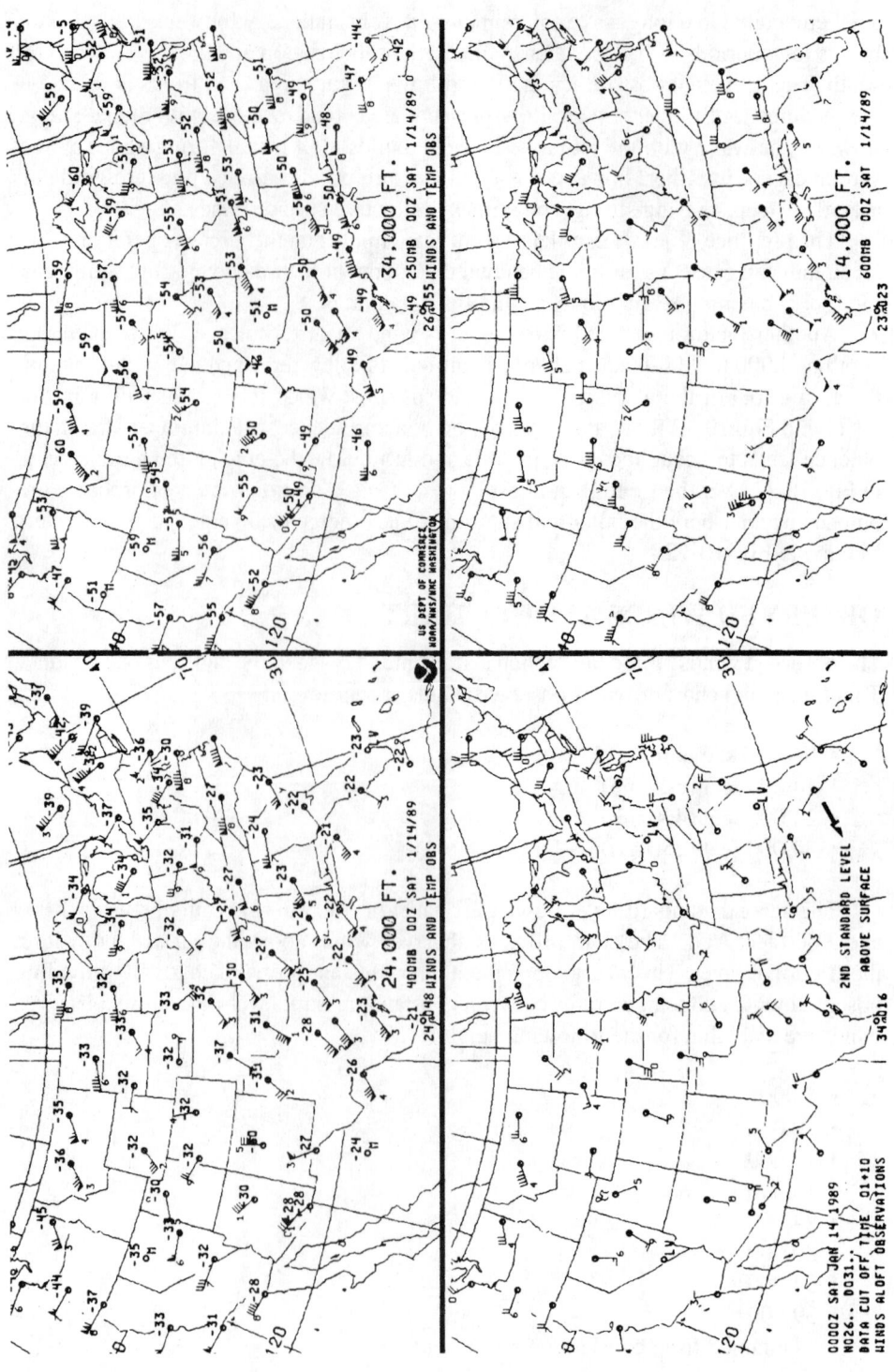

Fig. 10-13. *The observed winds aloft chart describes conditions that existed in the past. It should not be substituted for winds aloft forecasts.*

The chart portrays the state of the atmosphere in the past, approximately 2½ hours old by the time it becomes available, like constant pressure charts. Observed winds should not be substituted for winds aloft forecasts; however, if gross differences occur between observed and FD forecasts, an FSS or flight watch can be consulted. Both have direct access to NWS forecasters.

VORTICITY

Any nonmeteorologist pilot who wishes to better understand atmospheric phenomena will require a basic knowledge of vorticity. Although some pilot-meteorologists feel this subject is far too technical for pilots, I disagree. At the very least, pilots, especially those using DUATS, will come across this term in synopses, convective SIGMETs, and the convective outlook. (Vorticity is mentioned in chapter 4 with the discussion of WSTs and in chapter 5 with the synopsis.)

An understanding of vorticity will help relate the fact that not all weather occurrences can be attributed to pressure and frontal systems alone, as displayed on weather charts; however, for those "bottom-liners" who only wish to know the implications of vorticity, skip to the last few paragraphs of this section of the chapter.

Anything that spins has vorticity, which includes the Earth. Vorticity is a mathematical term that refers to the tendency of the air to spin; the faster that air spins, the greater its vorticity. A parcel of air that spins counterclockwise (cyclonically) has positive vorticity; a parcel of air that spins clockwise (anticyclonically) has negative vorticity.

The Earth's vorticity is always positive in the Northern Hemisphere because the Earth spins counterclockwise about its axis. An observer standing on the North Pole will have maximum vertical spin, one revolution per day. An observer's vertical spin will decrease when moving toward the equator, becoming zero at the equator. (Coriolis force is maximum at the poles and zero at the equator). The rate, or value, of the vorticity produced by the Earth's rotation is known as the *Earth's vorticity*.

Now consider the atmosphere, which is almost always in motion and generally will have its own vorticity relative to the Earth, or *relative vorticity*. The sum of the Earth's vorticity plus relative vorticity equals absolute vorticity. The value of absolute vorticity at middle latitudes almost always remains positive because of the Earth's rotation.

Air moving through a ridge, spinning clockwise, gains anticyclonic relative vorticity. Air moving through a trough, spinning counterclockwise, gains cyclonic relative vorticity; therefore, there tends to be downward vertical motion in ridge-to-trough flow, and upward vertical motion in trough-to-trough-to-ridge flow.

Absolute vorticity is analyzed at 500 mbs. Although normally not available to the pilot, 500 mb heights/vorticity charts are routinely transmitted over facsimile circuits. High values of absolute vorticity (greater than 16) have strong cyclonic rotation, indicating strong upward vertical motion, and low values (less than 6) have anticyclonic rotation, indicating strong downward vertical motion. Values between 7 and 15 produce weak or no vertical motion.

Vorticity is advected like other atmospheric properties; therefore, the pilot can expect to see the terms *positive vorticity advection* (PVA) and *negative vorticity advection* (NVA) (referring to absolute vorticity).

Positive vorticity advection indicates:

- A trough or low moving into an area.
- A ridge or high moving out of an area.
- Upward vertical motion probably occurring.
- Increasing cloud cover and precipitation.

Negative vorticity advection indicates:

- A ridge or high moving into an area.
- A low or trough moving out of an area.
- Downward vertical motion probably occurring.
- Decreasing cloud cover.

Regions of PVA are swept along within the overall flow. The PVA regions represent microsystems that can rotate around synoptic-scale highs and lows. These areas can be referred to as "A VORT MAX" or "VORT LOBE." If an area of PVA moves over a stationary surface front, a wave can form and a storm can develop. An area of PVA might be all that's required (a lifting mechanism) to trigger thunderstorms when moisture and instability are available. On the other hand, NVA might prevent thunderstorm development.

UPPER-LEVEL WEATHER SYSTEMS

This chapter's introduction mentions the fact that many aviation weather texts fail to adequately describe nonfrontal weather-producing systems. We've looked at warm- and cold-air advection in the lower troposphere and aloft, orographic lift (upslope/downslope), and convergence and divergence. We discussed the 500 mb wave and its significance. Short waves, vorticity, and the jet stream were described along with their effects on surface and high-altitude weather.

Upper-level weather systems tend to modify and direct surface weather. They can intensify or stabilize conditions at the surface, cause thunderstorms to occur, and enhance or retard the intensity of frontal zones.

Difluence, divergence aloft, develops when contours diverge or move apart as seen on the 300 mb chart to cause surface convergence and increased cyclonic vorticity. Surface lows can develop in this way. A perfect example occurred one afternoon with scattered thunderstorms forecast for northern California, northern Nevada, and eastern Oregon. No weather systems were depicted on either the surface analysis or 500 mb chart; however, thunderstorms did occur, right along a line of difluence. The difluence caused just enough surface convergence to trigger thunderstorms.

Figure 10-14 illustrates the effects of an upper-level weather system. The 0900Z surface analysis chart shows weak surface high pressure over the western United States; however, the 1000Z weather depiction chart discloses extensive areas of IFR and marginal VFR (gray shaded area), with rain and snow occurring throughout New Mexico, Texas, and Oklahoma. (The weather closed airports for days and was blamed for the deaths of dozens of people.) The 500 mb analysis reveals the culprit. A deep upper-level low along the southern Arizona and New Mexico border and an associated downstream trough produced devastating surface conditions. The synopsis read:

STRONG UPPER LOW OVER SOUTHWESTERN NEW MEXICO WILL MOVE TO NORTH CENTRAL TEXAS BY 22Z.

Lifting mechanisms have a cumulative effect. Upper-level troughs that are parallel to and behind a front intensify the storm. These fronts tend to be fast moving. Figure 10-15 shows a strong cold front moving through southern California. The satellite picture reveals a trough and low off the central California coast. These lows occur when cold air at the base of a trough is cut off from the cold air to the north. This closed circulation can lead to a circular jet stream. The weather in central California remains moist and unstable even though the surface front has passed.

With the approach of an upper-level trough, a period of 8 to 12 hours of poor weather can be expected. The surface front will precede the trough, usually bringing IFR weather; however, without a front, the upper trough or low might only bring marginal VFR conditions with localized areas of IFR. Under these conditions, VFR flight might be possible except in mountainous areas that remain obscured in clouds and precipitation.

Figure 10-16, an enhanced infrared satellite picture, shows an upper-level low off southern California. The area forecast for February 22, 1987, read:

SRN CA
CSTL SXNS...20-30 BKN/SCT 80 BKN LYRD 200 WITH WDLY SCT RW-/ISOLD
TRW-. CB TOPS TO 300.
INTR SXNS...80-100 SCT/BKN 120 BKN 200. ISOLD RW-.

These systems tend to form bands of weather, as can be seen in Fig. 10-16. The weather deteriorates as a band moves through, then improves, only to deteriorate with the next band. Notice that the area forecast cannot and does not take this into account. Under these conditions, a pilot must be careful not to get suckered by a temporary improvement.

Closed upper-level lows tend to remain stationary. And a closed low that is reflected vertically through the atmosphere tends to move erratically. These systems can cause poor weather and precipitation to linger for days. Off the West Coast, the systems can bring bands of moist unstable air from the Pacific.

During one episode, an upper low meandered over Red Bluff, CA, for five days. Pilots would call day after day wanting to know when the weather would clear. After awhile, the common response became, "The low is forecast to move east out of the area tomorrow, but that's what they said yesterday."

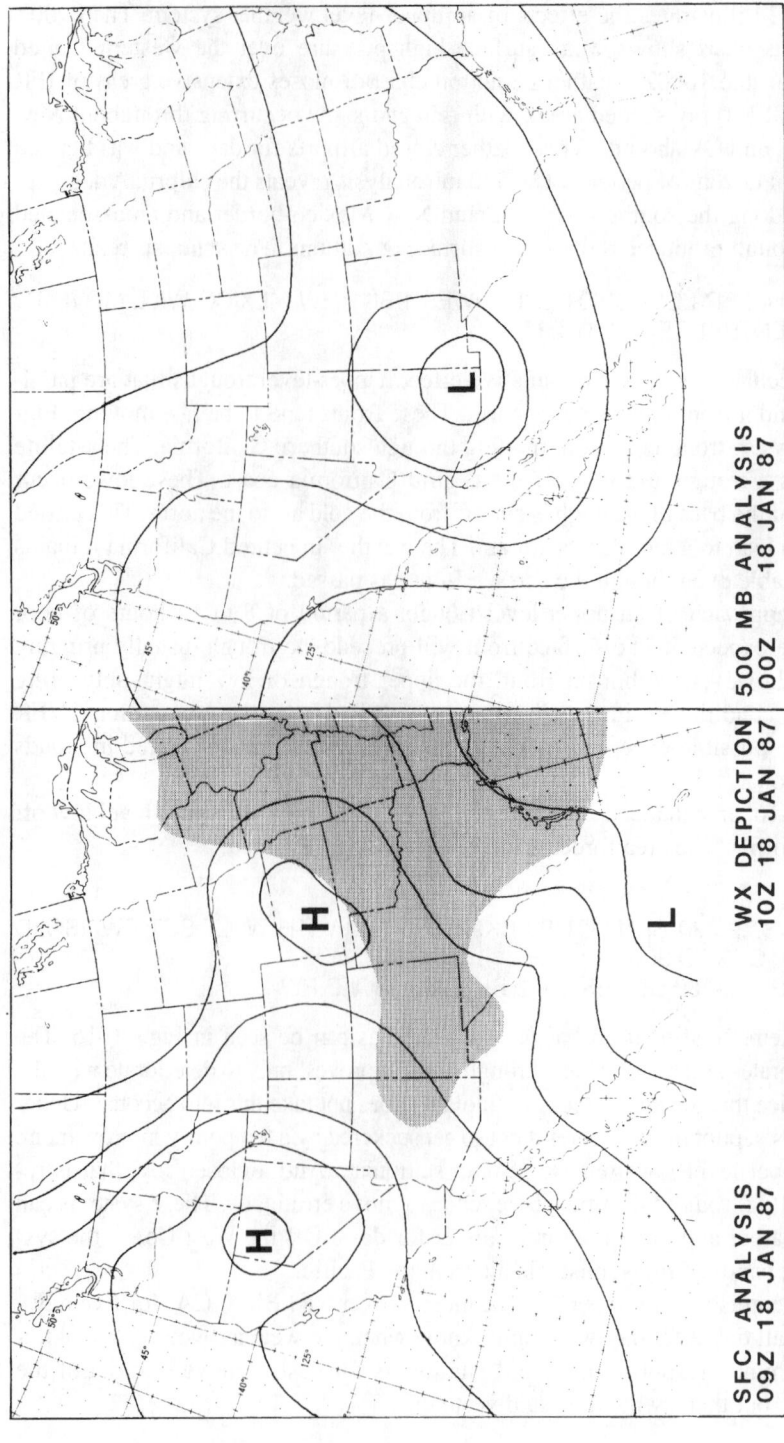

Fig. 10-14. Upper-level weather systems can cause devastating surface weather, often without a clue about the cause on the surface analysis chart.

`1946 16DE87 38A-2    01012 23162 WB1`

Fig. 10-15. *Strong surface cold fronts have support in the form of troughs parallel to and behind the front.*

During winter months, upper lows can form over the Great Basin of Nevada and Utah. Known as the Ely Lows, they can persist for days bringing snow and IFR conditions over extensive areas. Over the Midwest and eastern United States, deep upper lows can form over surface frontal systems. When this occurs, surface fronts tend to move slowly, bringing days or even weeks of snow and poor weather, closing airports for hours or even days, at a time. During summer months, closed lows aloft support the development of thunderstorms when surface lifting begins.

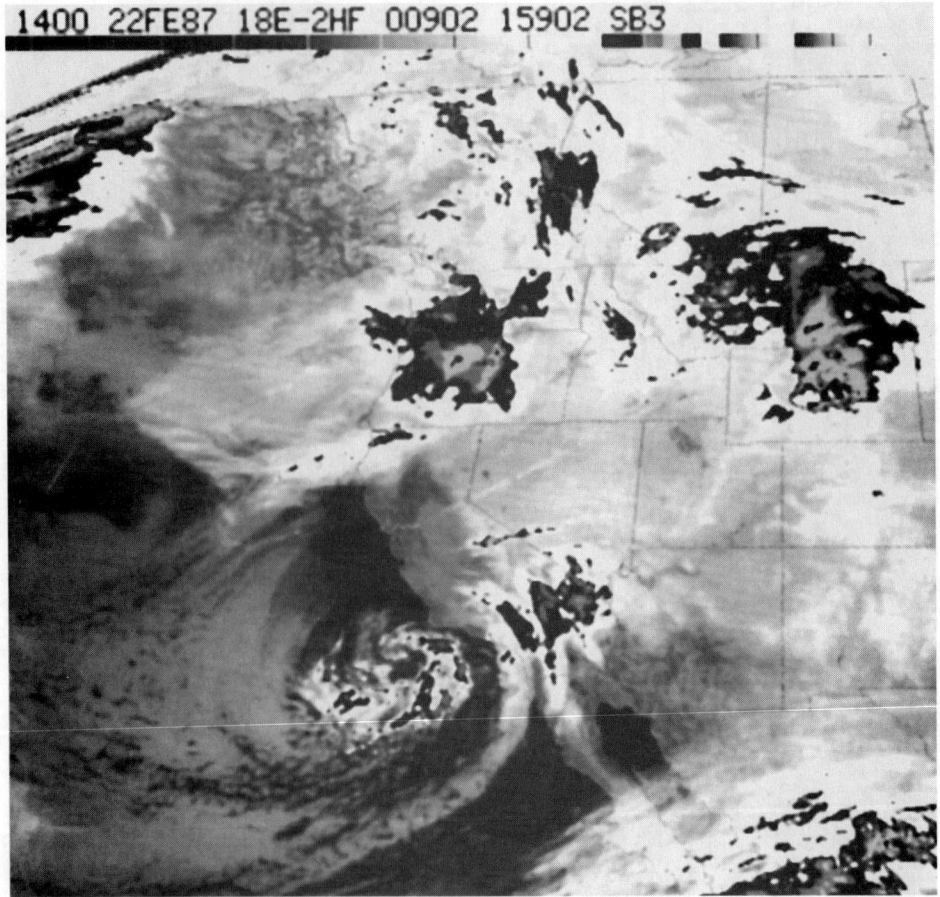

`1400 22FE87 18E-2HF 00902 15902 SB3`

Fig. 10-16. *Upper-level lows produce bands of weather. The weather deteriorates as a band moves through, then improves only to deteriorate with the next band.*

On the other hand, the absence of an upper-level trough will tend to weaken and slow a front's progress. A ridge aloft, even with a surface front, will not tend to produce thunderstorms or severe weather because the ridge prevents the vertical development required. Figure 10-17 shows a weak surface front moving into Idaho and Nevada. Notice the almost complete absence of clouds in California, even though these fronts often appear on the surface analysis chart. This front has minimal upper-air support. A temperature difference marks the boundary; there is insufficient moisture and lifting to produce clouds. It is not all that unusual to have cloud tops below 10,000 feet with weak fronts.

The examples provided illustrate how upper-level weather systems can cause se-

`1615 03OC87 28A-1    01734 22984 WA2`

Fig. 10-17. *Weak surface fronts have little upper-level support. Cloud tops are often below 10,000 feet. With insufficient moisture, there might be a complete absence of clouds.*

vere conditions at the surface, or dampen or cancel out the vertical motion required to produce weather. The point that not all weather is caused by frontal systems has been a theme of this chapter. In fact, nonfrontal weather-producing systems have considerable influence on surface conditions. The hurricane is the ultimate example.

Hurricanes produce just about every kind of nasty weather extending over thousands of square miles. Figure 10-18 is an infrared satellite photo of Hurricane Gilbert on September 13, 1988, when winds were reported to 140 knots, pressure in the eye was 26.66", tropical-storm-force winds extended over a diameter of 450 miles, and squalls with rainfall of 5 to 10 inches accompanied the storm. The NWS hurricane advisory: "Gilbert may still increase a little more in strength."

1801 13SE88 19E-4ZA 00911 15581 EC1

Fig. 10-18. *The hurricane is the ultimate example of a nonfrontal weather-producing system.*

11
Pilot briefings

LITTLE NEED EXISTED FOR METEOROLOGICAL INFORMATION IN THE EARLY days of aviation because all flights were local. Nor were Notices to Airmen (NO-TAMs) necessary because pilots departed and landed at the same field, assuming the engine didn't quit. By the spring of 1918, the U.S. Post Office Department began working on a transcontinental airmail route. A combination rail/air route between New York and Chicago was established by July and a month later extended to San Francisco. Authorization was granted in August 1920 for the establishment of 17 airmail radio stations.

Personnel originally were to load and unload mail; however, as traffic increased, the need for weather information became apparent. Airmail radio personnel soon began taking weather observations and developing forecasts. The information was relayed via radio telegraph to adjacent stations. In-flight weather reports were heavily relied upon for the weather briefing.

Postal personnel soon became involved in air traffic as well as postal services and in July 1927 were transferred to the Department of Commerce, Bureau of Lighthouses, along with their facilities, now known as *airway radio stations*. The stations were transferred in August 1938 to the Civil Aeronautics Authority, and they became *airway communication stations*. These facilities finally became flight service stations (FSSs) with the establishment of the Federal Aviation Agency in 1958. The Weather Bureau became increasingly responsible for the collection and distribution of aviation weather and forecasts. Pilots obtained briefings from the Weather Bureau and filed flight plans and acquired aeronautical information from the flight service station.

Due to the increase in air commerce and other factors, the Weather Bureau in 1961 began the certification of FSS personnel as pilot weather briefers. The Federal Aviation Agency and the Weather Bureau mutually signed a memorandum of agreement in 1965 delegating responsibility for pilot weather briefing to the FAA. FSS briefers had little in the way of guidelines during this period regarding the structure of the briefing; they basically read weather reports and forecasts verbatim as requested by the pilot.

An analysis group began a special evaluation of pilot-briefing services in 1975. The group cited deficiencies in the use of a standardized briefing format. (The standardized format had been taught at the FAA Academy for some time; however, it had yet to be incorporated in the FSS handbook.) Other areas identified were the reading of weather reports and forecasts verbatim as opposed to interpreting, translating, and summarizing data; a poor level of proficiency in reading, understanding, and employing facsimile charts; and briefers failed to obtain sufficient background information to tailor the briefing to the type of flight planned.

From this study came the agency's emphasis on an extremely rigid briefing format and an ambitious refresher training program. Unfortunately, the agency did little to inform pilots or other offices within the agency of this change in policy. This led to a good deal of friction between briefers and pilots.

Over the years, however, through mostly local efforts, pilots became acquainted with the standard briefing format. The refresher training was to be a continuing program conducted at least every 5 years; however, with few exceptions, this program has been abandoned presumably due to fiscal constraints.

The National Transportation Safety Board also conducted a special investigation into flight service station weather briefing inadequacies. In six of 72 accidents involving fatalities, the board determined that pertinent meteorological information was not passed to the pilot during the weather briefing. These deficiencies basically consisted of a failure to pass weather advisories and icing forecasts and downplaying forecasts of hazardous weather. The result of which led to what many pilots considered overdoing dissemination of these advisories.

A major change occurred in 1983 when the extremely rigid format was relaxed somewhat. Three types of briefings emerged:

- Standard briefing
- Abbreviated briefing
- Outlook briefing
- (Requirements for in-flight briefings were also specified)

Almost from the time the FAA took over pilot briefing responsibility in 1965, the Service A (weather) teletype system was obsolete; numerous proposals were made to update weather distribution. Even by the early 1980s, most flight service stations still used the 100-word-per-minute electromechanical teletype equipment. Briefers had to sift through mountains of paper to provide a briefing with weather reports as much as 1½ hours old by the time they were relayed and available. Figure 11-1 shows a typical teletype-era briefing position.

Fig. 11-1. *In the teletype era, briefers had to sift through mountains of paper, and weather reports were often one and a half hours old.*

The FAA approved what was termed the interim Service A system in November 1978 that had been tested at the Chicago FSS. Subsequently, the Service B teletype system for the transmission of flight plans and other messages was incorporated. Referred to as the Leased A and B System (LABS), it is in use at nonautomated FSSs. LABS was designed to update FSS Service A until a complete computer system could be developed and installed. LABS eliminates the need for the briefer to sort and post SPs, PIREPs, NOTAMs, and most amended forecasts. These housekeeping chores, which took considerable time, have been eliminated. With this system, most weather reports are available within 5 to 15 minutes of observation. Figure 11-2 shows a LABS briefing position.

Fig. 11-2. *With LABS, special reports and amendments became more timely. Weather reports are only 5–20 minutes old.*

Development of Model 1, a so-called completely computerized system, began in 1982 and came online in 1985. Automated flight service stations (AFSSs) use this equipment. Figure 11-3 shows a typical Model 1 briefing display.

According to the FAA, "The primary benefit of the (FSS automation) program is improved productivity through automation of specialist's access to detailed briefing information and flight-plan filing. To some extent the improved quality of pilot briefings reduces the need for multiple briefings as in the past." Model 1 is in the same evolutionary category as ARTCC flight data processing was in the early 1970s. It takes care of many of the data-processing functions, such as flight-plan transmission and tracking.

From a weather briefing point of view, however, Model 1 presents the same information as was available from teletype and LABS. Amendments are more timely

Fig. 11-3. *Much of the flight data processing at Model 1 facilities has been automated, but a weather briefing remains essentially the same.*

with Model 1, but this won't be directly obvious to the pilot. Model 1 does not necessarily improve pilot briefing productivity. Model 1 presents information in much the same way as commercial briefing services, which will be discussed at the end of this chapter.

An edited portion of FAR 91 regarding preflight action states: "Each pilot in command shall, before beginning a flight, become familiar with all available information concerning that flight. This information must include, for a flight under IFR or a flight not in the vicinity of an airport, weather reports and forecasts . . . and any known traffic delays of which the pilot has been advised by ATC."

Additional regulations specify fuel and alternate airport requirements; however, the regulations do not require that meteorological and aeronautical information be obtained from the FAA. A number of sources satisfy this requirement.

STANDARD BRIEFING

The standard briefing is designed for a pilot's initial weather rundown prior to departure. Standard briefings are not normally provided when the departure time is beyond 6 hours. Current weather may be omitted when the departure time is beyond 2 hours, unless specifically requested by the pilot. It is to the pilot's advantage to obtain a standard briefing or obtain an update to an earlier standard briefing as close to departure time as possible.

Background information

Before beginning a briefing, the specialist must gather background information that is pertinent and not evident or already known. The amount of information varies with the training and experience of the briefer, weather conditions, and the pilot's request. Pilots can assist the briefer and reduce delays by volunteering the following information.

The aircraft number or pilot's name. This is evidence that a briefing was obtained and is an indicator of FSS activity. In the absence of an aircraft number, the pilot's name is sufficient. Some pilots object to providing either a number or name. One briefer requested the pilot's aircraft number: "I don't have one." When asked for his name, the pilot again replied, "I don't have one." Most briefings are recorded and reviewed in case of an incident or accident, so the briefer can't get in trouble on that one. It's in the pilot's interest to get "on the record" as having received a briefing.

The aircraft type. Low-, medium-, and high-altitude flights present different briefing problems. This information allows briefers to tailor the briefing to a pilot's specific needs. For example, the briefing for a Learjet pilot will not match the briefing for a student pilot in a Piper Cherokee. By knowing the aircraft type, the briefer can probably estimate general performance characteristics such as altitude, range, and time en route.

The type of flight planned. Always advise the briefer if the flight is VFR-only, if an IFR flight is an option, or if an IFR flight is planned. Normally the briefer will assume a pilot is planning VFR unless stated otherwise. A student pilot should always identify himself or herself as a student to help the briefer provide information tailored for a student's needs. New pilots, low-time pilots, and pilots unfamiliar with a geographic area will receive better service if they so advise the briefer. This alerts the briefer to proceed more slowly with greater detail.

The departure airport. For some reason, some pilots are reluctant to specify their departure airport. They use generalities such as Los Angeles when their actual departure airport is Oxnard, more than 50 miles away. And of course there is always the ever popular departure point "Here." Pilots must be specific. They know the airport, but the briefer usually doesn't. This is important with FSS consolidation, toll-free telephone numbers, and in metropolitan areas.

The route of flight. The briefer will assume a pilot is planning a direct flight, unless otherwise stated. If not, a pilot must provide the exact route or preferred route, and any planned stops. The increased use of area navigational systems has quite logically increased the capability of direct flights; however, if the planned flight is more than 200 miles, the pilot should specify intermediate points. This will assist the briefer in providing weather for the planned route. It is often difficult to visualize reporting locations for a direct flight without studying a map. A pilot will receive a faster, more accurate briefing by specifying intermediate points. For instance, rather than "Oakland direct El Paso," the pilot could specify "Oakland, Las Vegas, Prescott, El Paso." Computerized systems do help by displaying all weather reporting stations, normally within 25 miles of the route.

The destination airport. Again, pilots must be specific; if not, a pilot might not receive all available weather and NOTAM information. A Piper Cub pilot on one occasion requested a briefing to Los Angeles. The briefer asked if his intended destination was Los Angeles International. It was! Another pilot obtained a briefing from Chino, CA, to his stated destination of Stockton, CA, and was told there were no NOTAMs for the route. At the end of the briefing, he matter-of-factly said he was actually going to Columbia, an airport about 20 miles east Stockton. Now there were a few NOTAMs! The airport would be closed during certain hours, a temporary tower was in operation, and aerobatic flight and parachute jumping were being conducted.

The estimated time of departure and estimated time en route. The estimated time of departure is essential, even if general, such as morning or afternoon. Many briefers can estimate time en route based on aircraft type. This information is needed to provide en route and destination forecasts. Total time en route is essential when stops or anything other than a direct flight is planned; for IFR flights, the estimated time of arrival is required to determine alternate requirements. Briefer: "When are you planning on departing?" Pilot: "Well, that depends on the weather." This response tells the briefer nothing. In such a situation a pilot could respond, "I'd like to go this afternoon, but I can put the flight off until tomorrow."

The proposed altitude or altitude range. This information is needed to provide winds and temperatures aloft forecasts. If an altitude range is specified, for example 8,000 to 12,000 feet, the briefer can provide the most efficient altitude for direction of flight.

This might seem like a lot of information, but it really isn't. The briefer must obtain this information before or during the briefing. Providing accurate background information will allow a briefer to do his or her job better, which is to provide the pilot with a clear and concise, well-organized briefing tailored to a pilot's specific needs.

Briefing format

All right, the background information has been provided. What can a pilot expect in return? The briefer is required to use all available weather and aeronautical information to provide a briefing in the following order. Pilots should be as familiar with this format as the pretakeoff mnemonic CIGAR (controls, instruments, gas, attitude, runup) or the IFR clearance format.

Adverse conditions. Any aeronautical or meteorological information that might influence the pilot to cancel, alter, or postpone the flight will be provided at this time.

Items will consist of weather advisories, major NAVAID outages, runway or airport closures, or any other hazardous conditions. The idea is to present the bad news first. Keep in mind that even though such phenomena as moderate turbulence, mountain obscurement, or IFR conditions might not necessarily be hazardous to an individual pilot or flight operation, the briefer is required to provide this information at this time.

Similar to the introduction of the CWA, it took the FAA three years to update the *Flight Services* handbook to reflect the deletion of "flight precautions." Pilots can expect to hear the terms "flight precautions," "weather advisories," or "AIRMETs" used interchangeably.

Only the adverse conditions that are pertinent to the intended flight should be provided. This is one reason why the pilot must provide the briefer with accurate and specific background information. The briefer should then only furnish those conditions that affect the flight. There is, unfortunately, some paranoia among briefers, and they will provide anything within 200 miles of the flight.

VFR flight is not recommended (VNR). Undoubtedly the VNR statement is the most controversial element of the briefing, nevertheless, the FAA requires the briefer to: "Include this recommendation when VFR flight is proposed and sky conditions or visibilities are present or forecast, surface or aloft, that in (the judgment of the specialist) would make flight under visual flight rules doubtful."

This leaves considerable leeway for the briefer; some use this statement more than others. The inclusion of this statement should not necessarily be interpreted by the pilot as an automatic cancellation, nor its absence as a "go-for-it" day. Notice that VNR applies to sky condition and visibility only. Such phenomena as turbulence, icing, winds, and thunderstorms of themselves do not warrant the issuance of this statement. And it is important to remember that this is a recommendation. Why then such a statement?

It's simple; every year, pilots insist on killing themselves and their passengers at an alarming and relatively constant rate by flying into weather where they have no business. This statement was instituted in 1974, presumably because the last person a pilot would talk to was usually the briefer. A logical although alarming result of this statement is the increasing number of pilots who, in the absence of VNR, ask, "Is VFR recommended?" So far, the answer remains that the decision as to whether the flight can be safely conducted rests solely with the pilot.

According to the *Flight Services* handbook, the *reason* for VNR must also be provided. For example, "VFR is not recommended into the Monterey area because of ceilings 200 feet and visibilities one-half mile, conditions are not expected to improve until around noon." Briefers have been known to use some exceedingly poor technique in this area. A briefer told a pilot "The San Fernando Valley is still VNR." Oh well.

Synopsis. The synopsis is extracted and summarized from FA- and TWEB-route synopses, weather advisories, and surface and upper-level weather charts. This element might be combined with adverse conditions and the VNR statement, in any order, when it would help to more clearly describe conditions. The synopsis should indicate the reason for any adverse conditions, and tie in with current and forecast weather.

Current conditions. Current weather will be summarized for the point of departure, en route, and destination. Current and relevant PIREPs and weather radar reports will be included. Weather reports will not normally be read verbatim, and might be omitted if the proposed departure time is beyond 2 hours, unless specifically requested by the pilot. The briefer might provide NOTAMs during this portion of the briefing. Because the SA database is normally reloaded between 3 and 6 minutes past the hour, obtain the latest reports by calling just after the hour.

En route forecast. The en route forecast will be summarized in a logical order (climbout, en route, and destination) from appropriate forecasts (FAs, TWEBs, weather advisories, and prognosis charts). The briefer will interpret, translate, and summarize expected conditions along the route.

Destination forecast. Using the terminal forecast where available or appropriate portions of an FA or TWEB forecast, the briefer will provide a destination forecast and any significant changes from 1 hour before until 1 hour after ETA. This contradicts a notion held in some circles that without an FT there is no destination forecast. A destination forecast is always available, although maybe not as detailed as one might prefer.

Winds aloft forecast. The briefer will summarize forecast winds aloft for the proposed route. Temperatures will normally only be provided on request. Large differences between altitudes or flight levels will be brought to the pilot's attention. On request, the briefer will provide the most favorable altitude for the proposed flight.

Notices to Airmen (NOTAMs). The briefer will review and provide applicable NOTAMs that are on hand for the proposed flight and not already carried in the *Notices to Airmen* publication. This information consists of NAVAID status, airport conditions, temporary flight restrictions, changes to instrument approach procedures, and flow-control information. In the briefing, the term NOTAMs is all-inclusive. I briefed a student one day and as is my practice informed him there were no NOTAMs for the route. There was a pause. I asked him if he knew what NOTAMs were; he didn't.

The United States Notam System (USNS) is computerized, and occasionally the system fails. When this occurs, briefers will include the statement: "Due to temporary NOTAM system outage, en route and destination NOTAM information may not be current. Pilots should contact FSSs en route and at destination to ensure current NOTAM information." (Chapter 13 is a detailed explanation of NOTAMs.)

Other services and items provided on request. At this point in the briefing, briefers will normally inform the pilot of the availability of flight plan, traffic advisory, and flight watch services, and request pilot reports. Upon request by the pilot, the specialist will provide information on military training route (MTR) and military operation area (MOA) activity, review the *Notices to Airmen* publication, check LORAN or GPS NOTAMs, and furnish additional specialized data.

It's not necessary to copy all the information provided because much is supplementary and is a background for other portions of the briefing. Pertinent information should be noted, and it's often advantageous to copy this data. The weather log in Fig. 11-4 is specifically designed to organize information in the FAA's briefing format. This form also serves as a checklist to ensure that all necessary areas have been covered.

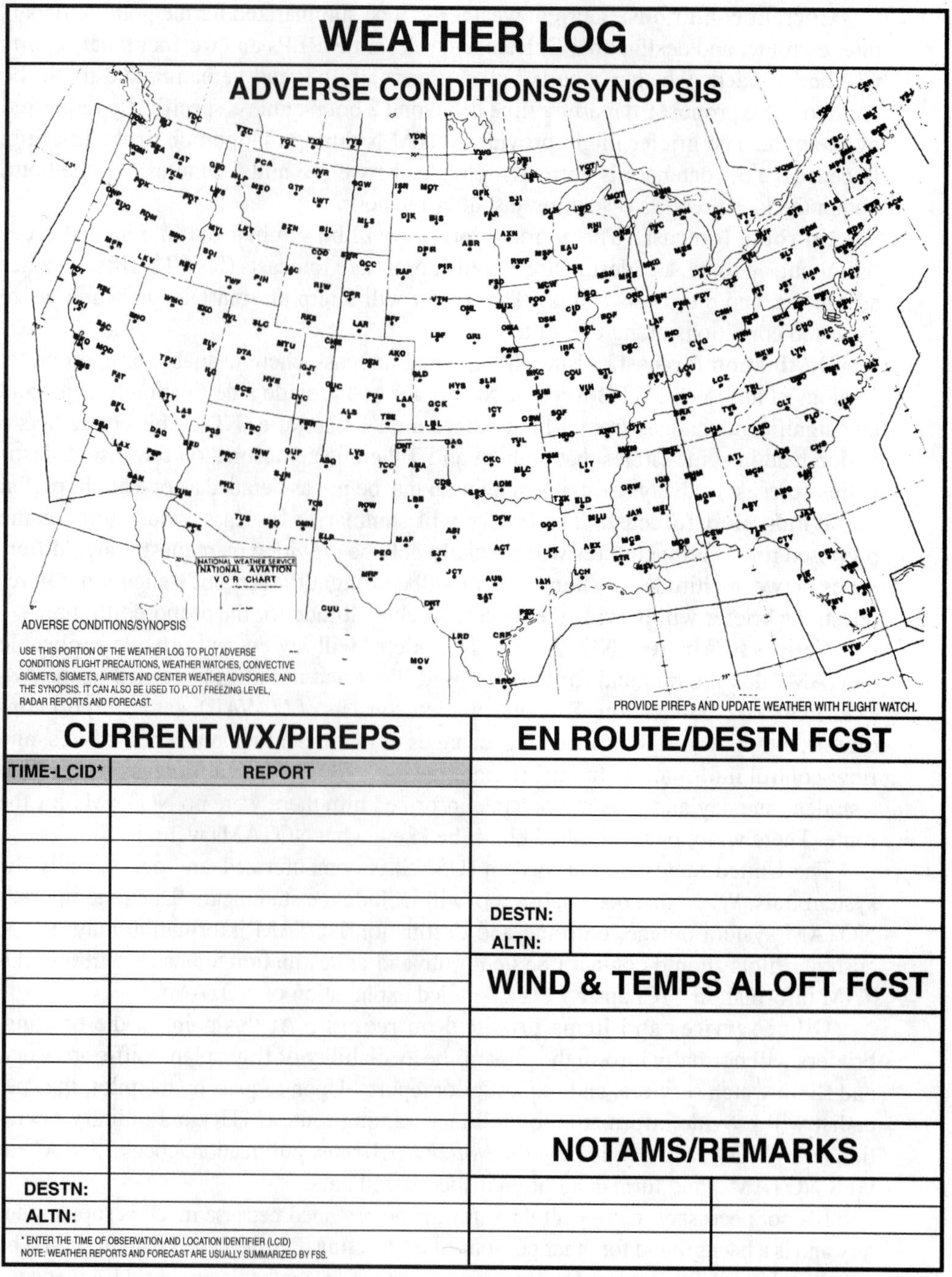
WEATHER LOG
ADVERSE CONDITIONS/SYNOPSIS
NATIONAL WEATHER SERVICE
NATIONAL AVIATION
V O R CHART
ADVERSE CONDITIONS/SYNOPSIS
USE THIS PORTION OF THE WEATHER LOG TO PLOT ADVERSE CONDITIONS FLIGHT PRECAUTIONS, WEATHER WATCHES, CONVECTIVE SIGMETS, SIGMETS, AIRMETS AND CENTER WEATHER ADVISORIES, AND THE SYNOPSIS. IT CAN ALSO BE USED TO PLOT FREEZING LEVEL, RADAR REPORTS AND FORECAST.
PROVIDE PIREPs AND UPDATE WEATHER WITH FLIGHT WATCH.
CURRENT WX/PIREPS
TIME-LCID*
REPORT
DESTN:
ALTN:
* ENTER THE TIME OF OBSERVATION AND LOCATION IDENTIFIER (LCID)
NOTE: WEATHER REPORTS AND FORECAST ARE USUALLY SUMMARIZED BY FSS.
EN ROUTE/DESTN FCST
DESTN:
ALTN:
WIND & TEMPS ALOFT FCST
NOTAMS/REMARKS

Fig. 11-4. *It's often helpful to note significant items on a form, such as this weather log, during a briefing.*

ABBREVIATED BRIEFING

Briefers provide abbreviated briefings when a pilot requests specific data, information to update a previous briefing, or supplement an FAA mass-dissemination system such as a transcribed weather broadcast, telephone information briefing service, or pilot's automatic telephone weather answering service.

Requests for specific information

When only specific information is required, a pilot should state this fact and request an abbreviated briefing. Because the briefer must normally input a computer request for each individual item, it's extremely helpful to request all items at the beginning of the briefing, which reduces delays. The briefer will then provide the information requested.

When using this procedure, the responsibility for obtaining all necessary and available information rests with the pilot, not the briefer. Pilots must realize that the briefer is still required to offer adverse conditions. Pilots sometimes become irritated when the briefer mentions weather advisories; however, this is a *Flight Services* handbook requirement.

Requests to update a previous briefing

Pilots requesting an update to a previous briefing must provide the time the briefing was received and provide necessary background information. The briefer will then to the extent possible limit the briefing to appreciable changes. Too often, when a briefer requests the time of the previous briefing, an alarming number of pilots respond, "I got the weather last night." Needless to say, this practice does not comply with FARs. These individuals should request a standard briefing.

Requests for information to supplement
FAA mass-dissemination systems

Again, the briefer must have enough background information and the time that the recording was obtained to provide you with appropriate supplemental data. The extent of the briefing will depend on the type of recording and the time received.

OUTLOOK BRIEFING

With a proposed departure time beyond 6 hours, an outlook briefing will normally be provided. The briefing will contain available information applicable to the proposed flight. The detail will depend on the proposed time of departure. The further in the future, the less specific. As a minimum, the outlook will consist of a synopsis and route/destination forecast.

SIGNIFICANT WEATHER PROGNOSIS CHARTS

Outlooks beyond FA, TWEB, and FT valid times are available using significant weather prognosis charts. (You occasionally might see it abbreviated as "sig wx progs.") Two charts are manually prepared at the National Meteorological Center to provide forecasts up to 42 hours from the time of reception at the FSS. Figure 11-5 shows an NMC meteorologist preparing a significant weather prog. These charts use standard weather symbols.

Fig. 11-5. *The significant weather prog is manually prepared by meteorologists at the National Meteorological Center outside Washington, D.C.*

The 12–24-hour low-level significant weather prog is issued four times a day, valid at 0000Z, 0600Z, 1200Z, and 1800Z depending on issuance time. By the time the chart becomes available, however, valid times are only 6 and 18 hours because it takes approximately 6 hours to prepare and distribute the chart. If a pilot calls just prior to the next issuance, only a 12-hour forecast would be available. The 12–24-hour prog consists of four panels. The two upper panels forecast significant weather from the surface to 400 mb (24,000 feet). Forecast weather categories (VFR, MVFR, and IFR) have similar definitions and limitations to the weather depiction chart. Areas of turbulence are depicted with dashed lines; turbulence symbols indicate intensity.

In Fig. 11-6 at 0600Z over the Rockies, moderate turbulence is forecast below 16,000 feet MSL. Over southeastern Canada, moderate to severe turbulence is expected below 24,000 feet. Note the turbulence symbols representing moderate and severe.

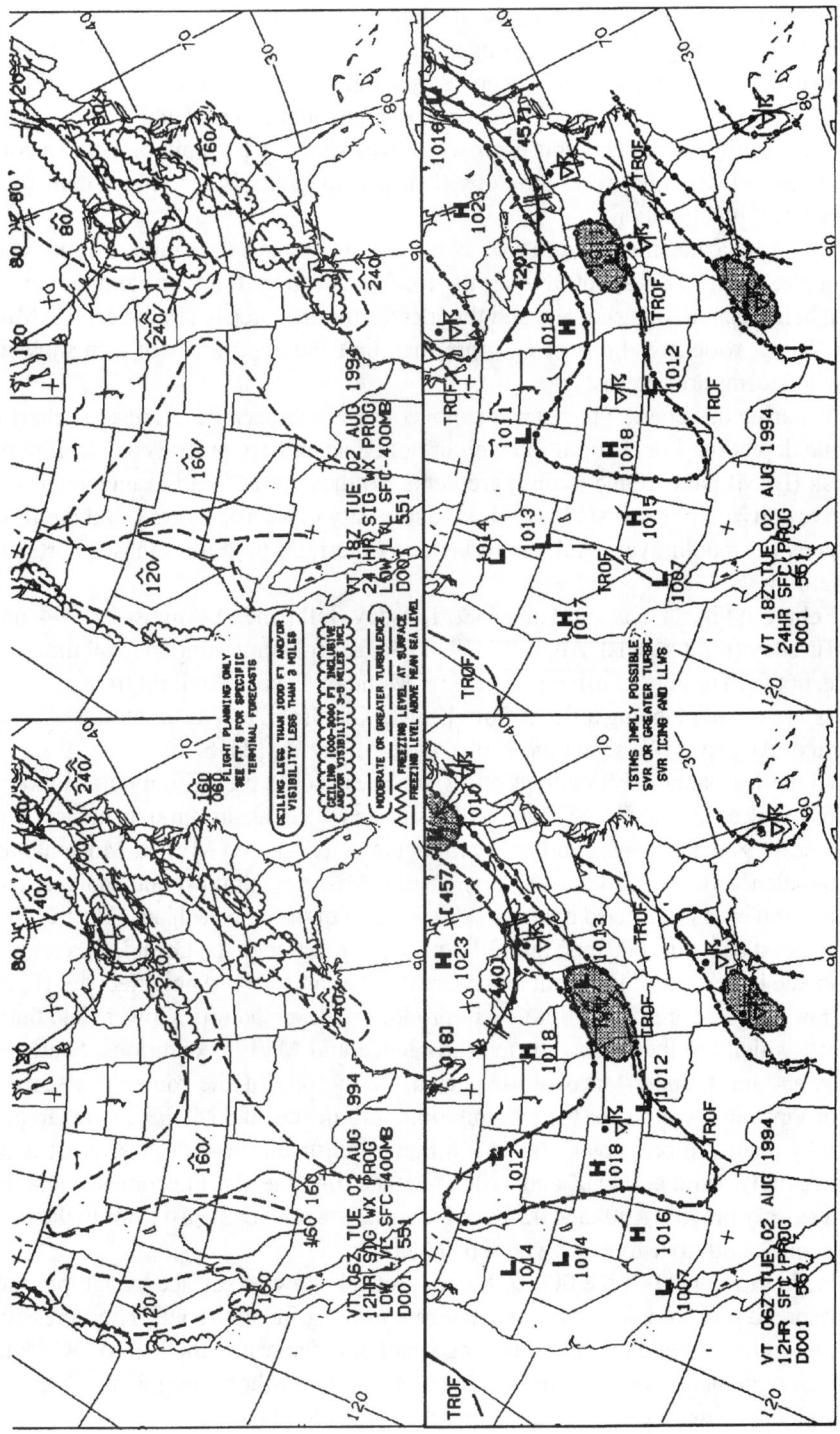

Fig. 11-6. *The 12-to-24-hour significant weather prog describes the location and movement of synoptic-scale weather systems. Small-scale and local events cannot be depicted.*

Although icing is not directly forecast, it's implied when flying in clouds and precipitation above the freezing level. The freezing level aloft is represented by short dashed lines, and the freezing level at the surface is denoted by a zigzag line and the contraction SFC.

The lower panels are surface progs that feature standard frontal and weather symbology. Alternating dashes and dots enclose areas of showery precipitation, and a continuous line encloses areas of continuous (two precip symbols) or intermittent (one precip symbol) precipitation.

Hatching indicates the phenomenon is forecast to cover half or more of the area. For example, in Fig. 11-6 at 0600Z over the Rockies, scattered, intermittent (covering less than half the area) rain showers and thunderstorms are forecast; in the central Mississippi Valley, widespread (covering more than half the area) areas of rain showers and thunderstorms are forecast.

In preparing this chart, forecasters cannot consider mesoscale features; the chart is a synoptic depiction. Local conditions might not be accurately portrayed. The limited fine mesh (LFM) data are the main ingredients, so these progs tend to underestimate convective activity in the west along with the intensity of Pacific storms, and the forecaster tends to smooth over local variations due to terrain; therefore, these charts are most useful in the Midwest and East.

The chart in Fig. 11-6 is valid at 0600Z Tuesday (left panels), August 2, 1994, and 1800Z Tuesday (right panels), August 2, 1994. Be careful converting to local time, for example, 0600Z Tuesday translates to 11 p.m. Monday, Pacific daylight time.

Let's say we're planning a flight from Kansas City, MO, to Washington, D.C., the afternoon of August 2. We select the 1800Z panels from Fig. 11-6.

From the top panel: VFR ceilings and visibilities should prevail through Missouri. Conditions will deteriorate to MVFR in Illinois, then VFR again through Indiana and Ohio. Over the Appalachians, conditions will again deteriorate to MVFR. Moderate-to-severe turbulence is expected for extreme eastern Missouri, Illinois, and Indiana, and moderate turbulence is expected below 16,000 feet east of the Appalachians. The freezing level is forecast above 12,000 feet; the 120 isotherm runs through the Great Lakes region.

From the lower panel: The front in the Great Lakes area should not affect our flight. But the low-pressure area over Illinois will produce widespread rain showers and thunderstorms, which are the reason for the turbulence and MVFR conditions. Scattered rain showers and thunderstorms are expected for the rest of the route. The surface trough in Virginia accounts for the precipitation, turbulence, and MVFR in that area.

Strictly a surface prognosis, the 36–48-hour significant weather prog chart is issued twice daily, valid at 0000Z and 1200Z. By the time the chart becomes available, valid times only provide a 30- and 42-hour forecast. If a pilot calls just prior to the next issuance, only a 30-hour forecast would be available.

Referring to Fig. 11-7, the 0000Z Monday panel shows a surface trough producing widespread rain showers and thunderstorms moving into Missouri. High pressure over New England should bring good flying weather to the rest of the KC-to-DC route. Pilots must remember that the scalloped lines on the 36–48-hour prog denote areas of overcast clouds with no reference to the height of the cloud bases.

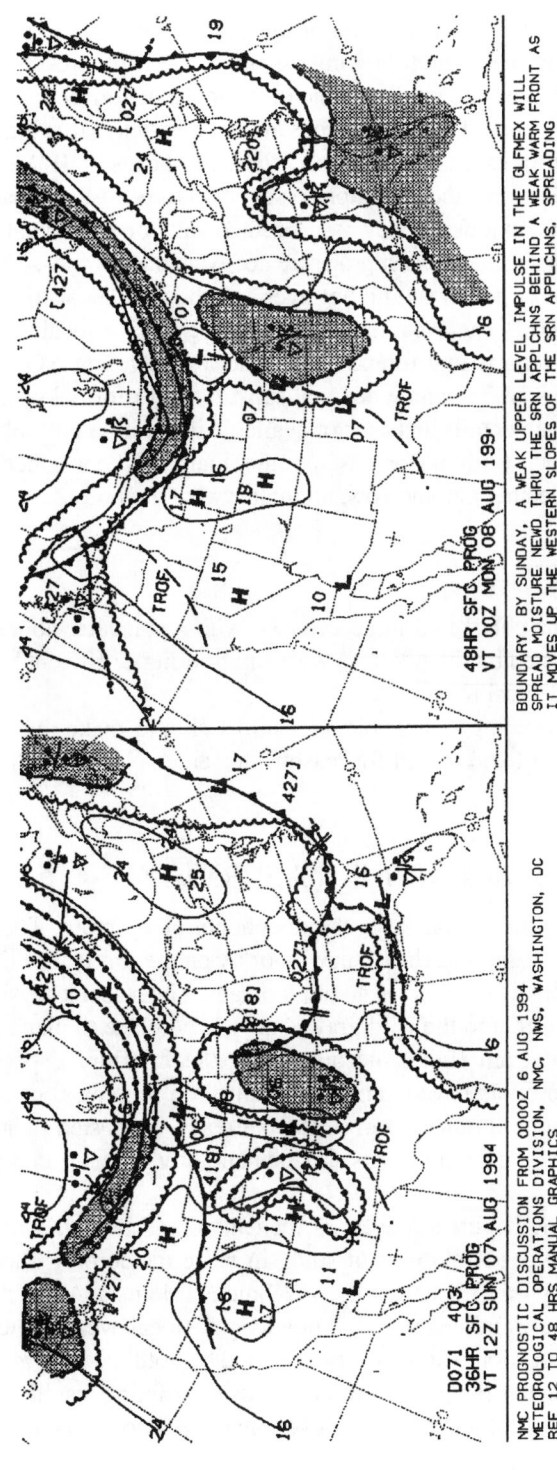

NMC PROGNOSTIC DISCUSSION FROM 0000Z 6 AUG 1994
METEOROLOGICAL OPERATIONS DIVISION, NMC, NWS, WASHINGTON, DC
REF 12 TO 48 HRS MANUAL GRAPHICS

COLD FRONT MOVING OFF THE NEW ENG AND MID ATLC COAST WILL BRING COOLER
AND DRIER AIR TO MOST OF THE NATION EAST OF THE MS RVR EXCEPT THE GULF
COAST AND LOWER SERN STATES. THE FRONT WILL QUICKLY BECOME NEARLY
STATIONARY IN THE LOWER MS AND TN VALLEY AND WILL SLOWLY RETURN AS A
WARM FRONT OVER THE NEXT TWO DAYS IN THE SRN PLAINS. EASTERN PORTION
OF THE FRONT WILL HANG UP ACROSS SERN NC WESTWARD THRU THE TN VALLEY
BEFORE BEGINNING TO WASH OUT TOWARD THE END OF THE PERIOD. AN EAST TO
WEST TROF WILL PERSIST THRU THE PERIOD ACROSS NRN FL WESTWARD INTO LA.
SCATTERED TO BROKEN MAINLY DIURNAL CONVECTION IN A VERY MOIST AND
UNSTABLE AMS WILL OCCUR ALONG AND NEAR THE TROF AND NEWD ALONG THE
ATCL SEABOARD AS FAR N AS SERN NC NEAR THE SLOW MOVING FRONTAL

BOUNDARY. BY SUNDAY, A WEAK UPPER LEVEL IMPULSE IN THE GLFMEX WILL
SPREAD MOISTURE NEWD THRU THE SRN APPLCHNS BEHIND A WEAK WARM FRONT AS
IT MOVES UP THE WESTERN SLOPES OF THE SRN APPLCHNS. SPREADING
SCATTERED SHOWERS AND THUNDERSTORMS NWD. THE UPPER LEVEL CLOSED LOW
NEAR NRN VANCOUVER ISLAND WILL SLOWLY DRIFT EASTWARD EJECTING
SHORTWAVES OVER THE WEAKENING WESTERN RIDGE. THESE SHORTWAVES WILL
DROP SEWD DOWN THE BACK SIDE OF THE WEAK ERN TROF INTERACTING WITH
GULF MOISTURE AND INSTABLITY SPREADING NWD THRU THE CENTRAL PLAINS
TODAY AND SUNDAY RESULTING IN AN AREA OF BROKEN CONVECTION FROM IA TO
ERN OK. IN THE WEST, SCATTERED MONSOONAL CONVECTION WILL CONTINUE TO
OCCUR DIURNALLY OVER THE SOUTHWEST AND SRN ROCKIES ROTATING AROUND THE
UPPER LEVEL RIDGE OVER AZ AND NM. BY SUNDAY, A NRN STREAM SHORTWAVE
WILL DROP SOUTHWARD DRIVING A COLD FRONT THRU THE SRN CANADIAN PLAINS
INTO THE NRN U.S. WITH SHOWERS AND THUNDERSTORMS NEAR THE FRONT.
ROSENSTEIN/WEATHER FORECAST BRANCH

Fig. 11-7. *On the 36-to-48-hour significant weather prog, scalloped lines enclose areas of overcast clouds with no reference to the height of cloud bases.*

What can be concluded about clouds, visibilities, turbulence, and icing? Drawing on our knowledge, we can make the following assumptions. High pressure generally means fair weather. Weak pressure gradients and the absence of convective activity indicate no significant turbulence. These conditions are not conducive to icing; however, conditions could be right for extensive areas of radiation fog causing IFR weather.

When using progs, remember the limitations on aviation forecasts, especially timing. Then when you request the outlooks, review issuance and valid times to obtain the latest forecasts. Keep in mind that these forecast progs are not available beyond 42 hours.

A last word on outlooks: pilots should not overlook the weather section of the local newspaper, *USA Today*, "The Weather Channel" on cable, or local TV weathercasts. Regional weather reports and forecasts are also available via NOAA weather radio transmitters. These sources cannot be legal substitutes for formal weather briefings, but they might contain details that are not available in aviation outlook products. Regardless of the source, the outlook-forecast axiom remains: "The weather tomorrow is going to be what the weather is tomorrow, no matter what anybody says."

IN-FLIGHT BRIEFING

Airborne briefings will be conducted in accordance with a standard, abbreviated, or outlook briefing as requested by the pilot. As with any briefing, sufficient background information must be made available.

Unless an airborne briefing is unavoidable, pilots should make every effort to complete all briefings prior to taking off for reasons that are explained in the next subsection.

FAA's PILOT WEATHER BRIEFING SERVICE

Briefings can be obtained in person, over the telephone, or by radio. The preferred methods are to obtain a weather briefing in person or by phone. Initial briefings by radio are discouraged, except where there is no other means. The reasons are simple. The cabin of an aircraft plunging into the wild gray yonder is no place to plan a flight; attention must be diverted from flying the aircraft to the briefing. Especially with marginal weather, certain pilots have a tendency to push on regardless of conditions. Finally, a lengthy briefing will unnecessarily tie up already-congested radio frequencies preventing any short and routine transmissions and any urgent communications between FSSs and pilots.

FSS specialist training begins at the FAA Academy, in Oklahoma City, with the equivalent of a collegiate academic year of study in basic meteorology and briefing techniques. The weather portion is taught by NWS meteorologists. At field facilities, "developmentals" receive training in "area knowledge" (local weather, terrain features, weather reporting locations) and must be certified by both the FAA and National Weather Service. Briefers at some point must get hands-on training just like pilots, and from time to time you the pilot will talk to a developmental who offers a briefing that might not be clear or concise; when this occurs, be as patient with him or her as a veteran briefer was patient with you when you planned your first solo cross-country.

The briefing is supposed to be presented in a clear, concise manner. Ambiguous terms, for instance "looks bad" or "scuzzy" or generalizations such as "VFR," are to be avoided. An FSS friend of mine received a one-liner briefing from Salinas to Sacramento, CA, "It's VFR." What does that mean? It could range from clouds almost to the ground and 1 mile visibility to clear-and-100.

The point is made that FSS briefers are not meteorologists—usually meteorologists make the point. These meteorologists call attention to the problem that when a forecast goes bad, the quality of the briefing falls apart and the pilot is left on the proverbial limb. This is not often the case. Briefers are trained to recognize forecast *variance*, which is the difference between the forecast for a given time and existing conditions. A briefer might suggest that the pilot wait for a new forecast or coordinate with a forecaster for resolution. In any case, the pilot is made aware of the problem.

Pilots will have less and less access to NWS personnel. The NWS for most practical purposes is out of the weather briefing business; however, where the NWS is accessible, telephone numbers are in the *Airport/Facility Directory* or are available from the local FSS. Pilots will not necessarily talk to a forecaster.

A rather new flight instructor brought his student into a National Weather Service office collocated with an FSS. The instructor explained that if the student wanted a really good briefing, always go to the NWS. Unfortunately for this young instructor, a rather crusty old "met tech" (meteorological technician) was on duty. The met tech quite unceremoniously admonished the instructor by explaining that many NWS specialists are not meteorologists nor engaged in aviation, and the FSS was the place to go for flying weather.

Many FSS specialists are excellent interpreters of the weather; they are familiar with local weather patterns and terrain and pass on their knowledge and experience to the pilot.

The biggest complaint about the FAA's pilot briefing service is delays. An Aviation Safety Reporting Service (ASRS) study states: "The inability to reach flight service by telephone was the complaint" in a number of incidents. "Reporters relate waits as long as 20 to 45 minutes on hold and then being disconnected. Reporters allege that because of the inability to reach flight service, many pilots in their area depart without preflight weather briefings or take off and contact en route flight advisory service." Flight Watch is not for an initial briefing; pilots who elect to use this procedure must call an FSS on the specific station's discrete frequency for an initial briefing.

It's no big mystery why delays are so lengthy. Let's take a large FSS that is responsible for a region with about 30,000 pilots. At any one time, there might be from four to eight briefers during peak periods. Guess what happens when more than eight pilots call? The longest delays occur when weather is marginal; during these periods, the average briefing might take 5 to 8 minutes, whereas briefings during good weather take an average of only two or three minutes. A final point to consider: After the briefer's initial checkout, the FAA provides little training to improve productivity.

Pilots complain about the deluge of superfluous information provided by certain briefers and the difficulty of getting information from other briefers. These complaints result from poor briefer training and perceived paranoia by the briefer about accident investigations.

Pilots are equally guilty of tying up briefing lines. Pilots inhibit the system by not being prepared for the briefing nor prepared to file a flight plan. Some pilots have unrealistic expectations; this begins with the flight instructor who fails to properly prepare a student for the briefing and ends with uncaring pilots who call to file an IFR flight plan, but haven't looked at the charts yet. These are usually the people that complain the loudest about delays. I have had many students call for a briefing and file a flight plan with zero knowledge of how to accomplish either. A practice briefing is an excellent option, but the instructor must still prepare the student and the call must be made during an off-peak period; ideally, the student will be polite enough to tell the briefer that it's for student practice and it would be easy enough to call back if the facility were too busy for the dry run. Instructors must ultimately take the responsibility to properly prepare their students.

Pilots need to be specific about the information they require. Ambiguous statements have no place in the pilot briefing environment: "Is it VFR?" "I'm looking for some soft IFR." "Where is it good?" "Just tell me what I need to know." "Is there anything significant?" (Is that VFR in controlled or uncontrolled airspace, above or below 10,000 feet, or special VFR?)

Try finding "soft IFR" in the *Pilot/Controller Glossary*. "Good" to one briefer could be "1,000-foot ceiling and 3 miles visibility." "Significant" falls into the same category as "good." Some pilots will still simply ask the briefer, "Are there any AIRMETs and SIGMETs for the route?" These individuals might miss significant information.

In the January 1989 issue of *AOPA Pilot*, Richard Collins wrote "Six Bits . . . of the Right Stuff." Collins discusses six virtues that make a good pilot. These virtues equally apply to a pilot's dealings with air traffic controllers and flight service station specialists.

Collins says "Patience is the only virtue and impatience the only sin." That's certainly true when holding on the phone for 20 or 30 minutes. There are very few reasons for departing without a briefing or requesting an initial briefing en route; there are many, many excuses. The pilot is as cool on the ground as in the air. These pilots realize the problems of other pilots, controllers, briefers, and forecasters. They are patient with training controllers and briefers because they remember when they were student pilots. These pilots would no more lose control with a controller or briefer than lose control of the aircraft.

A pilot with intuition seldom blunders into situations and occasionally cancels or discontinues trips because of the weather, according to Collins. A pilot with this virtue knows the limitations of weather reports and forecasts and plans accordingly. Based on their experience and the capabilities of the aircraft, they know when the weather answer is "no-go."

A pilot with the right stuff has his or her act together. The homework is done; routes have been reviewed, and a partial flight plan is completed. This pilot is ready for the briefing and ready to file a flight plan. He or she doesn't guess at the route, trying to file from memory. The pilot has the virtue of organization.

The decisive pilot uses information to make sound decisions. He or she always has more than one way out. When only one out is left, it's exercised. This might mean canceling a flight, circumnavigating weather, avoiding hazardous terrain, or an additional landing en route. The 180° turn is made before entering clouds. If the situation becomes uncertain, assistance is obtained before an emergency becomes an accident.

This pilot will never be caught on top or run out of fuel. Collins says "coordination deals with what is happening now, what is coming next, and the use of the brain." This pilot combines organization, cool, decisiveness, and coordination to update weather en route, devise a plan based on this information, and coordinate the action before the situation becomes critical.

Perhaps one more virtue: courtesy. A specialist that's been briefing for 4 to 6 hours on a marginal weather day is in no mood for pilot sarcasm. Briefers do not have to put up with obnoxious, rude, or profane pilots; that's the purpose of the telephone-release button. Courtesy is a two-way street, however, because pilots don't have to put up with obnoxious or rude briefers. If you don't think you're being treated in a courteous, professional manner, talk to the supervisor or facility manager, or call the FAA's toll-free hot line (800) FAA-SURE.

USING THE FAA's WEATHER BRIEFING SERVICE

The weather briefing is a cooperative effort between the pilot and FSS specialist. A pilot's preliminary planning should be complete, including a general idea of route, terrain, minimum altitudes, and possible alternates. Where available, obtain preliminary weather from one of the recorded services. From the broadcast, determine the type of briefing required: standard, abbreviated, or outlook.

During the briefing, try not to interrupt, unless the briefer is going too fast. Pilots too often interrupt with a question that is just about to be answered, which can cause the briefer to lose his or her train of thought, resulting in the inadvertent omission of information.

Here is a comment from the "things-that-bug-briefers-the-most" category. Some pilots unintentionally engage in a form of Chinese water torture; after every word the briefer says, the pilots interject "ah ha." This is terribly annoying and distracting.

I briefed a student one day and about 15 minutes later he called back. I recognized the aircraft number and said, "Didn't I just brief you?" "Well, I couldn't make heads or tails of my notes. This time I'm recording it." An outstanding idea! More students and instructors should adopt this practice. International student pilots who are training in the United States would probably find the technique very beneficial. An added advantage: The pilot can *listen* without taking his or her attention from the briefing and/or being forced to request repeated information from the briefer.

Briefers make mistakes, and many are not pilots. At the end of the briefing, don't hesitate to ask for clarification or additional information on any point you do not understand. If conditions are right for turbulence or icing and these phenomena were not mentioned, ask the briefer to verify that there are no weather advisories. Remember that forecasts for light to locally moderate icing do not warrant an advisory, nor does locally severe turbulence warrant a SIGMET. Forecasts for these conditions can be overlooked. On the other hand, don't expect the briefer to provide the freezing level on a clear day.

With this as a background and FSS staffing being further reduced **as a result of FSS consolidation**, the question becomes how can a pilot best use the services available?

Become familiar with recorded weather information in your area: TWEB, TIBS, and PATWAS. Figure 11-8 shows a typical FSS broadcast position. These programs have been established to help reduce delays. They provide a general weather picture, usually with enough information to determine if further checking is warranted. If the weather is IFR or beyond a pilot's limits, there's no need to tie up a briefer. Additionally, pilots can determine if on a particular day a flight to the coast, the desert, or the mountains would be best. Briefers can be on the telephone for 10 minutes or more with pilots looking for a place to fly. This could be eliminated if those pilots would use the recordings. This also applies to student pilots looking for suitable cross-country routes.

Fig. 11-8. *Recorded broadcasts in the form of TWEB, TIBS, and PATWAS provide a first look at the weather. The broadcasts can often be used to determine if further checking is warranted.*

Broadcasts are available over low frequency radio beacons, VORs, and by phone. TWEB and TIBS provide much of the information in a standard briefing. Some areas are served by another recorded service called PATWAS. Although summarized to a greater degree than TWEB, PATWAS contains basically the same information. The *Airport/Facility Directory* contains frequencies and phone numbers. Normally these services contain the synopsis, adverse conditions, route, and winds aloft forecasts through 12,000 feet, plus selected surface weather reports. Forecasts are normally available 24 hours, although surface weather reports might be suspended between 10 p.m. and 5 a.m. Depending on the broadcast, other information, such as terminal forecasts, NOTAMs, and military training activity, are not available. These broadcasts do not meet FAR requirements for IFR; however, often the information will be sufficient for a VFR flight. If any clarification or additional information is required, the FSS should be consulted.

When using TWEB, TIBS, or PATWAS, the weather log illustrated in Fig. 11-4 is ideal for jotting down information because the log is organized in the same order as the broadcast.

There are additional uses of the FAA's recorded weather systems. Knowing when the broadcasts are updated, pilots can obtain an outlook for the following day. Broadcast updates usually coincide with FA and TWEB forecast issuance times. Pilots can check with their FSS for broadcast update times; most publish these times in letters to airmen or pilot bulletins. Student and low-time pilots can use broadcast systems to learn aviation weather terminology. I always recommended this to students. In this way, they can become familiar with the terms and phrases used in weather briefings. Anything that they don't understand can be discussed with their instructor.

Pilots have a say in the content of these services. Although the FAA prescribes the general content and format, the exact items of information, such as individual weather reports, are left to the discretion of the facility. Facility managers are supposed to solicit comments from users, pilots, about their content. If pilots want a particular item on the broadcast, they should contact the facility and make a request. (A letter is better than a telephone call.)

Finally, pilots should know where to complain. As far as broadcasts go, this usually means it was unintelligible or read too fast. If there's a problem, contact the supervisor or manager, or use the FAA's hot line (800) FAA-SURE. If pilots don't make the FAA aware of a problem, who is at fault if the problem doesn't get fixed?

COMMERCIAL WEATHER BRIEFING SERVICES/DUATS

The FAA authorized Data Transformation and Contel in February 1990 to provide direct-user access terminal service (DUATS) to pilots within the contiguous United States. This computerized system, available at airports and through personal computers, allows direct access to weather briefing and flight plan services. DUATS and virtually all other commercial services use National Weather Service products, which contradicts the misconception that computer briefings somehow provide a different product than available through an FSS.

When using these services, it's essential to know what information is available. Pilots using a commercial system must check with the vendor to determine how its system handles aviation products. Certain advisories, for example CWAs, might not be available on some systems; none provide local NOTAMs. Know your service, and check with an FSS for any additional information required or to clarify anything that you don't understand—remember the disclaimer "Contact an FSS if you need more flight information assistance."

Let's review a DUATS aviation briefing, planning a flight in a Cessna 182 from Livermore, CA (LVK), to Salt Lake City, UT (SLC). The briefing filled nine complete pages and the weather was good. The briefing contains:

- The area forecast
- SIGMETs
- Convective SIGMETs
- Center weather advisories
- The AIRMET bulletin
- Surface weather observations
- Pilot reports
- Radar weather reports
- Terminal forecasts
- Winds aloft
- Notices to airmen

The pilot is presented with the same products available at the flight service station, except for charts and local NOTAMs. He or she must then decode, translate, and interpret the information to determine which reports and forecasts apply. This briefing took about 15 minutes to obtain and print, and an additional 10 minutes to analyze and apply. After a little practice, you should be able to scan the material as it is displayed and only print significant portions for further review.

A briefing provided by an FSS specialist would go something like this:

"There are weather advisories for moderate turbulence over California. There is a strong northerly flow aloft over California with an upper-level trough over the Rockies moving eastward. Livermore's reporting 15,000 thin broken, visibility two zero, wind calm. Over Stockton at 7,500 a Baron reports light turbulence with northerly winds 35 to 40 knots. En route, broken to overcast cirroform clouds and unrestricted visibilities becoming, by the Elko, Salt Lake City portion of the route, scattered clouds based around 6,000 to 7,000. Salt Lake surface winds zero two zero at one one, with standing lenticular altocumulus southeast through west. A Gulfstream II during climbout of Salt Lake southbound reports smooth, tops of scattered clouds 9,000 to 10,000. Conditions forecast to remain the same en route, with Salt Lake 6,000 scattered, a slight chance of ceilings 5,500 broken in light rain showers, surface winds three five zero at one five gusts two zero. Winds aloft forecast at 11,500, three three zero at two five knots. There are no NOTAMs for the route."

The advantages of computer briefings are the relatively prompt access and the capability of a personal copy. With these advantages come the responsibility to decode, translate, interpret, and apply information to a flight. The pilot will have to sift through the mountains of written data formerly reserved for the FSS specialist to determine if a particular flight is feasible under existing and forecast conditions and aircraft/pilot capability.

The sheer amount of information might be overwhelming, especially for long-distance flights. A pilot might have to study several pages for a single sentence that applies. Flight instructors and pilot examiners might wish to save briefings for training and flight tests. Remember that you have to contact the vendor with any service problems, and contact an FSS briefer with any content problems.

When obtaining a briefing from an FSS or other source, realize that it is only a *messenger*, not the message *writer*.

12
Updating information

A PILOT'S RESPONSIBILITY DOES NOT END WITH AN UNDERSTANDING OF forecast products and limitations and the means of collecting meteorological and aeronautical information. Due to the dynamic character of the atmosphere, data must be continually updated. Surprisingly, many pilots have not been taught or have not learned the importance of updating weather reports, forecasts, and NOTAMs en route. Failure to exercise this pilot-in-command prerogative can have disastrous results.

The importance of updating weather and NOTAMs en route cannot be overemphasized. The focal points for these services are the FAA's flight service stations via communications with and broadcasts from FSSs and Flight Watch.

FSS COMMUNICATIONS

With FSS consolidation, correct, concise, and accurate communications becomes more important. FSS frequencies are busier than ever with fewer specialists providing communications over larger areas. Correct communications technique, which seemingly is a simple task, will take on a greater significance. Towers, approach controls, and centers have specific frequencies for specific purposes (ground control, local control, clearance delivery, ATIS, etc.). Flight service stations also have different frequencies for specific services. Normally available are the common FSS, local airport advisory, facility discrete, and emergency frequencies. FSSs also have voice capability over many VORs. A pilot's first task is to select the appropriate frequency for the service desired.

FSS common frequency

The FSS common frequency is 122.2 MHz *simplex* (transmit and receive on the same channel) and is available at virtually every FSS. It is normally not published on aeronautical charts unless available at a remote site. When unsure of the appropriate frequency, 122.2 MHz can be used, although it is likely to be congested, especially over flat terrain or when used at high altitudes. If at all possible, *the FSS common frequency should never be used to obtain an initial weather briefing or file a flight plan.*

Local airport advisory frequency

The FSS's local airport advisory (LAA) frequency is 123.6 MHz (123.62 or 123.65 MHz at some locations). Used at nontower airports, this service provides wind, altimeter setting, favored or designated runway, and known traffic. Local weather conditions can also be included. At airports where part-time towers are collocated with an FSS, local airport advisory will be provided on the tower local control frequency when the tower is closed. VFR flights should monitor the frequency when within 10 miles of the airport. IFR flights will be instructed to contact the advisory frequency by the control facility.

The FAA's position remains that automated flight service stations will not provide LAA. This seems to be a waste of a valuable resource. And an FAA group has proposed a similar service for AFSSs. LAA for local or remote airports would be considered on an individual-airport basis. Wind and altimeter would be provided either from direct-reading instruments or the local weather report. If the local weather report is used, the time of observation would be included. The inclusion of time will alert the pilot that wind direction is true, rather than magnetic, as reported from direct-reading instruments.

Discrete frequency

Routine communications (weather information, flight plan services, position reports, etc.) should be accomplished on the station's discrete frequency. These frequencies are unique to individual facilities. Their use will usually avoid frequency congestion with aircraft calling adjacent stations.

FSS frequencies can be found on aeronautical charts as illustrated in Fig. 12-1 and in the *Airport/Facility Directory*. A heavy-line box indicates standard FSS frequencies, 122.2 MHz and the emergency frequency 121.5 MHz. Other frequencies, such as the station discrete and local airport advisory, are printed above the box. If a frequency is followed by the letter R (122.15R), the FSS has receive-capability-only on that frequency; the pilot *transmits* on that frequency. The pilot must *receive* transmissions from the FSS on another frequency, usually the associated VOR; this duplex communications arrangement means that the pilot has to make sure the volume is turned up on the VOR receiver.

FSS COMMUNICATIONS

HEAVY-LINE BOX indicates FSS.
Normally **122.2** and **121.5** are available.

122.35   122.5
```
┌─────────────────────────┐
│     HAWTHORNE  HHR       │
└─────────────────────────┘
```

122.35 (simplex) FSS primary discrete frequency.
122.5 (simplex) FSS secondary discrete frequency.

REMOTE COMMUNICATIONS OUTLET/NAVAID
(duplex) "FRIANT" is the name of the RCO and NAVAID. RANCHO MURIETTA is the controlling FSS.

122.1R
```
┌─────────────────────────┐
│         FRIANT          │
│   115.6   Ch 103   FRA  │
└─────────────────────────┘
```
RANCHO MURIETTA

122.1R (duplex) FSS has receiver ONLY. *Pilot must transmit on* **122.1** *and listen on the VOR frequency* **115.6.**

An underlined frequency (i.e., 109.4) indicates NAVAID only. NO FSS communications.

HEAVY-LINE BOX indicates FSS.
Normally **122.2** and **121.5** are available. A square inside the lower right corner indicates HIWAS, or a circle with a "T" indicates TWEB available on the VOR frequency.

122.6
123.65 VOR/DME
```
┌─────────────────────────┐
│         ARCATA          │
│    110.2   Ch 39 ACV    │
└─────────────────────────┘
```

122.6 (simplex) FSS discrete frequency.
123.65 (simplex) FSS local airport advisory.

REMOTE COMMUNICATION OUTLET
(simplex) "BURBANK" is the name of the RCO with a frequency of **122.35**.

122.35
```
┌─────────────────────────┐
│     BURBANK  RCO        │
└─────────────────────────┘
```
HAWTHORNE

HAWTHORNE is the controlling FSS.

Fig. 12-1. *FSS communication frequencies are depicted on aeronautical charts. Pilots should select the appropriate frequency for the service desired.*

For example, refer to the remote communications outlet/NAVAID portion of Fig. 12-1. Rancho Murietta FSS, noted below the box, has a remote receiver at the VOR site; the FSS's receiver is "listening" to 122.1 MHz, noted above the box. A pilot wishing to establish duplex communications would tune the airplane's transmitter to 122.1 MHz and select the Friant VOR frequency, 115.6 MHz, on the VOR receiver. The pilot must remember to turn up the volume on the VOR receiver because "Rancho Radio" will transmit on that frequency. A thin-line box indicates a remote communications outlet. The frequency or frequencies available are printed above the box with the name of the controlling FSS below.

After selecting the frequency for the service desired, correct communications technique must be used. By following the procedures below, pilots will realize faster, more efficient service and reduce the chances of errors and delays.

Monitor the frequency before transmitting. Monitoring the frequency before transmitting is paramount to effective communications. How many times have we heard someone transmit over someone else? We've all done it; select the frequency, and press the transmit button. This only adds to the congestion of already-crowded frequencies. This basic procedure should be followed when contacting any facility.

Use the complete aircraft identification. The FSS needs the full aircraft call sign. For example, "November one two three four Golf" or "Commander five six seven Alpha Delta."

Advise the FSS on which frequency you expect a response and your general location. Most FSSs monitor from 5–10 different frequencies. With FSS consolidation, this practice has become more and more important for efficient communications. Figure 12-2 shows the in-flight console at the McAlester, OK, AFSS. Each light on the panel represents one receiver.

Establish communications before proceeding with your message. The specialist might be busy with other aircraft on other frequencies or other duties. When the specialist replies with your aircraft number, listen carefully to the reply. A classical failure occurs when a pilot calls to file a flight plan. The specialist is already busy with another aircraft on another frequency and advises the calling pilot to "stand by," then the calling pilot proceeds with a flight plan. The specialist has no option but to mute the receiver and conclude the contact with the other aircraft. The pilot wishing to file might be somewhat miffed when the specialist says "Go ahead with your flight plan," but the specialist is not at fault.

EN ROUTE FLIGHT ADVISORY SERVICE (FLIGHT WATCH)

The objective and purpose of Flight Watch is to enhance aviation safety by providing en route aircraft with timely and meaningful weather advisories. This objective is met by providing complete and accurate information on weather as it exists along a route pertinent to a specific flight. The information is to be supplied in sufficient time to prevent unnecessary changes to a flight plan, but when necessary permit the pilot to make a decision to terminate the flight or alter course before adverse conditions are encountered.

Fig. 12-2. *Correct techniques are imperative for effective communication. Pilots should always advise the FSS on which frequency they expect a response.*

Flight Watch is not intended for flight plan services, position reports, and initial or outlook briefings, nor is it to be used for aeronautical information, such as NOTAMs, center or navigational frequencies, or for single or random weather reports and forecasts. The altimeter setting will normally only be provided on request. Pilots requesting services not within the scope of Flight Watch will be advised to contact an FSS.

Using all sources, Flight Watch provides en route flight advisories, which include any hazardous weather, presented as a narrative summary of existing flight conditions (real-time weather) along the proposed route of flight; the summary is tailored to the type of flight being conducted.

The purpose of Flight Watch is to provide meteorological information for that phase of flight that begins after climbout and ends with descent to land; therefore, the specialist can concentrate on weather trends, forecast variance, and hazards. Flight Watch is specifically intended to update information previously received and serve as focal point for system feedback in the form of PIREPs. Air traffic controllers do accept PIREPs, but weather is a secondary duty and unfortunately PIREPs aren't always passed along from ATC. If at all possible, PIREPs should be reported directly to Flight Watch. The effectiveness of Flight Watch is to a large degree dependent on this two-way exchange of information.

A Bonanza pilot approached an area of thunderstorms in California's Central Valley. The pilot received from Flight Watch the latest weather radar and satellite information as well as PIREPs and surface observations. The pilot safely traversed the area with minimum diversion or delay.

This is not a very exciting story, but that's the purpose of Flight Watch—to assist pilots in conducting uneventful flights. En route flight advisory service (EFAS) has been around for more than 20 years. In spite of this, its function and the best way to use this important service are misunderstood by many pilots.

EFAS was originally "en route weather advisory service"; no one could pronounce the radio call "EEE'waas." It began on the West Coast in 1972 originally as a 24-hour service; Flight Watch now normally operates from 6 a.m. until 10 p.m. local time. Flight Watch is not available at all altitudes in all areas. The service provides communications generally at and above 5,000 feet AGL. Although in areas of low terrain and close to communication outlets, service will be available at lower altitudes. Figure 12-3 shows the Los Angeles Flight Watch position circa 1976.

Fig. 12-3. *Flight Watch began on the West Coast in 1972 when specialists only had teletype reports and forecasts. Pilot reports are the mainstay of this service.*

The system expanded in 1976 and a network of 44 flight watch control stations became operational in 1979. The common frequency 122.0 MHz immediately became congested, especially from aircraft at high altitudes. A discrete high-altitude frequency was assigned Flight Watch stations in the Southwest in 1980 to help resolve the problem. With flight service station consolidation, Flight Watch responsibility has been assigned the FSSs associated with the air route traffic control centers (Oakland FSS-Oakland Center, Hawthorne FSS-Los Angeles Center, etc.). Figure 12-4 shows the Columbia, MO, AFSS Flight Watch console. A discrete high-altitude frequency, for use at and above Flight Level 180, has been assigned each flight watch control station to cover the associated center's area.

Fig. 12-4. *Flight Watch specialists in the 1990s have at their command a vast network of information that includes real-time weather radar and satellite imagery.*

Flight Watch procedures

Establishing communications is the first step. Because only one frequency is available for low altitudes, pilots must exercise frequency discipline. In addition to the basic communications technique already discussed, the following procedures should be used when contacting Flight Watch.

1. When known, call the name of the associated air route traffic control center (ARTCC) followed by Flight Watch: "Salt Lake Flight Watch." If the ARTCC is not known, simply calling "Flight Watch" is sufficient.

2. State the aircraft position in relation to a major topographical feature or navigation aid: "In the vicinity of Fresno," "Over the Clovis VOR," etc. Exact positions are not necessary, but the general location of the aircraft is needed. Flight Watch facilities cover the same geographical areas as the ARTCC. With numerous outlets on a single frequency, the specialist needs to know which transmitter serves the pilot's area. This will eliminate interference with aircraft calling other facilities, garbled communications, and repeated transmissions. Failure to state the aircraft position on initial contact is the single biggest complaint from Flight Watch specialists.

3. When requesting weather or an en route flight advisory, provide the specialist with cruising altitude, route, destination, and IFR capability, if appropriate. The specialist needs sufficient background information to provide the service requested.

Flight Watch specialists are required to continually solicit reports of turbulence, icing, temperature, wind shear, and upper winds regardless of weather conditions. This information and PIREPs of other phenomena are immediately relayed to other pilots, briefers, and forecasters. Together with all sources of information, the specialist has access to the most complete weather picture possible.

ARTCC controllers are helpful about relaying reports of turbulence and icing and providing advice on the location of convective activity, but the information is limited by equipment and usually to immediate and surrounding sectors. The ARTCC controller's primary responsibility is the separation of aircraft. On the other hand, Flight Watch has only one responsibility, weather. Flight Watch provides specific real-time conditions, as well as the big picture, with live National Weather Service weather radar displays, satellite pictures, and the latest weather and pilot reports. And Flight Watch specialists have direct communications with Center Weather Service Unit personnel and NWS aviation forecasters.

Getting ahold of Flight Watch is usually a simple matter, even for single-pilot IFR operations. ATC will almost always approve a request to leave the frequency for a few minutes, but don't wait until the last minute. Trying to find an alternate airport in congested approach airspace is no fun for anyone. I routinely request a temporary frequency change when air traffic and radio traffic are uncongested and have never been denied the request from en route controllers.

HIGH-ALTITUDE FLIGHT WATCH

The early days of airline flying were plagued by thunderstorms as well as icing, turbulence, widespread low ceilings and visibilities, and the limited range of the aircraft. Today's jets have virtually overcome these obstacles. More and more pilots of general aviation aircraft equipped with turbochargers and oxygen or pressurization are encountering the same problems as yesterday's airline captains. The only difference is a vastly improved air traffic and communication system. And among one of the FAA's best kept secrets is the implementation of high-altitude Flight Watch.

Continually updating the weather picture is the key to managing a flight in an aircraft with relatively limited range, especially at high altitude in aircraft without ice protection and storm detection equipment. Winds aloft can be a welcome friend eastbound or a terrible foe westbound. With limited range, even a small change in winds at altitude can have a disastrous result. At the first sign of unexpected winds, Flight Watch should be consulted if for no other reason than to provide a pilot report. A significant change in wind direction or speed is often the first sign of a forecast gone sour. A revised flight plan might be required. Flight Watch can provide needed additional information on current weather, PIREPs, and updated forecasts upon which to base an intelligent decision.

A primary reason for high-altitude flying is to avoid frontal, mountain wave, and mechanical turbulence; however, the flight levels have their own problem: wind shear, or clear air turbulence. When problems are encountered, Flight Watch can help find a smooth altitude or alternate route. If the pilot elects to change altitude, an update of actual or forecast winds aloft is often a necessity.

Icing is normally not a significant factor in the flight levels, except around convective activity or in the summer when temperatures can range between 0°C and −10°C; however, icing can be significant during descent, especially when the destination temperature is at or below freezing. Flight Watch can provide information on tops, temperatures aloft, reported and forecast icing, and current surface conditions.

Many aircraft are equipped with airborne weather radar and lightning detection equipment; however, these systems are plagued by low power, attenuation, and limited range. A pilot might pick his or her way through a convective area only to find additional activity beyond. With RRWDS, Flight Watch has the latest NWS weather radar information. Well before engaging any convective activity, a pilot should consult Flight Watch to determine the extent of the system, its movement, intensity, and intensity trend. Armed with this information, the pilot can determine whether to attempt to penetrate the system or select a suitable alternate. ATC prefers issuing alternate clearances rather than handling emergencies in congested airspace and severe weather.

High-altitude Flight Watch frequencies for individual ARTCCs are provided in Fig. 12-5. Frequencies and outlets can also be found on the inside back cover of the *Airport/Facility Directory*. And although not easy to find, Jeppesen high-altitude charts carry the frequencies. The standard frequency 122.0 MHz can be used when a pilot is unsure of the discrete frequency.

Finally, destination and alternate weather should be obtained from Flight Watch. The preflight briefing provided current and forecast conditions at the time of the brief-

High-altitude flight watch
(for use at and above FL180)

Seattle	135.92	Chicago	134.87
Oakland	135.70	Indianapolis	134.82
Los Angeles	135.90	Memphis	133.67
Salt Lake City	133.02	Cleveland	135.42
Denver	124.67	Atlanta	135.47
Albuquerque	127.62*	Jacksonville	134.17
Minneapolis	135.67	Miami	132.72
Kansas City	128.47	Boston	133.92
Fort Worth	133.77	New York	134.72
Houston	126.62	Washington	134.52

*134.82 in northern Arizona

Fig. 12-5. *Discrete high-altitude Flight Watch frequencies are being commissioned. The use of these channels will eliminate much frequency congestion for aircraft at low altitudes.*

ing. This information should be routinely updated en route; airlines do it, often through Flight Watch. Are the updated reports consistent with the forecast? If not, why? Flight Watch specialists through their training are in an excellent position to detect forecast variance. Whether the forecast was incorrect or conditions are changing faster or slower than forecast, the pilot needs to know and plan accordingly. A knowledge of forecast issuance times is often helpful. Forecasts might not be amended if the next issuance time is close. Flight Watch is in the best position to provide the latest information and suggest possible alternatives.

Updates must be obtained far enough in advance to be acted upon effectively. This must be done before adverse weather is encountered or fuel runs low. Hoping a stronger-than-forecast headwind will abate or arriving over a destination that has not improved as forecast is folly. At the first sign of unforecast conditions, Flight Watch should be consulted and, if necessary, an alternate plan developed. This might mean an additional routine landing en route, which is eminently preferable to at best a terrifying flight or at worst an aircraft accident.

13
Notices to Airmen (NOTAMs)

FAILURE TO CHECK NOTICES TO AIRMEN HAS LED MANY A PILOT INTO AN embarrassing and potentially hazardous situation. Increased use of DUATS and other commercially available briefing systems means that interpreting and understanding NOTAMs will take on a greater significance.

The FAA advertises the status of components or hazards in the National Airspace System (NAS) through aeronautical charts, the *Airport/Facility Directory*, other publications, and the National Notices to Airmen System. Changes are normally published on charts, in the directory, or appear in the *Notices to Airmen* publication. Published NOTAMs are sometimes referred to as Class II, which is the international term used to identify NOTAMs that appear in printed form for mail distribution. The need for current charts and publications cannot be overemphasized.

A pilot called flight service and requested a briefing from Bishop to Santa Cruz, CA. The briefer explained that the airport was closed. "Oh, I must be using an old chart," said the pilot. Indeed, the airport had been closed for 2 years.

Aeronautical information not received in time for publication is distributed on the FAA's telecommunications systems; the information would include unanticipated or temporary changes or hazards when the duration is for a short period or until published. Unpublished NOTAMs are not necessarily given during abbreviated or outlook briefings, but are routinely provided as part of an FSS standard briefing.

NOTAMs are issued for the commissioning and decommissioning of facilities and restrictions to landing areas (runways and waterways), including closures, braking action, and problems due to snow, ice, slush, or water. Lighting aids that affect landing areas (airport beacons, VASIs, wind-T lights, obstruction lights, pilot-controlled lighting, etc.) or are part of an instrument approach procedure receive NOTAM distribution. NAVAID status, air traffic control facility hours of operation, controlled airspace, and services (EFAS, HIWAS, AWOS, etc.) receive NOTAM distribution.

Unpublished NOTAMs are divided into three groups:

- NOTAM (D)
- NOTAM (L)
- FDC NOTAMs

FSS specialists are responsible for classification, format, and transmission of NOTAM (D)s and NOTAM (L)s. The country is divided into *flight plan areas*; a designated FSS has NOTAM responsibility for airports and NAVAIDs within a respective flight plan area. The tie-in FSS that is responsible for a facility can be determined from the *Airport/Facility Directory*.

TERMINOLOGY

Times used on NOTAMs are UTC. The day begins at 0000Z and ends at 2359Z. The absence of a date/time group means the condition is in effect and will continue until further notice (UFN); however, UFN in the NOTAM text is not transmitted. To allow automated NOTAM processing, one of the following terms will be used: EFF (effective), TIL (until), or THRU (through). The term will be followed by a time or a date and time group describing the effective period.

Runways are identified by magnetic bearing (12-30, 12, or 30). If magnetic bearing has not been established, the runway is identified by the nearest eight points of the compass (NE-SW, N 200 N-S RY). A solidus (/) is used to indicate "and" (17-35/1-19 CLSD means Runways 17-35 and 1-19 are closed). The plus sign (+) is used to indicate "additional" or "also."

NOTAM (D)

NOTAM (D)s contain information that might influence a pilot's decision to make a flight, or require alternate routes, approaches, or airports. They are considered "need-to-know" and issued for certain landing area restrictions, lighting aids, special data, and air navigation aids that are part of the National Airspace System. Table 13-1 contains NOTAM (D) issuance criteria.

Table 13-1. NOTAM criteria.

NOTAM (D)	NOTAM (L)
Landing area	
Commissioning or decommissioning all or a portion of a landing area.	Runway information that does not restrict or preclude the use of a runway.
Conditions that restrict or preclude the use of any portion of a runway or waterway.	Conditions pertaining to taxiways and ramps.
	Bird activity.
Braking action fair, poor, or nil.	Personnel and equipment on or adjacent to runway.
Airport closure.	
Lighting aids	
Runway lights.	Taxiway lights.
Approach lighting system.	Airport beacon.
Pilot controlled lighting.	VASI/REIL/obstruction lights.
NAVAID/Communications/Services	
NAVAID which is part of NAS.	NAVAID which is not part of NAS, ATIS/TWEB.
Communication outlet or service.	Class C airspace services.
Special data	
Weather reporting service.	Fuel jettisoning.
Restricted areas.	Changes to SIDs/STARs.
Airshow-high speed aircraft.	Airshow, aerobatic demonstrations, glider operations, balloon meets.
Parachute jumping.	
Aerial refueling.	

NOTAM (D) information is given distant dissemination on the FAA's weather telecommunications system. NOTAM (D)s are prepared and transmitted using standard contractions and abbreviations that can be found in appendix A.

!OAK 03/005 OAK VORTAC OTS TIL 2200

This Oakland NOTAM was issued in the third month of the year and is the fifth NOTAM (D) issued during the month for the OAK NOTAM file (03/005). The Oakland VORTAC is scheduled to be out of service until 2200Z (OAK VORTAC OTS TIL 2200).

NOTAM (D)s are issued for public-use airports listed in the *Airport/Facility Directory*. Depending on the system, the pilot might have to determine the NOTAM file for a particular facility. This has become more complex with FSS consolidation. The appropriate *Airport/Facility Directory* contains this information. For the following discussion, refer to Fig. 13-1. Note that the NOTAM files for the Goffs and Gorman VORTACs are Riverside (RAL) and Bakersfield (BFL) respectively (GOFFS-NOTAM FILE RAL; GORMAN-NOTAM FILE BFL); therefore, information on the status of these NAVAIDs would be found in the respective NOTAM files RAL and BFL.

For the Grass Valley, Nevada County Airpark, airport NOTAMs are contained in the Rancho Murieta NOTAM file (RIU); however, NAVAID NOTAMs appear in the Marysville NOTAM file (MYV). This is because the tie-in FSS for Nevada County is the Rancho Murieta FSS. The NAVAID serving the airport is Marysville and its NOTAM file is MYV. The directory indicates that the Visalia Municipal Airport has both airport (listed under COMMUNICATIONS-FRESNO FSS) and NAVAID (listed under RADIO AIDS TO NAVIGATION) NOTAMs assigned to the Visalia (VIS) NOTAM file. Pilots planning either a VFR or IFR flight to Visalia would need to check VIS for both airport and NAVAID NOTAMs.

SAC 12/005 SAC PRIRA 30 NMR MYV045005 OTS TIL 03092359
TVL 01/035 SWR TACAN AZM OTS

The SAC NOTAM was the fifth NOTAM issued for the SAC NOTAM file during the 12th month of the year (12/005). The Sacramento primary radar (SAC PRIRA) within a 30 nm radius of the Marysville VOR 045 radial at 5 miles (30 NMR MYV045005) will be out of service until March 9 at 2359Z (03092359). This alerts pilots that radar services to aircraft not equipped with transponders will be unavailable. This might be significant for a nontransponder aircraft if an approach required radar to execute the procedure.

The second NOTAM advertises the fact that the Squaw Valley (SWR) TACAN azimuth is out of service. This would normally only affect military aircraft.

Here are a few NOTAM (D)s that illustrate common usage.

SFO 10L-28R CLSD EFF 170700-1500

San Francisco Runway 10 Left/28 Right will be closed on the 17th between 0700Z and 1500Z, which are noted as UTC dates and times that require conversion; the runway is scheduled to be closed between 11 p.m. on the 16th until 7 a.m. on the 17th, Pacific Standard Time. Pilots must use care converting UTC to local times.

SJC 12R-30L CLSD 0800-1400 DLY TIL 271400

CALIFORNIA

GEORGETOWN   (Q61)   2 NW   UTC-8(-7DT)   N38°55.26'@120°51.88'   SAN FRANCISCO
   2588   B   FUEL 100LL   TPA—3388(800)   L-2G
   RWY 16-34: H2980X60 (ASPH)   S-22   MIRL
      RWY 16: Trees.   RWY 34: Trees
   AIRPORT REMARKS: Attended 1600-0100Z‡. Fuel not avbl on Tues. Touch and go landings prohibited on Rwy 34.
      ACTIVATE MIRL RWY 16-34—CTAF.
   COMMUNICATIONS: CTAF/UNICOM 123.05
      RANCHO MURIETA FSS (RIU) TF 1-800-WX BRIEF. NOTAM FILE RIU.
   RADIO AIDS TO NAVIGATION: NOTAM FILE RIU.
      HANGTOWN (L) VOR/DME 115.5   HNW   Chan 102   N38°43.48'W120°44.96'   318°13.0 NM to fld.
      2602/17E.

GILLESPIE FLD   (See SAN DIEGO (EL CAJON))

GNOSS FLD   (See NOVATO)

GOFFS   N35°07.87'W115°10.59'   NOTAM FILE RAL.   LOS ANGELES
   (L) VORTAC 114.4   GFS   Chan 91   020°22.9 NM to Searchlight, NV. 4000/15E.   H-2B, L-3D, 5B
   VORTAC unusable:
      200°–235° beyond 30 NM below 8700'   290°–320° beyond 20 NM below 9500'
      235°–260° beyond 25 NM below 7400'   320°–010° beyond 30 NM below 8500'
      260°–290° beyond 25 NM below 8000'
   RCO 122.05R 114.4T (RIVERSIDE FSS)

GORMAN   N34°48.24'W118°51.68'   NOTAM FILE BFL.   LOS ANGELES
   (L) VORTAC 116.1   GMN   Chan 108   081°32.1 NM to General Wm. J. Fox Airfield. H-2B, L-3B, 5A
   4920/16E.
   VORTAC unusable:
      190°–220° beyond 27 NM below 9500'   255°–280° beyond 20 NM below 10,500'
      220°–255° beyond 8 NM below 9500'   280°–300° beyond 15 NM below 8000'
   RCO 122.1R 116.1T (BAKERSFIELD FSS)

GRASS VALLEY

NEVADA CO AIR PARK   (O17)   3 E   UTC-8(-7DT)   N39°13.44'W121°00.14'   SAN FRANCISCO
   3151   B   S4   FUEL 100LL, JET A, MOGAS   TPA—4151(1000)   L-2G
   RWY 07-25: H3925X50 (ASPH)   S-35   MIRL   2.1% up NE   IAP
      RWY 07: VASI(V4L)—GA 3.0° TCH 25'. Trees
      RWY 25: VASI(V2L)—GA 3.25° TCH 30'. Thld dsplcd 204'. Road
   AIRPORT REMARKS: Attended Nov-Apr 1600-0100Z‡, May-Oct dalgt hours. Watch for air tankers July-Oct. Rwy
      07-25 slopes downhill to west, recommend take off Rwy 25. Separation distance between Rwy 07-25 and
      parallel taxiway centerlines less than 150'. Due to rwy gradient width and crosswinds use of Rwy 07-25 not
      recommended for student solo cross-country flights. ACTIVATE MIRL Rwy 07-25—CTAF.
   COMMUNICATIONS: CTAF/UNICOM 123.0
      RANCHO MURIETO FSS (RIU) TF 1-800-WX-BRIEF, NOTAM FILE RIU.
   ® **SACRAMENTO APP/DEP CON** 125.4
   RADIO AIDS TO NAVIGATION: NOTAM FILE MYV.
      MARYSVILLE (T) VOR/DME 110.8   MYV   Chan 45   N39°05.92'W121°34.38'   058° 27.7 NM to fld. 60/16E.

GRAVELLY VALLEY   (See UPPER LAKE)

GROVELAND

PINE MOUNTAIN LAKE   (Q68)   3 NE   UTC-8(-7DT)   N37°51.70'W120°10.71'   SAN FRANCISCO
   2930   B   S4   **FUEL** 80, 100LL   TPA—3930(1000)   L-2F
   RWY 09-27: H3625X50 (ASPH)   S-12   MIRL
      RWY 09: VASI(V2L)—GA 4.5° TCH 16'. Trees.   **RWY 27:** PAPI(P2L)—GA 4.0° TCH 37'. Tree. Rgt tfc.
   AIRPORT REMARKS: Attended 1600-0100Z‡. Fuel available by request ctc-UNICOM. Arpt advisories unavailable
      through UNICOM. Be alert deer on and in vicinity of arpt especially Nov-Apr. Rwy 27 PAPI unusable byd 7
      degrees either side of centerline. ACTIVATE MIRL Rwy 09-27 and PAPI Rwy 27—CTAF. Fee for overnight
      parking. Transient parking avble. Ldg fee.
   COMMUNICATIONS: CTAF/UNICOM 123.05
      RANCHO MURIETA FSS (RIU) TF 1-800-WX-BRIEF. NOTAM FILE RIU.
   RADIO AIDS TO NAVIGATION: NOTAM FILE MOD.
      MODESTO (H) VOR/DME 114.6   MOD   Chan 93   N37°37.64'W120°57.47'   052° 29.7 NM to fld. 90/17E.

Fig. 13-1. *The* Airport/Facility Directory *can be used to determine NOTAM files and the tie-in FSS for airports and NAVAIDs.*

CALIFORNIA

VISALIA

SEQUOIA FLD (Q31) 8 N UTC-8(-7DT) N36°26.91'W119°19.14'   **SAN FRANCISCO**
  313 S2 TPA-1113(800)   **L-2E, 5A**
  RWY 13-31: H3012X60 (ASPH) S-30, D-50 MIRL
    RWY 13: Thld dsplcd 210'.Fence.
  AIRPORT REMARKS: Unattended.
  COMMUNICATIONS: CTAF/UNICOM 122.9
    FRESNO FSS (FAT) TF 1-800-WX-BRIEF. (1400-0600Z‡). NOTAM FILE FAT.
    RANCHO MURIETA FSS (RIU) TF 1-800-WX-BRIEF. (0600-1400Z‡).
  RADIO AIDS TO NAVIGATION: NOTAM FILE VIS.
    VISALIA (T) VOR/DME 109.4 VIS Chan 31 N36°22.04'W119°28.93' 042° 9.3 NM to fld. 260/16E.
      VOR/DME without voice when FSS clsd.

- -

VISALIA MUNI (VIS) 4 W UTC-8(-7DT) N36°19.12'W119°23.57'   **SAN FRANCISCO**
  292 B S4 **FUEL** 100LL, JET A TPA—1292(1000) ARFF Index Ltd.   **H-2A, L-2E, 5A**
  RWY 12-30: H6556X150 (ASPH)-PFC S-60, D-100, DT-160 MIRL   **IAP**
    RWY 12: REIL. VASI(V4L)—GA 3.0° TCH 50'. Thld dsplcd 275'. Railroad.
    RWY 30: MALSR. PAPI(P4L)—GA 3.0° TCH 52'. Tree.
  AIRPORT REMARKS: Attended Sep-May 1500-0300Z‡, Jun-Aug 1500-0400Z‡. PPR for air carrier operations with
    more than 30 passenger seats call arpt manager 602-738-3201 for ARFF svc. Remaining overnight tiedown fee.
    MIRL Rwy 12-30 preset on low ints dusk-0700Z‡ ACTIVATE higher ints—CTAF. After 0700Z‡ ACTIVATE MIRL
    Rwy 12-30—CTAF. ACTIVATE MALSR Rwy 30—CTAF.
  WEATHER DATA SOURCES: AWOS-3 119.925 (209) 651-2418.
  COMMUNICATIONS: CTAF/UNICOM 123.05
    FRESNO FSS (FAT) TF 1-800-WX-BRIEF. (1400-0600Z‡). NOTAM FILE VIS.
    RANCHO MURIETA FSS (RIU) TF 1-800-WX-BRIEF. (0600-1400Z‡).
    RCO 122.1R 109.4T (FRESNO FSS)
  ® **FRESNO APP/DEP CON** 118.5
  AIRSPACE: CLASS E svc effective 1400-0800Z‡ other times CLASS G.
  RADIO AIDS TO NAVIGATION: NOTAM FILE VIS.
    (T) VOR/DME 109.4 VIS Chan 31 N36°22.04'W119°28.93' 108° 5.2 NM to fld. 260/16E. VOR/DME
      without voice when FSS clsd.
    VILIA NDB (LM) 220 VI N36°15.29'W119°18.88' 299° 5.4 NM to fld. Unmonitored 0500-1500Z‡.
    ILS 108.5 I-VIS Chan 31 Rwy 30 LOM VILIA NDB. ILS and LOM unmonitored 0500-1500Z‡.

WARD FIELD (See GASQUET)

WASCO-KERN CO (L19) 2 NW UTC-8(-7DT) N35°37.18'W119°21.22'   **LOS ANGELES**
  313 B **FUEL** 100LL TPA—1113(800)   **L-2E, 3B, 5A**
  RWY 12-30: H3380X60 (ASPH) S-6 MIRL
    RWY 12: Thld dsplcd 455'. Road.   **RWY 30:** TRCV(TRIL). Thld dsplcd 20'. Road
  AIRPORT REMARKS: Attended irregularly. Fuel avbl by phone call only 805-758-5102 or 758-2507. Rwy 12-30 2734'
    lighted between dsplcd thlds. Crop dusting ops prohibited except by arpt manager 805-393-7990. ACTIVATE
    MIRL Rwy 12-30 and TRCV Rwy 30—CTAF.
  COMMUNICATIONS: CTAF/UNICOM 122.8
    BAKERSFIELD FSS (BFL) LD 805-399-1787. NOTAM FILE BFL.
  RADIO AIDS TO NAVIGATION: NOTAM FILE BFL.
    SHAFTER (H) VORTACW 115.4 EHF Chan 101 N35°29.07'W119°05.84' 289° 14.9 NM to fld. 550/14E.
      HIWAS.

Fig. 13-1. *Continued.*

This San Jose runway closure will occur daily (DLY) from 0800Z through 1400Z
until the 27th at 1400Z, which is daily from midnight local through 6 a.m., until the
27th at 6 a.m.

```
BUR 7-25 CLSD TKOF 12500/OVR
BUR 7-25 E 254 CLSD
```

Burbank Runway 7-25 is closed for takeoffs to aircraft with gross weights of 12,500 pounds and over probably because the eastern 254 feet of the runway is closed, which the second NOTAM reveals.

PAE PTCHY THN IR
CLM 8-26 THN SNW

There is patchy thin ice on the runway at Paine, WA. Port Angeles' Runway 8-26 is covered with thin snow. Often the depth of ice, snow, and slush will be indicated: 3/4 SLR (three-quarters of an inch of slush on the runway).

MIT 12-30 RY LGTS PCL CMSND/MED INTST CONT UNTIL 0800 OTRW KEY 122.8 FIVE TIMES WI 5 SEC MED INTST

At Minter Field, Shafter, CA, pilot-controlled lighting (PCL) has been commissioned for Runway 12-30. Lights are on medium intensity continuously until 0800Z, then must be activated by keying 122.8 five times within five seconds for medium intensity.

UKI LOC DME 15 OTS

The Ukiah localizer DME for Runway 15 is out of service. Be careful because the NOTAM is stating that only the DME is out, not the localizer. A Runway 15 localizer instrument approach may still be executed to the appropriate minimums without the DME.

SBA ILS DME UNMON 0700-1400 DLY

The Santa Barbara ILS DME is not monitored (UNMON) between the hours of 0700Z and 1400Z daily. This does not mean the DME is off or out of service, only that if the facility fails, a NOTAM would not be issued immediately; a pilot report would probably be the first indication of an outage. (Most NAVAIDs within the NAS are monitored.)

MQO VOR 290-090 UNUSBL BYD 7 BLO 7000

The Morro Bay VOR is unusable between the 290° and 090° radials beyond seven nm below 7,000 feet MSL. Any airway, approach segment, or fix that is dependent on the Morro Bay VOR signal and is within the specified "unusable" area could not be used for navigation.

DAG VORTAC UNMON/VOICE OTS

The Daggett VORTAC is in an unmonitored status, but not out of service. Most likely the monitor line is down because voice communications through the VOR are also out of service. Weather advisory broadcasts will not be available.

PDT A/C 1330-0630 DLY

Pendelton, OR, approach control services operate daily between 1330Z and 0630Z.

ONP CESA HRS 1600-0300 DLY

The Newport, OR, Class E surface airspace is effective daily between 1600Z and 0300Z daily.

Loran and global positioning system (GPS) NOTAMs are only available from the FSS on request. DUATS also provides loran and GPS NOTAMs. The FAA issues flow control information as NOTAM (D)s; the service is properly called *special traffic-management-program advisory messages*. For example:

JFK 07/033 JFK TMPA SEE ATCSCC MSG EFF 1900-2300

There is a traffic management program advisory in effect for JFK between 1900Z and 2300Z. These NOTAMs alert the pilot that flow control is in effect, without the details of the restriction. Pilots will need to check with an FSS for specific information contained in the ATC Systems Command Center advisory.

Flight service stations normally only retain military operations area (MOA) and military training route (MTR) activity within 100 miles of the facility. Pilots should not expect the status of areas or routes beyond this limit. AFSSs have the capability of retrieving MOA and MTR activity by MOA name and MTR route number; therefore, pilots requesting the status should provide the briefer with the MOA name or MTR route.

NOTAM (L)

Only distributed locally, NOTAM (L)s advertise conditions or hazards that do not meet the criteria for a NOTAM (D). Landing area information that does not restrict or preclude the use of the runway is issued as a NOTAM (L): cracks and soft edges, people and equipment on or adjacent to the runway, nonstandard runway markings, and bird activity.

NOTAM (L)s also pertain to items such as ATIS, TWEB, and UNICOM, or a single frequency outage when there is more than one frequency available. For example, if a tower's ground control frequency were out of service, the information would be issued as a NOTAM (L) because the local control frequency was normal. Table 13-1 describes conditions issued as NOTAM (L)s.

The FAA defines *local dissemination* as the area affected by the aid, service, or hazard being advertised—within the region covered by the issuing FSS and, when necessary, the appropriate towers or centers.

FDC NOTAMs

FDC NOTAMs contain regulatory information: chart amendments, changes to instrument approach procedures, and temporary flight restrictions. FDC NOTAMs are transmitted over the FAA's telecommunications system. Nonautomated FSSs normally only retain those FDC NOTAMs that pertain to their *flight service area*, which is an area within 400 miles of the facility. Automated FSSs will provide FDC NOTAMs for the route, regardless of distance.

During a standard briefing, FDC NOTAMs that are pertinent, on hand, and not yet published are provided. When published in the *Notices to Airmen* publication, they are only available on request. This is extremely important to National Ocean Service (NOS) chart users because NOS charts are not updated as often as other commercially available charts. NOS users should routinely check the *Notices to Airmen* publication.

FDC 9/1769 SMO FI/T SANTA MONICA, SANTA MONICA, CA.
NDB-B ORIG PROC NA.

This FDC was issued in 1989 and it was the 1,769th FDC issued during that year (FDC "9"/"1769"). It pertains to the Santa Monica Airport (SMO). The information contained in the NOTAM is temporary: FI/T (flight information of a temporary nature). The NDB-B original issuance approach procedure is not authorized.

FDC 4/4047 STS FI/T SONOMA COUNTY SANTA ROSA, CA.
ILS RWY 32 AMDT 15 . . . VOR RWY 32 AMDT 18 . . . VOR/DME RWY 14 AMDT 1 . . . CHANGE NOTE TO READ: WHEN CDSA NOT IN EFFECT, EXCEPT OPERATORS WITH APPROVED WEATHER REPORTING SERVICE, USE TRAVIS AFB /SUU/ ALTIMETER SETTING AND INCREASE ALL DH'S AND MDA'S BY 390 FEET.

This FDC changes a note on the three approaches listed to increase the decision heights and minimum descent altitudes by 390 feet when the Class D airspace (CDSA) is not in effect, unless the pilot has access to an approved weather reporting service. When Class D airspace is not in effect (the tower is closed), the altimeter setting is not available. Pilots without approved weather reporting service must use the Travis AFB (SUU) altimeter. The use of a remote altimeter setting requires an increase in minimums. (AWOS solves this limitation.)

THE *NOTICES TO AIRMEN* PUBLICATION

Published every 14 days, the *Notices to Airmen* publication (Fig. 13-2) is divided into three parts. Part 1 has three sections:

- Section 1: Airway FDC NOTAMs
- Section 2: Airport, facility, and procedure NOTAMs
- Section 3: FDC NOTAMs of a general nature (flight restrictions over foreign countries, changes to regulations, and the like)

Part 2 contains revisions to minimum en route IFR altitudes and changeover points, which were formerly published as a separate document. Part 3 contains graphic notices; information in part 3 varies widely, but is included because of its impact on flight safety. FDC NOTAMs for temporary flight restrictions are not contained in the *Notices to Airmen* publication.

Figure 13-3 is an example from part 1, section 2. Most are FDC NOTAMs, such as Santa Ynez where the NDB Runway 8 approach procedure is NA (not authorized). Other entries, such as South Lake Tahoe, advertise a change in VOR name and identification. The vertical bar in the margin indicates information that is new with this issuance (Santa Monica).

A letter to the editor of a national publication requested assistance in translating FDC NOTAM SANTA MONICA VOR-A. DME REQUIRED. The intersections (for the approach) are determined by either the SMO DME or radials off the LAX VOR. This

U.S. Department
of Transportation

**Federal Aviation
Administration**

NOTICES TO AIRMEN

November 10, 1994

Next Issue
November 24, 1994

SPECIAL NOTICE

Beginning with this publication, "PART 95 - Revisions to Minimum En Route IFR Altitudes and Changeover Points" will be included as Part 2. Graphic Notices are now published as Part 3. **Effective February 2, 1995, the Part 95 revisions will cease to be published as a separate document.** All current subscribers of the Part 95 Amendment will automatically receive the Notices to Airmen Publication.

NOTE

Items included in the Notices to Airmen publication are **NOT** *given during pilot briefings unless specifically requested by the Pilot.*

Airspace–Rules and Aeronautical Information Division (ATP–200)

Fig. 13-2. *Pilots are responsible for information contained in the* Notices to Airmen *publication. Information in this document will only be provided on request during FSS weather briefings.*

CALIFORNIA

RIVERSIDE

Riverside Muni

FDC 8/574 /RAL/ FI/T RIVERSIDE MUNI, RIVERSIDE, CA.
VOR-A AMDT 4. VOR RWY 9 AMDT 9. PROCS NA.

SACRAMENTO

Sacramento Executive ATCT

ATCT HRS 0600-2100 LCL. (2/88)

Sacramento Metropolitan

RY 16R ALSF2 OTS INDEFLY.(03/88)

SAN FRANCISCO

San Francisco Intl

FDC 8/796 /SFO/ FI/T SAN FRANCISCO INTL. SAN
FRANCISCO, CA. ILS RWY 19L AMDT 17 GLIDE SLOPE
UNUSABLE BELOW 140 FEET MSL.

SAN LUIS OBISPO

San Luis Obispo EFAS

HIGH ALT EFAS FREQ 135.9 CMSND. MONITORED BY HHR
FSS. (3/88)

San Luis Obispo County ATCT

ATCT HRS 0700-1900 LCL. (3/88)

SANTA ANA

John Wayne Arpt-Orange County

FDC 8/714 /SNA/ FI/T JOHN WAYNE AIRPORT-ORANGE
COUNTY, SANTA ANA, CA. ILS RWY 19R AMDT 11...DYERS
FIX MINIMUMS...LOC BC RWY 1L AMDT 10...NDB RWY 19R
ORIG... CIRCLING MDA 680/HAA 626 ALL CATS. NDB RWY
19R ORIG...CIRCLING CAT C VIS 1 3/4. TMPRY 375 FT MSL
CRANE 3/4 MI NE RWY 19R THR.

SANTA BARBARA

Santa Barbara Muni ILS/DME RY 7

(I-SBA) ILS UNMONITORED WHEN FSS CLSD. (2/88)

Santa Barbara Muni

FDC 7/1079 /SBA/ FI/T SANTA BARBARA MUNI, SANTA
BARBARA CA. ILS RWY 7 AMDT 1...VOR RWY 25 AMDT
4...ALTERNATE MINIMUMS NA 0700Z-1230Z.

SANTA MONICA

Santa Monica NDB

FDC 8/977 CA FI/P CORRECT US GOVT VFR TERMINAL AREA
CHART & AREA CHARTS -US LOS ANGELES CA 16TH
EDITION DATED 10 MAR 88. SANTA MONICA NDB (SQQ) IS
STILL A NON-OPERATIONAL FACILITY. DISREGARD SIGNALS
FROM NAVAID UFN. COORDINATES ARE LAT 34-00-50N
LONG 118-27-20W.

SANTA ROSA

Sonoma County

FDC 7/220 /STS/FI/T SONOMA COUNTY, SANTA ROSA CA.
IFR DEPARTURE PROCEDURE: RWY 1 TURN LEFT, RWY 14
TURN RIGHT, RWY 19 CLIMB STRAIGHT AHEAD, RWY 32
CLIMB STRAIGHT AHEAD TO 1800, THEN CLIMBING LEFT
TURN, DIRECT STS VOR. INTERCEPT AND CLIMB
SOUTHBOUND ON STS R-202 WITHIN 15 MILES, TO

CALIFORNIA

RECROSS STS VOR AT OR ABOVE MEA FOR DIRECTION OF
FLIGHT, OR COMPLY WITH SONOMA COUNTY SIDS.

FDC 7/11 /STS/FI/T SONOMA COUNTY. SANTA ROSA CA. ILS
RWY 32 AMDT 14...VOR RWY 32 AMDT 18. CHANGE NOTE TO
READ: WHEN CONTROL ZONE NOT IN EFFECT. EXCEPT FOR
OPERATORS WITH APPROVED WEATHER REPORTING
SERVICE: 1. USE TRAVIS AFB ALTIMETER SETTING. 2.
INCREASE ALL DH AND MDAS BY 180 FEET.

SANTA YNEZ

Santa Ynez

FDC 6/1617 /IZA/ FI/T SANTA YNEZ SANTA YNEZ, CA. NDB
RWY 8 ORIG PROC NA.

SOUTH LAKE TAHOE

Lake Tahoe VORTAC

(LTA) EFFECTIVE 5 MAY 88 NAV IDENTIFIER TO BE CHANGED
TO SWR & NAME TO BE CHANGED TO SQUAW VALLEY. (1/
88)

STOCKTON

Stockton Metropolitan

FDC 7/497 /SCK/ FI/T STOCKTON METROPOLITAN,
STOCKTON CA. ILS RWY 29R AMDT 18...VOR RWY 29R AMDT
17...NDB RWY 29R AMDT 14. RAISE CIRCLING MDA CATS A/
B/C TO 580. REASON: 275 MSL CRANE 4900 FT NW APCH
END RWY 11L.

COLORADO

AKRON

Akron FSS

FSS DCMSND. (3/88)

DENVER

Jeffco

RY 29R REIL DCMSND.(2/88)

FDC 8/436 /BJC/ FI/T JEFFCO, DENVER, CO. ILS RWY 29R
AMDT 10. TCH 53 GS OM CROSSING HEIGHT 7195.

Stapleton Intl

FDC 8/115 /DEN/ FI/T STAPLETON INTL, DENVER, CO. ILS/
DME 1 RWY 8R AMDT 4...CONVERGING ILS/DME 2 RWY 8R
AMDT 1. MISSED APCH HOLDING ALT 10000 FT. ILS/DME 1
RWY 17L AMDT 5...CONVERGING ILS/DME 2 RWY 17L AMDT
1. TERMINAL ROUTE DEN 15.0 DME TO DEN 13.0 DME ALT
9000 FT. DEN 13.0 DME TO TARGS INT/DEN 5.6 DME/TOT
NDB ALT 7000 FT. MISSED APCH HOLDING ALT 9000 FT.
LOC/DME RWY 18 AMDT 1. TERMINAL ROUTE WENNY INT
TO CHOSE/I-UGT 12.5 DME ALT 9000 FT, CHOSE INT TO
LAKEE/I-UGT 5.5 DME ALT 7000 FT. NDB RWY 26L AMDT
37...NDB RWY 26R AMDT 7. TERMINAL ROUTE IOC VORTAC
TO WATKI INT ALT 10000 FT. MISSED APCH HOLDING ALT
10000 FT. ILS RWY 26L AMDT 45. MISSED APCH HOLDING
ALT 10000 FT. ILS RWY 35L AMDT 27. TERMINAL ROUTE IOC
VORTAC TO SKIPS INT/I-SPO 17.7 DME ALT 10000 FT, SKIPS
INT TO DEBIT INT/I-SPO 13.4 DME ALT 9000 FT., DEBIT INT
TO LENDI OM/INT/I-SPO 7.7 DME ALT 7500 FT. PROC TURN
NA. ILS RWY 35R AMDT 10...ILS RWY 35R CAT II AMDT
10...ILS RWY 35R CAT III AMDT 10. TERMINAL ROUTE IOC
VORTAC TO SEDAL INT/I-RRV 18.5 DME ALT 10000 FT,
SEDAL INT TO ENGLE INT/I-RRV 14.2 DME ALT 9000 FT,
ENGLE INT TO GANDI OM/INT/I-RRV 8.5 DME ALT 7500 FT.
PROC TURN NA. MISSED APCH HOLDING ALT 10000 FT. LDA/
DME RWY 35R AMDT 1. MISSED APCH HOLDING ALT 10000

Fig. 13-3. *Pilots need to become familiar with reading and interpreting information in the* Notices to Airmen *publication. Anytime that doubt exists about an entry, clarification should be obtained from an FSS.*

FDC NOTAM was issued during the time that the LAX VOR was out of service for an extended period; thus, DME was required to make the approach. This emphasizes the point that pilots need to become familiar with reading and interpreting information of this nature.

USING THE NOTAM SYSTEM

Let's say we're planning a flight from Hayward, CA, in the San Francisco Bay Area to Agua Dulce Airpark north of the Los Angeles Basin. As part of the standard briefing, we receive any pertinent NOTAMs that are on hand. We can expect to receive NOTAMs on the status of NAVAIDs, airway changes, and airspace restrictions—NOTAM (D)s and FDC NOTAMs unless there is a temporary NOTAM system outage. In such a case, we will have to check with FSSs en route and at the destination to ensure receipt of current NOTAMs. Information contained in the *Notices to Airmen* publication will only be provided on request.

The *Airport/Facility Directory* shows that the tie-in FSS for Agua Dulce is the Riverside (RAL) FSS. Local NOTAMs will only be available from the destination airport's tie-in FSS; however, any local NOTAMs should not restrict or preclude our landing at Agua Dulce. An option would be to contact Riverside Radio prior to landing to obtain this information.

Bottom line: Request a standard briefing. NOTAMs are not necessarily provided during abbreviated or outlook briefings. If the specialist doesn't mention NOTAMs, ask. Briefers are human and NOTAMs are easy to overlook. Keep in mind that NOTAMs are not normally available on TWEB, TIBS, or PATWAS. A check of the *Airport/Facility Directory* should be a standard part of flight planning; significant information might be published. The option remains to request local NOTAMs from the tie-in FSS prior to descent and landing. Finally, if you don't have access to the *Notices to Airmen* publication, ask the briefer to check.

Pilots using DUATS or other commercially available systems will have to decode and translate NOTAMs. Remember the DUATS disclaimer, "FDC NOTAMs that are not associated with an affected facility identifier will now be presented unless you specifically choose to decline . . .." The notices that "are not associated with an affected facility identifier" include temporary flight restrictions and airway changes. DUATS does not provide NOTAM Ls or information contained in the *Airport/Facility Directory* and the *Notices to Airmen* publication. The contents of these documents will still remain the responsibility of the pilot. If any doubt exists about the meaning or intent of a NOTAM, consult a flight service station for clarification.

Pilots should be aware that prominent events (major sporting events, parades, ceremonies, etc.), disasters (earthquakes, flooding, forest fires, oil spills, train wrecks, etc.), and presidential visits invariably evoke temporary flight restrictions. It remains the pilot's responsibility to obtain these NOTAMs, and a call to the flight service station might be the easiest way.

The procedures discussed in this chapter apply equally to VFR and IFR flights. Only by understanding the system can pilots ensure that they meet their obligation of obtaining all available information.

It's like going to the restroom before a flight: We know we should obtain all available information and visit the restroom, but sometimes it's just not convenient. Both oversights can lead to a very uncomfortable flight!

14
VFR flight planning

IF YOU ASK SEVERAL PILOTS WHY THEY FILE VFR FLIGHT PLANS, YOU'RE guaranteed to hear some interesting answers. "My instructor told me to file a flight plan." "To let my boyfriend know when to pick me up at the airport." "To alert search and rescue in case of an accident." The primary purpose of a VFR flight plan is to alert search and rescue; however, secondary uses include requesting United States Customs for international flights, alerting medical personnel on life-guard missions, requesting special handling from air traffic control, or letting some-one know when you'll be arriving. It's all possible, but a pilot must know how to use the flight plan system for anything to work properly.

Most pilots take VFR flight plans for granted. Unfortunately, certain flight in-structors were never taught the purpose of flight plans or how to use them; therefore, those instructors are unable to properly teach students. The following classic example is all too often repeated.

The student pilot calls the FSS and says, "I would like to file a flight plan." The specialist responds, "Go ahead." There is silence. The specialist advises the pilot to just read off the information on the flight plan form, to which the student responds, "I don't have a form. My instructor told me to call flight service and file a flight plan." FSS specialists will walk pilots through the flight plan, if necessary. One pilot complained about a lengthy phone delay, then told the briefer he did not have a flight plan form; after taking the flight plan, the briefer explained that the initial delay was caused by the previous caller's lack of a flight plan form.

It's an appalling fact that many flight plans submitted to flight service stations contain errors or are in some violation of FARs. Most errors are small or technical and corrected by the FSS specialist. The time required, however, significantly contributes to delays in reaching the FSS. It is to every pilot's advantage to know and understand the FAA's VFR flight plan service.

A few pilots are in for a rude awakening with DUATS. DUATS flight-plan filing has not been available for DVFR, border crossings, or international trips. Before accepting the flight plan, the computer checks for errors and omissions, with the slightest fault resulting in a reject.

FLIGHT PLAN INFORMATION

OK, let's say we've decided to file VFR, obtain a copy of FAA Form 7233-1 (8-82) illustrated in Fig. 14-1. (The reverse side of the form is for military flight plan use.) Many commercially available flight planning organizers provide a flight plan form. The preflight planner in Fig. 14-2 is the other side of the weather log used in previous chapters and contains a flight plan form. This facilitates the transfer of information from the navigation log to the flight plan.

Check VFR, then enter the full aircraft identification, including the N, followed by the aircraft type and special equipment code. Pilots pay a lot of money for their airplanes and are understandably proud. One pilot filed an aircraft type as "a P-A Thirty-Two T slant T slant alpha." Upon inquiry it seemed he had a T-tailed, turbocharged Lance (PA32). Selecting correct type designators is required with DUATS. Appendix A contains a listing of type designators for general aviation aircraft. Pilots should become familiar with the designators for the aircraft they routinely fly.

Special equipment codes may be included on VFR flight plans. Below is a list of the codes.

- X No transponder; no DME
- T Transponder; no altitude encoding; no DME
- U Transponder with altitude encoding; no DME
- D DME only
- B Transponder; no altitude encoding; no DME
- A Transponder with altitude encoding; with DME
- W RNAV only (area navigation including loran, GPS, etc.)
- C Transponder; no altitude encoding; with RNAV
- R Transponder with altitude encoding; with RNAV
- G Flight management systems; electronic flight information systems

An added advantage of today's air traffic control computerized radar system is helping locate missing aircraft and saving lives. This technique involves the use of

US DEPARTMENT OF TRANSPORTATION FEDERAL AVIATION ADMINISTRATION **FLIGHT PLAN**	(FAA USE ONLY) ☐ PILOT BRIEFING ☐ VNR ☐ STOPOVER				TIME STARTED	SPECIALIST INITIALS

1 TYPE	2 AIRCRAFT IDENTIFICATION	3 AIRCRAFT TYPE SPECIAL EQUIPMENT	4 TRUE AIRSPEED	5 DEPARTURE POINT	6 DEPARTURE TIME		7 CRUISING ALTITUDE
VFR IFR DVFR			KTS		PROPOSED (Z)	ACTUAL (Z)	

8 ROUTE OF FLIGHT

9 DESTINATION (Name of airport and city)	10 EST TIME ENROUTE		11 REMARKS
	HOURS	MINUTES	

12 FUEL ON BOARD		13 ALTERNATE AIRPORT (S)	14 PILOTS NAME ADDRESS & TELEPHONE NUMBER & AIRCRAFT HOME BASE	15 NUMBER ABOARD
HOURS	MINUTES			
			17 DESTINATION CONTACT/TELEPHONE (OPTIONAL)	

17 COLOR OF AIRCRAFT	CIVIL AIRCRAFT PILOTS FAR Part 91 requires you file an IFR flight plan to operate under instrument flight rules in controlled airspace. Failure to file could result in a civil penalty to exceed $1,000 for each violation (Section 901 of the Federal Aviation Act of 1958 as amended.) Filing a VFR flight plan a recommended as a good operating practice. See also Part 99 for requirements concerning DVFR flight plans.

FAA Form 7233-1 (8-82)

CLOSE VFR FLIGHT PLAN WITH _____ FSS ON ARRIVAL

Fig. 14-1. *Every pilot should keep handy a copy of FAA Form 7233-1 Flight Plan. Forms are available at all FSSs and other locations and contained on many commercially available flight planning forms.*

recorded radar data that can be played back to determine the point at which an aircraft disappeared. The data retrieved enables rescuers to all but pinpoint a crash site and pick up the surviving pilot. (The *Flight Services* handbook no longer requires specialists to obtain special equipment codes for VFR flights, but it doesn't hurt to provide one anyway.)

Continuing across the form, enter true airspeed in knots. Next is departure point and proposed departure time. The proposed departure time should be UTC (Zulu). Forms such as the preflight planner contain time-conversion tables, but the FSS specialist will help if you aren't sure how to convert local to Z. It's better to provide local time and let the FSS specialist do the conversion than to give an incorrect UTC time. And please don't use the 24-clock format for local time; state "a.m." or "p.m." as appropriate. The proposed initial cruising altitude is next; because it's the "initial" altitude, if you select a different altitude while en route, an amendment to the flight plan

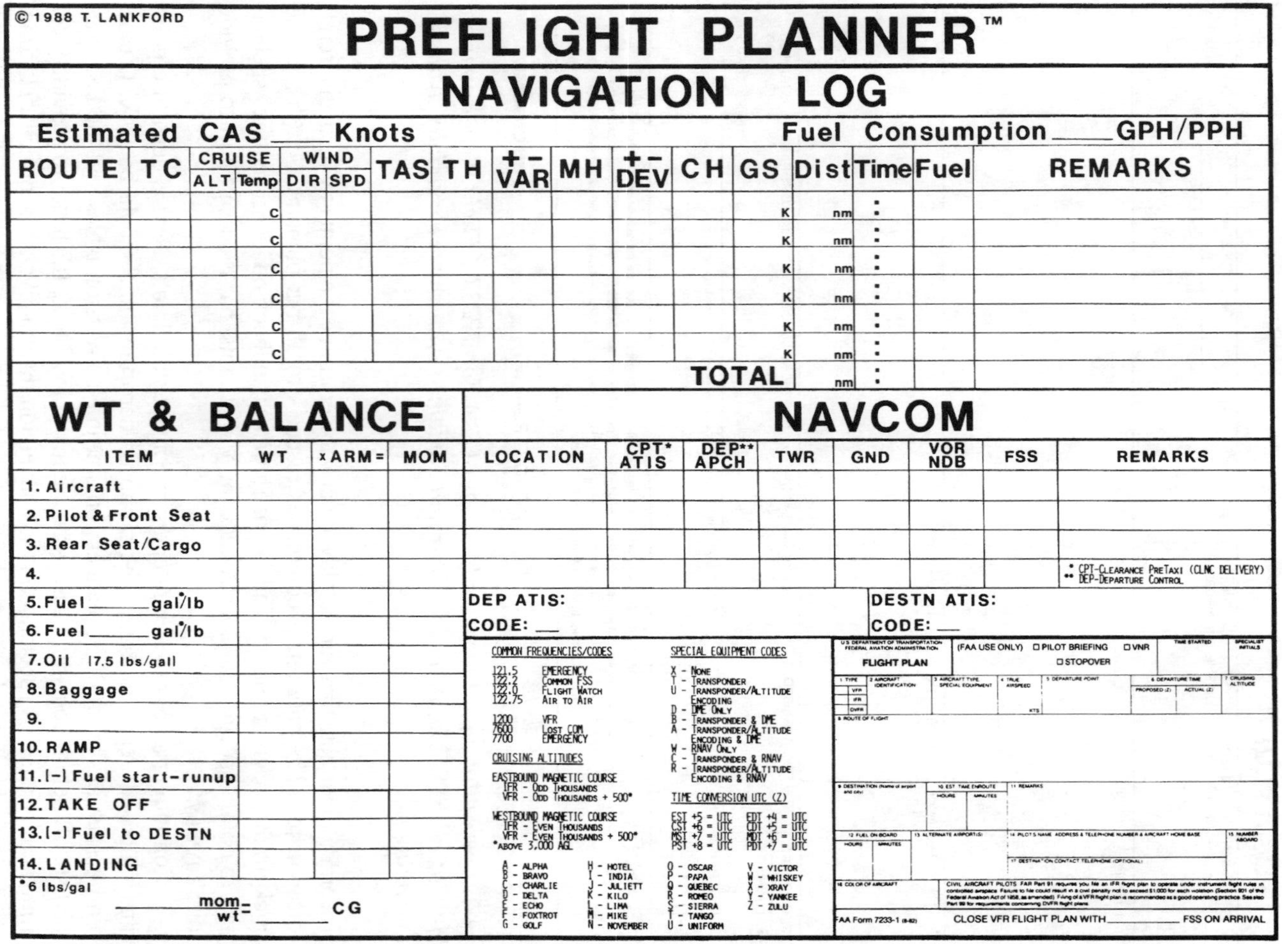

© 1988 T. LANKFORD

PREFLIGHT PLANNER™

NAVIGATION LOG

Estimated CAS ____ Knots — Fuel Consumption ____ GPH/PPH

ROUTE	TC	CRUISE ALT	CRUISE Temp	WIND DIR	WIND SPD	TAS	TH	+− VAR	MH	+− DEV	CH	GS	Dist	Time	Fuel	REMARKS
			C									K	nm	:		
			C									K	nm	:		
			C									K	nm	:		
			C									K	nm	:		
			C									K	nm	:		
			C									K	nm	:		
												TOTAL	nm	:		

WT & BALANCE

ITEM	WT	x ARM =	MOM
1. Aircraft			
2. Pilot & Front Seat			
3. Rear Seat/Cargo			
4.			
5. Fuel ____ gal*/lb			
6. Fuel ____ gal*/lb			
7. Oil (7.5 lbs/gal)			
8. Baggage			
9.			
10. RAMP			
11. (−) Fuel start-runup			
12. TAKE OFF			
13. (−) Fuel to DESTN			
14. LANDING			

*6 lbs/gal

mom/wt = ____ CG

NAVCOM

LOCATION	CPT* ATIS	DEP** APCH	TWR	GND	VOR NDB	FSS	REMARKS
							* CPT-CLEARANCE PRETAXI (CLNC DELIVERY) ** DEP-DEPARTURE CONTROL

DEP ATIS:
CODE: ___

DESTN ATIS:
CODE: ___

COMMON FREQUENCIES/CODES

121.5 EMERGENCY
122.2 COMMON FSS
122.0 FLIGHT WATCH
122.75 AIR TO AIR

1200 VFR
7600 LOST COM
7700 EMERGENCY

CRUISING ALTITUDES

EASTBOUND MAGNETIC COURSE
IFR - ODD THOUSANDS
VFR - ODD THOUSANDS + 500*

WESTBOUND MAGNETIC COURSE
IFR - EVEN THOUSANDS
VFR - EVEN THOUSANDS + 500*
*ABOVE 3,000 AGL

SPECIAL EQUIPMENT CODES

X - NONE
T - TRANSPONDER
U - TRANSPONDER/ALTITUDE ENCODING
D - DME ONLY
B - TRANSPONDER & DME
A - TRANSPONDER/ALTITUDE ENCODING & DME
W - RNAV ONLY
C - TRANSPONDER & RNAV
R - TRANSPONDER/ALTITUDE ENCODING & RNAV

TIME CONVERSION UTC (Z)

EST +5 = UTC EDT +4 = UTC
CST +6 = UTC CDT +5 = UTC
MST +7 = UTC MDT +6 = UTC
PST +8 = UTC PDT +7 = UTC

A - ALPHA, B - BRAVO, C - CHARLIE, D - DELTA, E - ECHO, F - FOXTROT, G - GOLF
H - HOTEL, I - INDIA, J - JULIETT, K - KILO, L - LIMA, M - MIKE, N - NOVEMBER
O - OSCAR, P - PAPA, Q - QUEBEC, R - ROMEO, S - SIERRA, T - TANGO, U - UNIFORM
V - VICTOR, W - WHISKEY, X - XRAY, Y - YANKEE, Z - ZULU

FLIGHT PLAN (FAA USE ONLY) □ PILOT BRIEFING □ VNR □ STOPOVER

FAA Form 7233-1

CLOSE VFR FLIGHT PLAN WITH ____ FSS ON ARRIVAL

Fig. 14-2. *The preflight planner organizes a pilot's preflight calculations and contains time conversions, special equipment codes, the international phonetic alphabet, and a copy of the flight plan form. The form lets the pilot transfer preflight data to the flight plan for FSS or DUATS filing.*

is not required. It's perfectly acceptable to enter "VFR" to indicate that the trip will be flown at various VFR altitudes.

The proposed route of flight and destination should be as accurate as possible referring to radio navigation aids, airways, airports, towns, or any prominent geographical landmarks. Remember, if search and rescue becomes necessary, this route will be searched first. The destination city and airport name must be complete. Many towns and cities have more than one airport; therefore, the specific airport name is required.

Pilots using DUATS will have to specify appropriate location identifiers (LCIDs). These can be obtained from charts, the *Airport/Facility Directory*, and through the DUATS location encode-decode function. In addition to providing identifiers, DUATS includes weather report types (SA, FT, etc.), tie-in FSS and center, and latitude-longitude for those with coordinate-capable navigation equipment. Pilots filing with an FSS are not required to provide LCIDs; specialists are normally familiar with LCIDs within their flight service area; however, it's often helpful if the pilot can provide the identifiers for distant locations. A word of caution: If you provide an LCID, make sure it's correct. I had a DC-8 pilot file for MKC (Kansas City Downtown Airport). I asked if he meant the Kansas City Mid-Continent International Airport (MCI). He did.

The estimated time en route (ETE) in hours and minutes seems straightforward but under certain circumstances can cause considerable confusion. A number of considerations in addition to forecast winds aloft must be included when determining ETE: landings en route, a round-robin flight plan, ground time, planned delays, and the like. The omission of ground time has all too often led to the unnecessary initiation of search and rescue. Additional time is required when a flight plan will be closed after arrival; consider the time necessary to taxi to a parking place, tie down the aircraft, and find a telephone. DUATS requires a four-digit ETE, representing hours and minutes. Be careful because 0010 represents 10 minutes, 0100 is one hour.

The remarks section of a VFR flight plan is used to indicate intended landings en route, United States Customs notification, or a pilot's intention to close the flight plan with other than the destination's tie-in FSS. Remarks can be used to indicate radio trouble, for instance "no transmitter" or "no radio" or to identify the source of a weather briefing (NWS PB, TWEB, TIBS, PATWAS, DUATS, etc.). Items of a personal nature, such as "Please call grandma," are not accepted. "Student pilot" or "training flight" are unnecessary and serve no purpose.

The amount of fuel on board is entered in hours and minutes. Pilots have been known to enter 25 gallons or full, which is worse than the gallon amount. In one instance, a student pilot called to file a flight plan, and when it came to fuel on board, he stated "Full." The specialist, who was also a flight instructor, asked for fuel in hours and minutes. The student had no idea. Because the specialist was familiar with the aircraft, he was able to assist the pilot in determining the aircraft's endurance. It is unconscionable that an instructor would send a student on a cross-country flight so poorly prepared. If search and rescue becomes necessary, this information would be used to conduct an extended search. A search for the "full" aircraft would have to start at the airplane's maximum range and backtrack; accurately stating hours and minutes will help searchers plot a search area that saves valuable time, which a downed pilot does not have to spare. Be careful with a DUATS-filed flight plan. A pilot who filed 0010 en route also filed 0040 for 40 minutes of fuel instead of the intended entry of 0400, which would have accurately represented the 4 hours of fuel on board.

An alternate airport is not required on VFR flight plans; however, if the destination is below VFR minimums, but forecast to improve, a pilot might want to indicate an alternate. If the destination is at the limit of fuel range, a safe alternate might be appropriate. An alternate might be helpful if the pilot changes destination and fails to advise the FSS or close the flight plan.

According to portion of FAR 91 regarding the VFR flight plan, information required specifies "The full name and address of the pilot in command," unless otherwise authorized by ATC. The *Flight Services* handbook, however, instructs specialists only to obtain the "pilot's name, telephone number, aircraft's home base, ..." The FAA in effect authorizes this deviation from the FARs. Although not required by FARs, the telephone number—especially a destination contact—is often helpful if the pilot fails to close the flight plan.

The total number aboard, including the pilot and crew, is entered—no pets, please—and finally state the predominant, generic colors of the aircraft.

FLIGHT PLAN FILING

Flight plans can be filed in person, over the telephone, by radio, and through DUATS and other private services. File on the radio only when there is no other alternative because this practice often unnecessarily congests already crowded frequencies.

Due to FSS consolidation, the opportunity to file in person is becoming rare. Flight plan information is typed directly into the computer at AFSS locations. A "mask" of the flight plan form is displayed on the computer monitor, and the specialist tabs to each "cell" on the form to enter appropriate information.

The telephone is the most common method for filing. Most areas are served by local or toll-free numbers. Many of the larger FSSs have a recorded flight plan filing system known as *Fast-File*. Telephone numbers and facilities where Fast-File is available can be found in the *Airport/Facility Directory*. The name Fast-File might seem misleading because the service is actually nothing more than a telephone answering machine.

Some pilots think Fast-File means to speak fast and that will result in instant processing; however, depending on the location and specialist priorities, it might take up to an hour for a flight plan to be copied and processed. It is extremely important to speak slowly, distinctly, and directly into the telephone. Fast-File for the pilot? Yes, but "slow retrieval" at the FSS is possible. Failure to provide complete and accurate information will only lead to confusion and delay.

Whichever method a pilot selects, numerous tasks that are completed prior to calling will expedite and simplify the process. Having the form filled out and ready is paramount. Merely read the information starting with block 1 and continuing through block 16. It's not necessary to say "Block 1 V-F-R, Block 2 . . .," and the like. Just remember the specialist has to write or type the information. (This is where devilish pilots get even for those lickety-split weather briefings, but you are not a devilish pilot.) Please spell out personal names, and do it slowly enough to be accurately heard.

Flight plans filed through DUATS are checked by a computer for completeness and accuracy. Correct aircraft types and LCIDs are required. Pilots may file, amend, or cancel the flight plan up to 1 hour before proposed departure time. Amendments or cancellations have to be made through a compatible computer terminal. One hour before proposed departure time, a flight plan proposal is transmitted to the departure airport's tie-in FSS. The proposal will not contain the entire flight plan, only information necessary to transmit a flight notification message upon activation. Incorrect aircraft identification or location identifiers result in lost flight plans. For example, if the pilot filed a departure point of BFI (Boeing Field, Seattle) instead of BFL (Bakersfield, CA), which are both perfectly good LCIDs, the flight plan would be sent to Seattle. Bakersfield would have no information on the flight plan nor any way of obtaining the information.

During the accident investigation, no record could be found of the pilot receiving a weather briefing. Several weather advisories were in effect. Because of this single incident, when a pilot files a flight plan with an FSS, the specialist is required to ask if the pilot has received any adverse-condition reports for the route.

FSS FLIGHT PLAN PROCEDURES

The FSS will hold the flight plan in abeyance until a departure report is received or until about one hour after the proposed departure time, when the flight plan is canceled and archived in the facility history file; therefore, if a flight will be delayed by more than an hour, the pilot should revise the departure time with the FSS.

The best way to open or activate a flight plan is by radio. A specific request is required. The FSS cannot assume that a radio call or even a departure time is a request to open a flight plan. It's always a good idea to provide the point of departure and destination. The specialist will provide any pertinent updates and ensure appropriate weather advisories have been received. For nonradio aircraft and in areas with poor or no radio coverage, a pilot can request an "assumed departure," and the FSS will activate the flight plan based on the pilot's proposed departure time; however, if the pilot is delayed or decides to cancel, the FSS must be advised. This procedure should only be used when direct radio contact is not possible.

Occasionally a pilot calls to open the flight plan and is told the FSS cannot find the flight plan. After a number of questions, the FSS determines that the pilot filed a different aircraft identification. If a pilot changes aircraft, he or she must advise the FSS about the swap before departure or upon activation. A swap can occur when the "filed" plane experiences mechanical difficulties or rental planes are shuffled by an FBO. Pilots must remember to provide any other changes, such as aircraft type or color. Pilots using DUATS who depart before the FSS has received the flight plan proposal will be required to refile.

With FSS consolidation and toll-free phone numbers, filing with one FSS and opening with another has become routine; however, it takes time to forward the flight plan proposal. The FSS receiving the flight plan forwards the proposal to the departure FSS. This is done automatically and almost instantaneous at AFSSs; it might take 30 minutes or more for a nonautomated facility to forward a flight plan proposal.

Upon activation, the departure FSS becomes responsible for search and rescue. That FSS transmits a flight notification message consisting of the aircraft identification, type, destination, ETA, and any pertinent remarks to the destination tie-in FSS. When the destination FSS acknowledges receipt of the flight notification message, it becomes responsible for search and rescue. The message is "suspended," either written on a flight progress strip or stored electronically in the "inbound" file of the computer, until the flight plan is closed or search and rescue is initiated.

A portion of FAR 91 requires that "When a flight plan has been activated, the pilot in command, upon canceling or completing the flight under the flight plan, shall notify an FAA flight service station or ATC facility." It's best to close with the destination's tie-in FSS. If a pilot intends to close the flight plan with other than the destination's FSS, that intention should be indicated in the remarks of the flight plan, although a flight plan can be closed with any FSS. Towers and centers are also required to accept VFR flight plan cancellations; however, this should be avoided because ATC is often busy with higher priority duties, and closures get lost.

Certain pilots prefer to close their flight plans in the air by radio, while others would rather wait until they are on the ground. It's often convenient to close in the air before switching to approach or the tower, or before descending below a flight service station's radio reception range. Sometimes pilots can become so involved with ATC and the completion of the flight that the flight plan is completely forgotten.

Two axioms apply to closing flight plans:

- There are many excuses but few reasons.
- There are those who have forgotten and those who will forget.

SEARCH AND RESCUE

An aircraft is considered overdue when it cannot be located and the pilot has not canceled or revised the flight plan within 30 minutes after the ETA. The FSS will attempt to locate an overdue aircraft by checking the destination and adjacent airports. Please note: The FAA looks for aircraft, not people. A request for complete flight plan information is sent to the departure FSS or the FSS holding the flight plan (if filed with other than the departure station) or the DUATS contractor.

When the flight plan has not been closed or the aircraft has not been located within one hour after the ETA, an *information request* (INREQ) is transmitted. This message is addressed to all FSSs, flight watch control stations, and ARTCCs along the flight-planned route, as well as the Rescue Coordination Center (RCC) at Langley AFB, Virginia.

When replies to the INREQ are negative and the aircraft has not been located within 2 hours after the ETA, an *alert notice* (ALNOT) is issued. An ALNOT requires an extended communications search of all airports within 50 miles of the flight-planned route from the last known position to the destination. This is a primary reason for frequent position reports, which reduce the search area, if a search becomes necessary.

RCC is kept informed and is responsible for the physical search for the aircraft. These procedures are time-consuming and expensive. They often involve dozens of FAA, law enforcement, and military agencies. Pilots must make every effort to close flight plans or revise ETAs when the flight will be late by 30 minutes or more.

It's unfortunate that FSSs must take action on about 10 percent of aircraft on a VFR flight plan. If a pilot fails to close, the FSS can file an incident report with the appropriate flight standards district office, which usually leads to the counseling of the pilot; however, depending on the circumstances or if there have been prior incidents, stronger measures such as a letter of warning or certificate suspension can result. Most often, the FSS locates the aircraft, and the pilot never hears a word. This regrettably leads to pilot complacency when it comes to closing VFR flight plans.

It's the pilot's prerogative to take advantage of the FAA's VFR flight plan service, but the pilot must adhere to certain rules and procedures. Procedures described here will take the maximum advantage of and the increased safety afforded by the VFR flight plan.

USING VFR FLIGHT PLAN SERVICE

My first FSS facility was in Lovelock, NV. My wife remained in the Los Angeles area, and for about six months I commuted weekly in my Cessna 150. I would normally file VFR from Lovelock to Van Nuys with a landing at Bishop, CA. In those days, the only way to contact flight service at Bishop was a long distance phone call or by radio; there were no procedures for separate VFR flight plans. (If the flight were today, I would have to file two separate VFR flight plans at the departure point: one from Lovelock to Bishop and another from Bishop to Van Nuys. This procedure would be a distinct advantage if search and rescue became necessary.)

The purpose of my VFR flight plan was to provide search and rescue protection over the sparsely populated sections of California and Nevada. Upon landing at Bishop, I would contact Tonopah Radio and advise them that I had landed. This in effect was a position report. I would also advise them of my departure. If search and rescue became necessary, this would have narrowed the search area to the second half of the route. I made additional position reports at approximately one-hour intervals.

My wife would contact Los Angeles FSS, request the flight data position at the station, and ask the person working that position for an ETA for my aircraft N number. Remember the FAA only has aircraft numbers, not names. If you want someone to be able to check on your flight, you must file and open a VFR flight plan; give the aircraft number to the person calling flight service to check on the progress of your flight. (If the Van Nuys area was IFR . . ., please refer to chapter 15 regarding IFR flight plans and flight planning.)

The decision to use the FAA's VFR flight plan service rests solely with the pilot. It is my opinion that a VFR flight plan for short legs (less than an hour) probably is not necessary, unless the filing is for student training or the route is over water, desolate, or unpopulated areas. But it's mighty comforting on long flights, especially over sparsely inhabited regions of the country.

Normally the FSS will not accept flight plans with en route delays of more than an hour. Pilots have been known to file flight plans of 6 to 10 hours en route with as many as half a dozen stops. This procedure defeats the purpose of a VFR flight plan because search and rescue would not start until 30 minutes after the ETA at the final destination. If search and rescue becomes necessary, the search area might extend for more than 1,000 miles, which decreases the odds of a timely rescue. Additionally, pilots on these flight plans often lose track of time, which causes the unnecessary initiation of search procedures. A far better idea is to file individual flight plans for each leg. This can be done with the original departure FSS.

Under certain circumstances, a *round-robin* flight plan, which accounts for a planned landing en route and a return to the original departure point, is the most practical procedure; however, a round-robin flight plan should only be used when filing separate legs is not practical and ground time will not exceed an hour. Make sure that all ground time is included in the ETA. Certain flight instructors have students file round-robin flight plans for all cross-country flights. Unfortunately, this denies the student the practice of filing, opening, and closing flight plans, which should be an essential part of training. Unexpected delays and losing track of time often occur.

Additionally, the closing and filing of separate flight plans assures that the student will obtain the latest weather information. One flight instructor called the FSS and inquired if his student had left Santa Barbara. (The student had filed a round-robin flight plan from the LA Basin to Porterville to Santa Barbara and back.) The FSS informed the instructor of the pilot's ETA, which had not yet expired. The instructor asked again if the pilot had left Santa Barbara. He was informed that because the student had filed a round-robin flight plan this could not be determined and that the FSS would not take any action because the aircraft was not overdue. If the instructor had taught the student to file separate legs, the FSS could have determined the status of the flight.

Revise an ETA with any FSS. Transmit the original point of departure, destination, and the revised estimated time of arrival. Model 1 equipment is unable to automatically process extensions; therefore, FSSs are required to obtain revised ETAs only. If a pilot wishes to revise a flight plan, he or she will have to provide the revision in the form of a revised ETA.

A destination change while en route can be accomplished without submitting a complete flight plan. The pilot will have to provide the FSS with aircraft identification, type, original departure point, original destination, new destination, new route, and new ETA. The FSS will notify the original destination and the new destination and provide search and rescue coverage until the new destination acknowledges receipt of the changes.

SPECIAL VFR

Basic VFR weather minimums within controlled airspace are designed to allow pilots to fly visually. This requires a visual horizon or contact with the ground and enough visibility to see and avoid terrain, obstructions, and other aircraft. Basic VFR weather minimums were developed in the 1930s when aircraft speeds averaged between 50 and 150 knots; the minimums do not take into account the increased speeds of aircraft in the 1990s. And properly realize that the FAR requirements are minimums; "minimum" does not necessarily equate to "safe."

Pilots have been known to become disoriented and lose control of the aircraft in conditions with visibilities as far as 5 miles. The pilot in command is still responsible to determine if the flight can be safely conducted based on his or her experience and capabilities. I recommend that low-time pilots obtain information from a competent instructor in operations under special VFR and ideally log some dual instruction in actual special VFR weather conditions.

Aircraft can be safely flown visually in less than basic VFR conditions required for controlled airspace. Special VFR allows pilots to operate in this environment, which is a visual horizon or visual contact with the ground and enough visibility to avoid terrain and obstructions. Under such conditions, air traffic control must ensure separation from other aircraft. Although ATC provides separation from other aircraft, it is still the pilot's responsibility to maintain terrain and obstruction clearance.

These operations must be conducted in accordance with the special VFR weather minimums in FAR 91. Special VFR applies only to operations within low-density surface-based Class B, C, D, and E airspace. Special VFR is prohibited within high-density Class B airspace, which is indicated on aeronautical charts by the phrase "NO SVFR" in the airport data block. Weather minimums for an airplane operating under special VFR are clear of clouds and 1 statute mile visibility. To operate at night under the provisions of special VFR, the pilot and airplane must be equipped and certified for IFR.

Prior to departure or before entering less-than-basic VFR weather conditions, the pilot must obtain a clearance from the ATC facility with jurisdiction over the airspace. IFR operations have priority over special VFR.

In one instance, a rather shaky voice called Van Nuys and requested taxi instructions. The visibility was less than basic VFR and the controller asked what type of clearance the pilot would like. The pilot replied, "I don't have one of those," presumably referring to an instrument rating. The controller responded, "That's OK, I've got plenty."

Special VFR must be *requested by the pilot*. ATC is not allowed to suggest the procedure.

USING SPECIAL VFR

It might seem that special VFR is of little practical use; however, this is not the case because special VFR has a specific and practical application. Special VFR is intended to allow a pilot to depart or enter surface-based controlled airspace when conditions are less than basic VFR but safe enough for contact flying.

This often occurs in metropolitan areas where surface visibilities are reduced to less than 3 miles but remain above 1 mile in haze, smoke, and fog. Visibilities usually improve significantly at a few thousand feet AGL. This procedure can also be used to allow a pilot to depart controlled airspace for uncontrolled airspace, which has less stringent VFR requirements.

Pilots must be careful because a special VFR can be a clearance to nowhere. Notice that the provisions of special VFR do not relieve the pilot from maintaining FAR 91's requirements for minimum safe altitudes over certain areas. For example, Fresno was reporting ceiling 400 overcast and visibility 1 mile; a pilot departed special VFR and flew 15 miles before tangling with some high tension wires.

Another example: Departing Long Beach, CA, one morning with a surface visibility of 2½ miles, I requested and received a special VFR clearance. I was cleared out of the Class D airspace to the north; "Climb and maintain VFR conditions. Report VFR on top or leaving Class D airspace." As soon the airplane topped the haze, flight visibility was almost unlimited. I called ATC and reported on top, and I was cleared to leave the frequency.

Pilots must report on top or leaving the surface-based controlled airspace. ATC is providing separation, so when the pilot fails to report, the airspace must be sterilized until the aircraft is located. This causes extensive and unnecessary delays for all aircraft in the area.

Arrivals are conducted the same way. Over the airport or other prominent landmark, while maintaining VFR conditions that are better than the visibility restriction, the pilot contacts the control facility (tower, approach control, or center) and requests a special VFR clearance.

Special VFR might be more efficient than an IFR approach. In one instance, Santa Barbara was reporting 500 broken, 10 miles visibility. Even though the ceiling was less than VFR, I could see the runway. I requested special VFR and made a straight-in approach rather than a 15-mile round-trip, which would have been required to execute the ILS approach.

On another occasion, the destination was Crescent City, CA, and the VOR was out of service, eliminating an IFR approach. The weather was ceiling 500 feet, 1 mile visibility. I found a hole by the coastline and requested a special VFR clearance from the Crescent City FSS. Because of the low ceiling and visibility, I slowed the Mooney to approach speed, which was about 100 knots. There was no sane reason to be blasting along in these conditions at 160 knots. I knew the tops were at 1,500; if the ceiling or visibility dropped, I could have climbed through the clouds to on top. ATC was providing separation, so there shouldn't have been any other aircraft in that airspace. Ceiling and visibility remained unchanged and a routine landing was accomplished.

I was flying from Ontario to my home airport Whitman Airpark in the Los Angeles Basin one rainy afternoon. I had obtained clearance through the Burbank Class C airspace when the controller advised visibility was 2½ miles in rain and fog and asked my intentions. I requested and received a special VFR clearance out of Class C airspace to the northwest. I reported leaving Class C airspace and landed at my destination, which was in Class G airspace.

Delays, extensive at times, should be anticipated. A pilot can never count on making it in special VFR; an absolute VFR alternate should always be within reach.

FLIGHT ASSISTANCE SERVICE

FAA air traffic controllers and flight service specialists are trained in various techniques to locate disoriented pilots and assist with navigational or mechanical difficulties; however, none of this training or resources are of use to a pilot unwilling to take advantage of this service. All too often pilots get hung up on semantics. What constitutes an emergency? When is a pilot lost?

As a brand-new private pilot, I was flying from Leeds to Oxford in England. Soon, nothing on the chart looked like anything on the ground. I was not lost. I was in the center of England. I was disoriented. Before things got out of hand, I requested assistance and reoriented myself.

Most pilots think an emergency is an engine failure or an in-flight fire. An emergency is simply a situation involving distress, the need to resolve uncertainty, or a means of alerting those who are in a position to help. Distress or uncertainty can result from mechanical failure, pilot incapacitation, insufficient fuel, penetration of IFR conditions by a VFR pilot, or a pilot unable to locate his or her position.

The FAA says an emergency exists when an emergency is declared by the pilot, by FAA facility personnel, or by officials responsible for the operation of aircraft.

FAA facilities use all available resources to assist aircraft in difficulty: radars, direction finders (DF), NAVAIDs (VORs and NDBs), and landmarks. When an aircraft is "located," the pilot determines the best course of action. This might be nothing more than continuing to destination, or the execution of an emergency DF or radar approach if the pilot is on top of a cloud deck and unable to proceed to a VFR airport.

Thorough flight planning, understanding weather reports and forecasts, and frequent weather updates should alleviate most problems. Mechanical and electronic systems fail and, being human, we can find ourselves in need of assistance. When a pilot encounters an uncertain situation, the five Cs apply:

- Confess
- Communicate
- Climb
- Comply
- Conserve

Confess and communicate the difficulty; certain pilots compound problems by waiting until they have only a few minutes of fuel. A pilot who requests assistance before the situation becomes critical has many more options. Anytime the outcome of a flight situation becomes uncertain, whether due to unknown position, marginal fuel or weather, navigational issue, or mechanical problem, contact an FAA facility for assistance. Resolve the situation before a simple flight assist becomes an accident.

Climb if possible to improve communications, radar, and DF capability. Comply with instructions. One pilot requested assistance and was instructed to squawk 7700. The pilot replied, "I don't think it is that much of an emergency." After some coaxing, code 7700 was selected. Within 60 seconds, the FSS was able to relay the flier's exact position, which had been determined by a local radar facility. This pilot proceeded to the planned destination without further difficulties.

Don't blindly follow instructions that will take you through clouds. Keep in mind that you are still the pilot in command. If penetrating clouds is unavoidable, inform ATC immediately.

Conserve the remaining fuel, if possible, by reducing to a maximum-endurance power setting. Pilots should be familiar with this setting for each aircraft that they fly.

Most parts of the country are covered by some kind of air-to-ground communications, whether military or FAA. Pilots already in contact with a facility when a problem arises should use that frequency. The facility might change the frequency if it feels that better handling will result; otherwise, the international VHF emergency frequency 121.5 MHz can be used. Some pilots believe that the only time this frequency should be used is when the aircraft is going down in flames. That's not true. Frequency 121.5 is simply a clear channel set aside for aircraft in difficulty. It can and should be used anytime a pilot encounters a situation of uncertainty or distress.

A major deterrent to requesting assistance is the pilot's ego. The perceived embarrassment and possible repercussions of a request have in many instances turned a bad situation into an impossible one. For example, a pilot attempting to locate an uncontrolled field in reduced visibility became critically low on fuel and landed in an open field. He never requested assistance during his hour-and-a-half futile search for the airport. Several days later under similar conditions, another pilot called the local FSS and advised them of his situation. Using VOR and DF positioning, the FSS led the pilot and his family to another airport in the area and obtained a special VFR clearance for the pilot. He landed safely. Both pilots were invited to have a little chat with the flight standards district office about the circumstances. One pilot also talked to his insurance company.

FAR 91 specifies the responsibility and authority of the pilot in command. Under this rule, the pilot is responsible and the final authority for the operation of the aircraft. The pilot in an emergency may deviate from any rule to meet that emergency. The pilot might be required to send a written report of that deviation.

Pilots who request flight assistance usually don't deviate from FARs; however, their actions might be reviewed by the FAA. With few exceptions, this results in counseling from the accident prevention specialist, unless there is evidence of excessive pilot deficiencies. Flight assistance service cannot be continually used by a few pilots who refuse to obtain or maintain basic proficiency.

When a student pilot requires assistance, the FAA wants to talk to the flight instructor. I know because I've been in that position. We ironed out the problems, I became a better instructor and, in my favor, my students knew to contact the FSS for assistance.

Students should be thoroughly educated about DF steers and practice the technique before they begin solo cross-country flights. The worst time to receive a first DF steer is during an actual flight assist; a training flight with the instructor onboard is the ideal time to perfect the technique. Facilities are usually quite willing to provide practice DF steers and approaches where commissioned.

Then there was the pilot who called and requested a practice DF steer. When told the facility was too busy, the pilot said, "Well, how about the real thing?"

As stated in the *Airman's Information Manual*, "When you are in doubt of your position or feel apprehensive for your safety, do not hesitate to request assistance. The FAA's air traffic service facilities are ready and willing to help."

PAN PAN PAN

It was the second day of the trip back from the Oshkosh fly-in. I had departed Gillette, WY, for Pocatello, ID, in a Bonanza, and I was planning to use VOR and pilotage navigation through Jackson Hole and Idaho Falls. The weather was not a factor, but visibility was restricted by the smoke of forest fires. In this part of the country, even at 12,500 feet, I was still below the mountain peaks. After passing what I identified as the Grand Tetons, I turned southwest toward Pocatello.

You guessed it. I couldn't get any VOR or establish communications with any facility. Continuing to fly down what I thought was the Snake River Valley, things didn't seem quite right. Following an old aviation axiom, I followed a river and road that normally should bring a pilot to a town and hopefully an airport.

Even with two and a half hours of fuel, I decided it was time to resolve the issue of position. Because I was unable to establish communications on standard frequencies, I selected 121.5. I was not in distress, but there was a sense of urgency; therefore, as outlined in the *Airman's Information Manual*, I broadcast "PAN PAN PAN" followed by the aircraft identification. Almost immediately a military air evacuation flight responded. Based on the assumed position, the Snake River Valley, the air evacuation pilot provided me with a frequency for Salt Lake Center.

After several tries, however, I was unable to establish communication with Salt Lake. By this time, I had come across a small town with a good-sized airport. Unfortunately, there was no name on the airport. Because the transponder was being interrogated, I knew someone had the airplane on radar. I selected 7700 and again broadcast "PAN PAN PAN." The air evacuation pilot responded again, and I requested that he ask the center where it was painting 7700. In a few moments, the evac pilot provided us with another center frequency.

Calling the center on this frequency, the controller immediately responded: "Your position is 6 miles east of Big Piney." After a few moments shuffling the chart to find Big Piney, I was on my way again, although not by the originally planned route. As Maxwell Smart said in the 1960s TV series "Get Smart!," "Missed it by *that* much!" Well, it wasn't very much on a WAC chart.

15
IFR flight planning

AN IFR FLIGHT PLAN ALLOWS AN APPROPRIATELY RATED AND CURRENT PILOT with a properly equipped aircraft to operate in weather conditions that are less than VFR. The IFR flight plan also provides separation between known aircraft and alerts search and rescue in case of an accident. When less than VFR conditions exist in controlled airspace, FARs require the pilot to file an IFR flight plan and obtain a clearance. An IFR flight plan is also required in Class A airspace where aircraft speeds require positive, ground-controlled separation.

The proper filing of an IFR flight plan can make the difference between an on-time departure or a lengthy delay. The more a pilot knows about the filing and processing of an IFR flight plan, the better he or she will be able to make maximum use of DUATS, the FAA's air traffic control system, and his or her flying activity.

In the early days of aviation, locations were abbreviated by two letters: NK for Newark. As the number of NAVAIDs and airports increased, three-letter identifiers came into use: ATL for Atlanta. International identifiers use four letters: CYVR for Vancouver, British Columbia, Canada. Some publishers have placed the four-letter international identifiers on domestic charts; use only the three-letter identifiers for operations in the United States.

Airway intersections initially were given local names, for example, Twin Lakes. The names were abbreviated with number-and-letter identifiers: 4TW for Twin Lakes. NAVAIDs in the vicinity of airports were often given the same name and identifier as the nearby airport. Computerization has impacted LCIDs two ways. All intersections are now identified by five letters (EUGEN). In our computerized ATC system, to ensure accurate flight data processing and eliminate ambiguity, NAVAIDs not collocated with airports will have different LCIDs. Some still remain, but all will eventually be changed.

FLIGHT PLAN INFORMATION

With a few important differences, the IFR flight plan is completed in much the same way as the VFR flight plan in chapter 14; type (IFR), full aircraft identification, aircraft type, and special equipment codes are similar. Aircraft identification must include the N. Approved lifeguard (L) and air-taxi (T) operators should prefix the N with L or T, as appropriate ("LN"XXX or "TN"XXX).

Operators with approved three-letter call signs can use the call sign: AMF123 (Amflight one twenty-three). DUATS users must use appropriate aircraft type designator abbreviations contained in appendix A.

The *Airman's Information Manual* (AIM) recommends that pilots file the maximum transponder or navigation capability for their aircraft, as described in chapter 14. This will provide ATC with the necessary information to utilize the system to its maximum. Pilots flying aircraft with a *collision-avoidance system* (TCAS) should prefix the type of aircraft with a "T" (T/BE20/R).

Pilots should file /R (or other appropriate area navigation and transponder code) when the aircraft is equipped with any FAA-approved-and-certified area navigation (RNAV) system, including VORTAC RNAV, inertial, omega, loran, or GPS systems.

True airspeed, point of departure, proposed time of departure, and initial cruising altitude are similar to the VFR flight plan. The departure point must be an aeronautical fix: airport, NAVAID, radial-distance, or latitude-longitude. For IFR in controlled airspace, the pilot can file for any appropriate altitude. This might be advantageous because of aircraft ceiling limitations, icing, or passenger comfort. ATC will normally approve altitudes based on traffic. On top (OTP) is also a perfectly good altitude, but the pilot should specify the requested and appropriate on-top altitude: OTP/175 (on top at 17,500 feet).

Information is computer processed, so route of flight on an IFR flight plan must be accurate, even the smallest route or fix error will result in a reject. For example, the IFR en route chart in Fig. 15-1 shows V485 almost directly over the San Jose VOR (SJC); however, the airway is actually a radial from the Sausalito VOR (SAU). Figure 15-2 is an example of a flight plan route filed for SJC V485. The computer-generated error message specifies that the fix SJC is not on route V485. Close does not count. The route is corrected by inserting the intersection LICKE (Fig. 15-1). The computer responds with an acknowledgment message: R (roger message number 001).

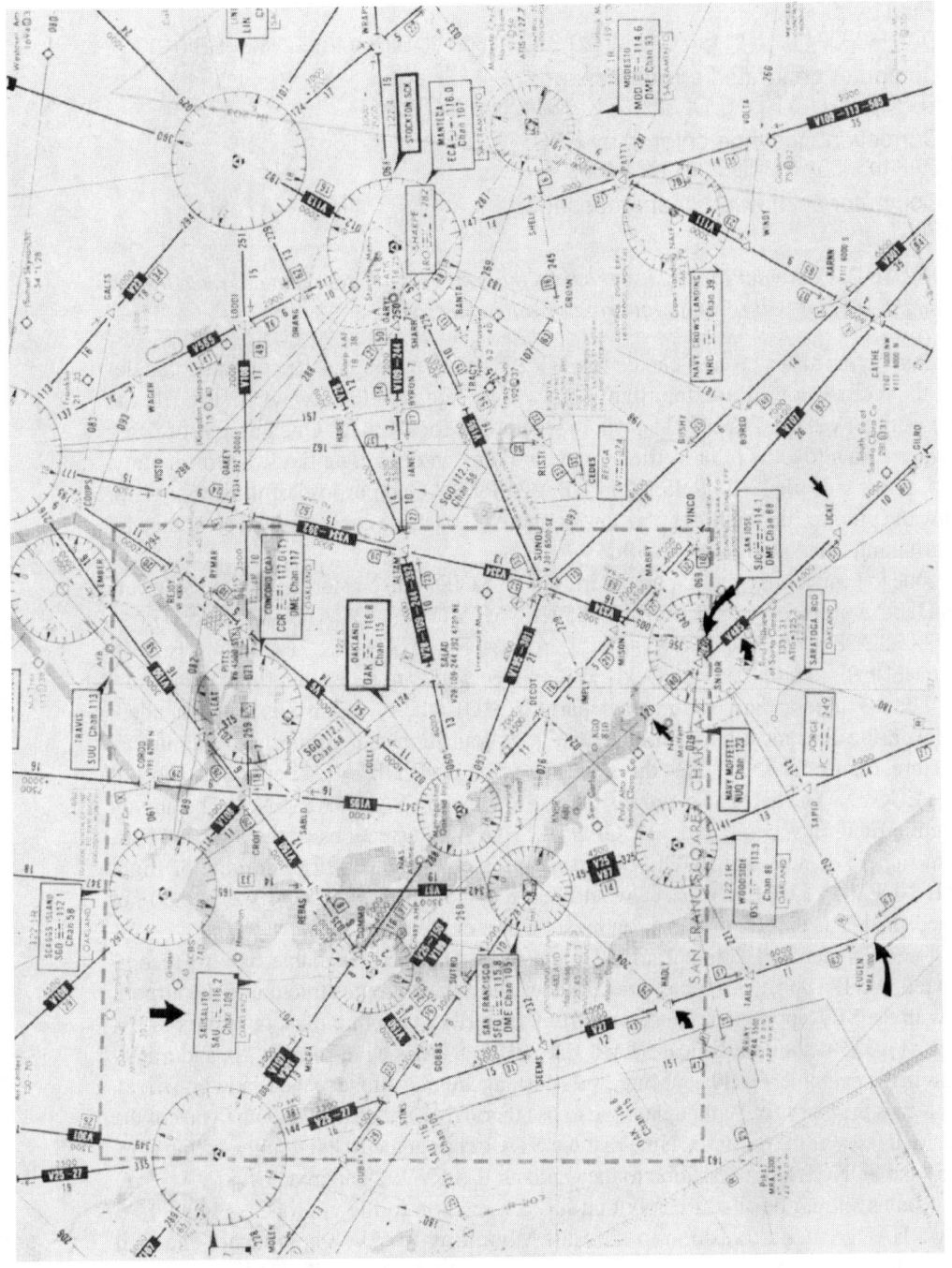

Fig. 15-1. *Pilots must be careful specifying routes on IFR flight plans. A thorough review of the charts will prevent errors.*

Filed by pilot:

OAK1425001 FP N1115R C150/T 90 SJC P1800 90 SJC.V485.ROM..PRB/0100

Computer generated error message:

ERROR 001 10 RTE SJC.V485. FIX NOT ON ROUTE

Correction message entered by FSS:

OAK1430001 CM SJC..LICKE.V485.

Computer acknowledgment message:

R001

Fig. 15-2. *Computer processing requires that routes and fixes connect. Even the slightest error will result in an error or reject message.*

Airway-to-airway routes can be filed (V28 V334), as long as the computer can define the intersection; if not, the flight plan will be rejected. It's always best to file the intersection (V28 ALTAM V334), which is required for DUATS. One purpose of filing airways is to allow the pilot to file a route and omit intermediate fixes. Unfortunately, a few pilots still file "V25 SNS V25 PRB V25 RZS." One purpose of the airway is to allow the pilot to file the flight plan easier and ATC to more easily issue a clearance by omitting intermediate fixes: "V25 RZS."

Due to limitations of ATC computers, not all LCIDs can be stored. A pilot using certain DUATS protocols can file ROM V485 HENER V186 . . .. DUATS will respond Posting FIX REDDE for ZOA ARTCC adaption purposes. Route of Flight...? ROM V485 REDDE V485 HENER.... This allows the Oakland Center computer to accept the flight plan.

The use of a standard instrument departure (SID), where available, is recommended; however, the SID must have an intersection or transition that connects to the airway structure. Figure 15-3 illustrates this point. The SID is the NUEVO4 departure. The exit fix is EUGEN. The SID has two transitions, SHOEY and SALINAS (SNS). The airway structure filed must begin at the EUGEN or SHOEY intersections or the Salinas VOR. From Fig. 15-1, note that EUGEN intersection is on V-27; therefore, a route of flight OAK NUEVO4 EUGEN V27 . . . would be accepted; OAK NUEVO4 V27 would not be accepted. The exit fix or transition must be part of the route.

Use of a standard terminal arrival route (STAR), where available, is also recommended. A STAR provides routes and altitudes from the en route structure to the airport. As with the SID, appropriate transitions from the airway structure must be used. For example, Fig. 15-4 shows a typical STAR. Like SIDs, STARs have an entry fix and transitions. In the example, Coaldale, Mina, and Mustang are transitions to the Madwin Arrival. Manteca is the entry fix. A pilot planning to use this arrival should file a route to one of the transition fixes or the entry fix. Note that the STAR ends at the SUNOL intersection (Fig. 15-5) and SUNOL has a transition to the Oakland ILS RWY 29 approach.

Routes should be filed to a fix that has a transition to the approach. Figure 15-5 shows HADLY has a transition to the Half Moon Bay RNAV-A approach. (The Half Moon Bay RNAV-A approach has been canceled; it is used for illustration purposes only.) Referring to Fig. 15-1, HADLY is also an intersection on V27; therefore, the route V27 HADLY HAF could be filed.

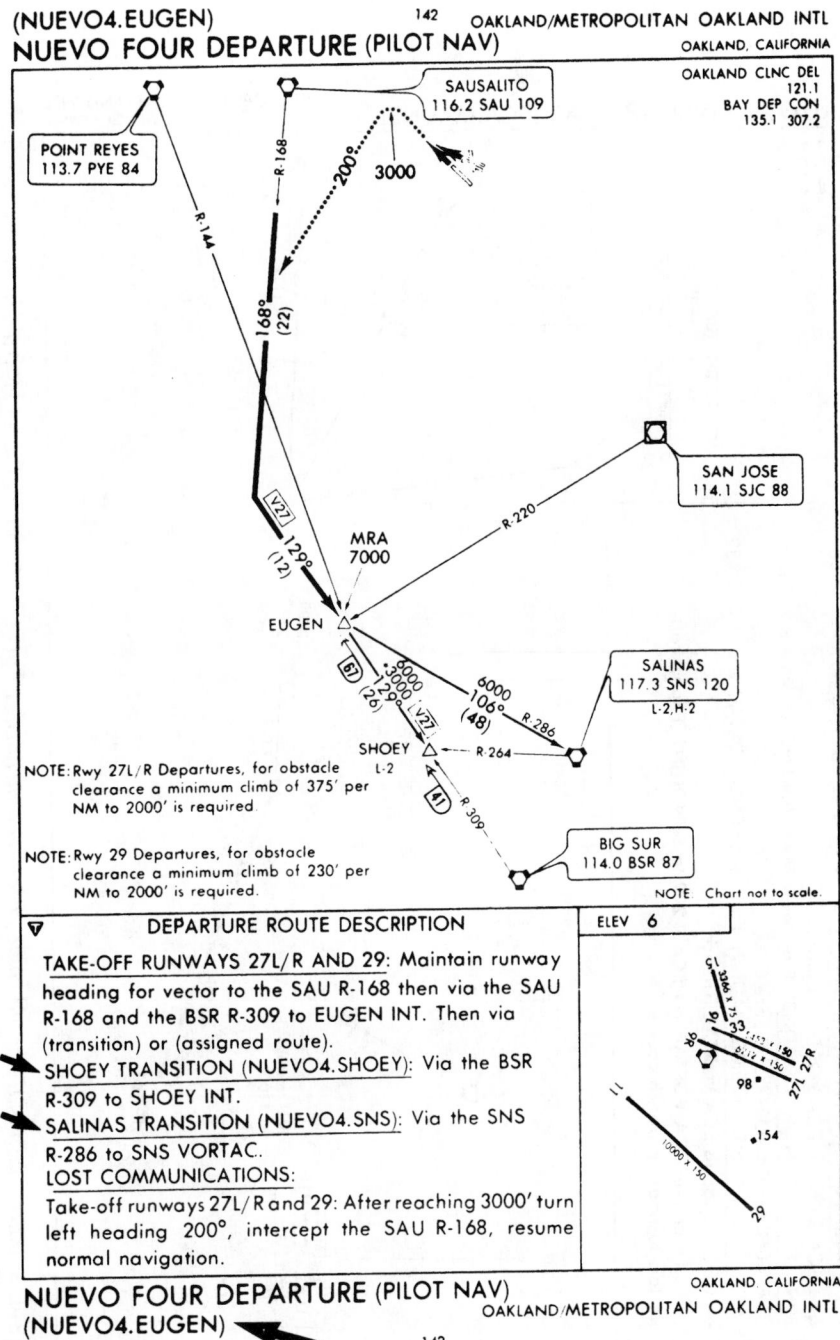

(NUEVO4.EUGEN) 142 OAKLAND/METROPOLITAN OAKLAND INTL
NUEVO FOUR DEPARTURE (PILOT NAV) OAKLAND, CALIFORNIA

OAKLAND CLNC DEL
121.1
BAY DEP CON
135.1 307.2

POINT REYES
113.7 PYE 84

SAUSALITO
116.2 SAU 109

3000

200°

R-168

R-144

168°
(22)

V27 129°
(12)

MRA
7000

EUGEN

SAN JOSE
114.1 SJC 88

R-220

SALINAS
117.3 SNS 120
L-2,H-2

6000
106°
(48)

R-286

61

3000
129°
(26)

6000

V27

SHOEY
L-2

R-264

41

R-309

BIG SUR
114.0 BSR 87

NOTE: Rwy 27L/R Departures, for obstacle
clearance a minimum climb of 375' per
NM to 2000' is required.

NOTE: Rwy 29 Departures, for obstacle
clearance a minimum climb of 230' per
NM to 2000' is required.

NOTE: Chart not to scale.

▽ DEPARTURE ROUTE DESCRIPTION ELEV 6

TAKE-OFF RUNWAYS 27L/R AND 29: Maintain runway
heading for vector to the SAU R-168 then via the SAU
R-168 and the BSR R-309 to EUGEN INT. Then via
(transition) or (assigned route).
SHOEY TRANSITION (NUEVO4.SHOEY): Via the BSR
R-309 to SHOEY INT.
SALINAS TRANSITION (NUEVO4.SNS): Via the SNS
R-286 to SNS VORTAC.
LOST COMMUNICATIONS:
Take-off runways 27L/R and 29: After reaching 3000' turn
left heading 200°, intercept the SAU R-168, resume
normal navigation.

NUEVO FOUR DEPARTURE (PILOT NAV) OAKLAND, CALIFORNIA
(NUEVO4.EUGEN) OAKLAND/METROPOLITAN OAKLAND INTL
142

Fig. 15-3. *Standard instrument departures should be filed. As with any route segment, exit fixes and transitions must connect with the filed en route structure.*

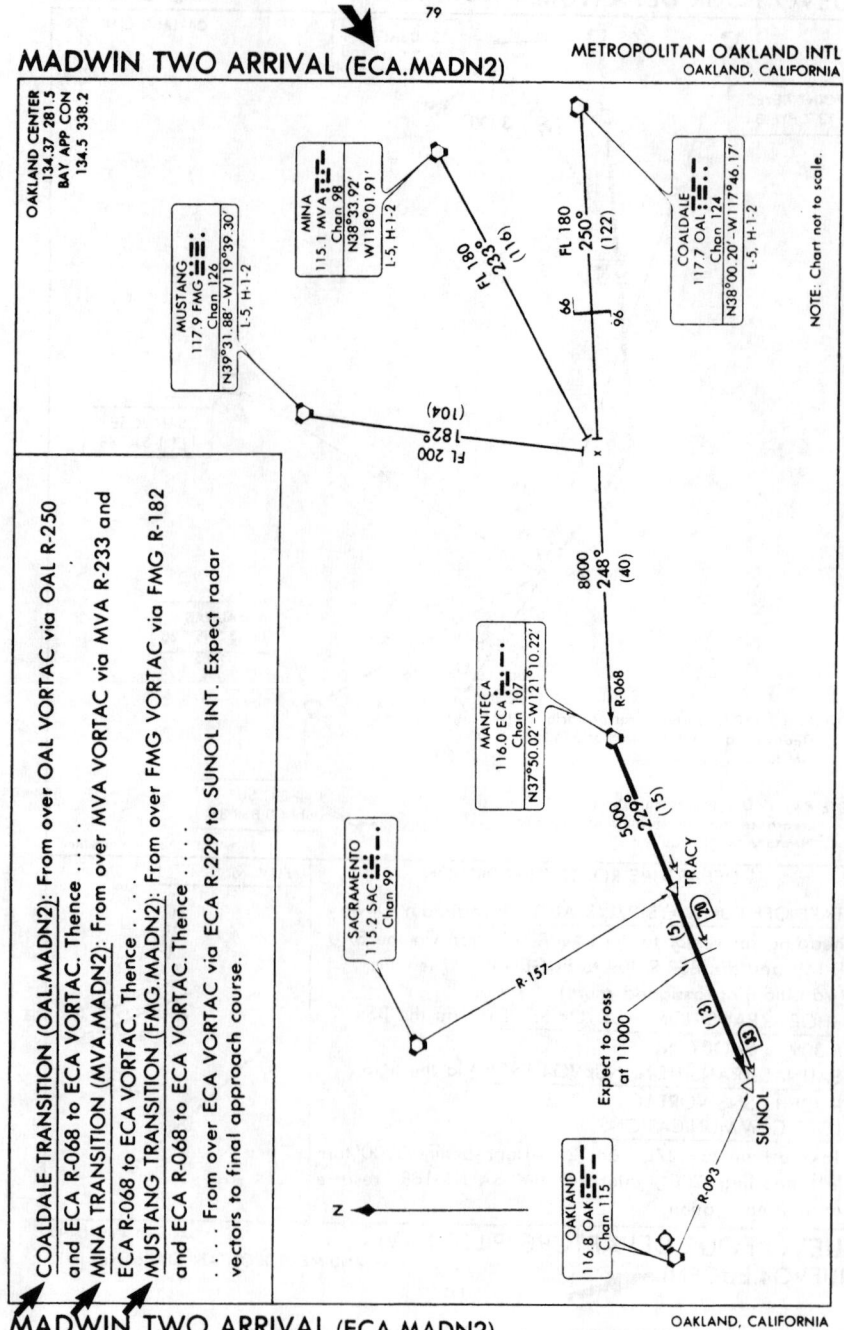

Fig. 15-4. *Standard terminal arrival routes are recommended. Pilots should file to the STAR transition or entry fix.*

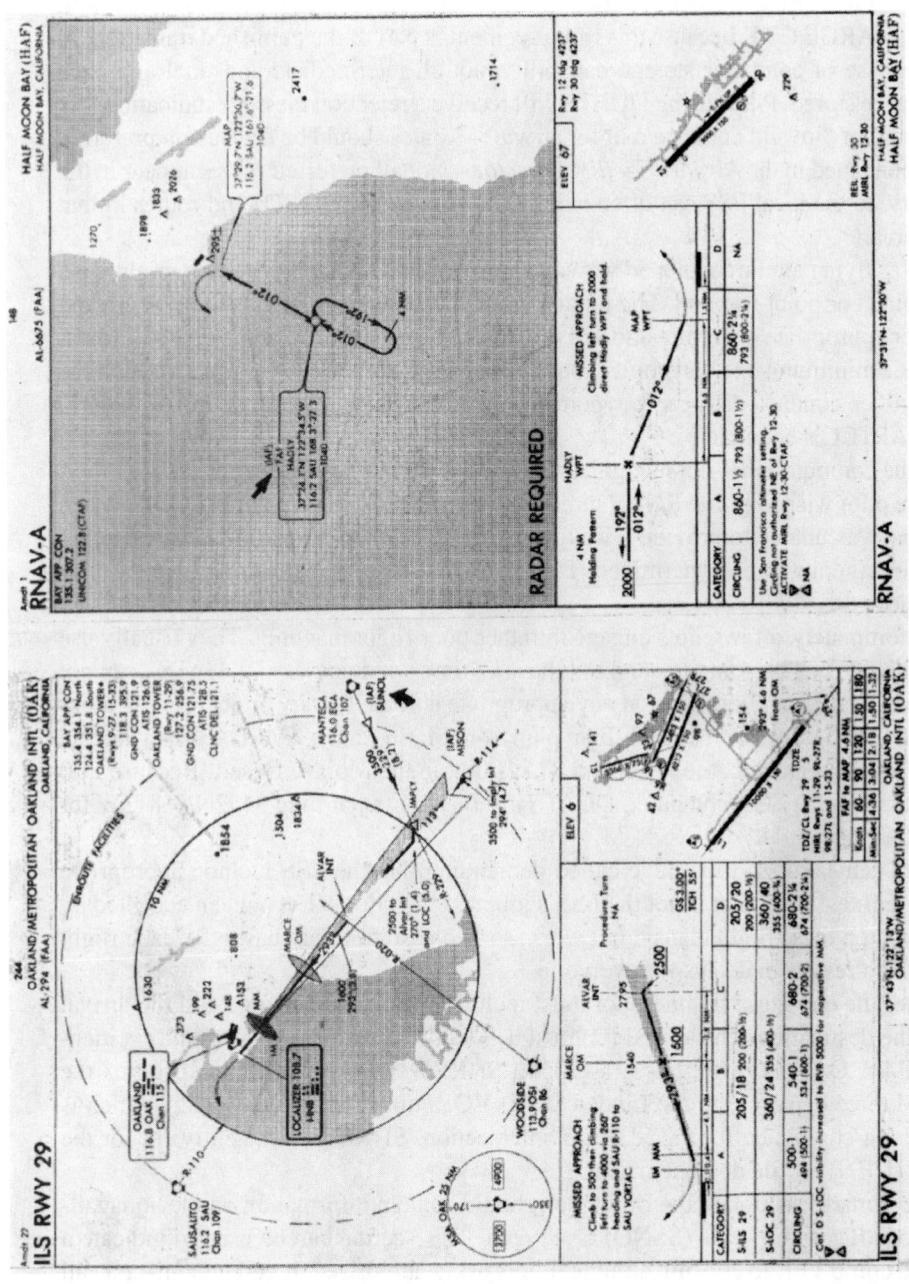

Fig. 15-5. *Pilots should review approach charts carefully and file to an intersection with a transition to the approach. Although the Half Moon Bay RNAV-A approach has been canceled, it serves to illustrate this point.*

This is important if radio communications are lost. Certain pilots think they must file to the final approach fix (FAF); the FAF for the Oakland ILS RWY 29 approach (Fig. 15-5) is the MARCE outer marker. It is not necessary to file, for example, SUNOL IMPLY MARCE OAK because this route segment is part of the published transition. In fact, because of computer storage capability, not all intermediate and final approach fixes can be stored. Pilots using DUATS will receive a rejection message indicating "fix not stored" or "invalid entrance/exit for airway." Routes should be filed using preferred routes contained in the *Airport/Facility Directory*. Not all preferred routes appear in the directory, so the local FSS can often help pilots with preferred SIDs and routes in that FSS's area.

Pilots flying aircraft equipped with an approved area navigation system might wish to file direct or point-to-point. The random-route portion of the flight should begin and end over appropriate departure and arrival fixes. The use of SIDs and STARs is advisable. As a minimum, one waypoint should be filed for each ARTCC through which the flight will be conducted. These waypoints should be located within 200 nm of the preceding ARTCC's boundary.

If the computer does not recognize the fix, the flight plan will be rejected. For example, a pilot wished to file RNAV direct from Hollister, CA, to John Day, OR. When asked, he was unable to provide a waypoint. The specialist pulled out a chart and acceptable waypoints were determined. The flight plan, as filed by the pilot, would have been rejected.

Unfortunately, a few pilots engage in rather poor flight planning. They usually use the excuse, "They're going to send me the way they want anyway." DUATS does not discriminate against pilots filing at any appropriate altitude for any direct route. For example, DUATS will accept an IFR flight plan from Marin Ranch, CA (CA35), in the San Francisco Bay area, direct to Rosmond, CA (L00), in the Mojave Desert. Because L00 is not stored in the ZOA computer, DUATS inserts the L00 lat-long 3452N/11813W for the destination.

Irate center supervisors have called demanding that the FSS include appropriate departure fixes. Well, that is not the FSS's job, it is the pilot's. Even when accepted by the computer, these flight plans don't work; someone somewhere has to fix each flight plan, which results in delays for everyone.

After the departure fix, the pilot must include each course change and the arrival fix for the destination. The arrival fix must be identified by both lat-long and fix identifier: OAK OAK3 OAK LIN 3910/10542 SHREW DEN. This route specifies the Oakland three departure to the Linden (LIN) VOR, direct to "3910/10542," which is the lat-long coordinate for the SHREW intersection. SHREW is the entry fix for the Denver (DEN) profile descent arrival.

The remarks section of the flight plan should contain information on the nonavailability of SIDs and STARs (SSNO) or oxygen. This section can be used to indicate a desire to climb VFR to cruising altitude. Pilots using authorized three-letter air-taxi call signs should write out the radio telephony in remarks: "Amflight" for AMF. Enter only those remarks that are pertinent to ATC or to clarify other flight plan information. Requests for specific approaches, multiple approaches, or altitude changes en route should be made directly with the controller.

When required, an alternate airport should be entered in block 13. Remember that it is the pilot's responsibility to determine the requirement for an alternate. Pilots will from time to time ask the FSS specialist if an alternate is necessary. Although the specialist can often be helpful selecting an appropriate alternate, he or she is not responsible for this determination.

At an accident prevention meeting some years ago, an FAA FSDO representative insisted that pilots file separate flight plans from the destination to alternate in case of radio failure and a missed approach. *Do not file a separate flight plan* because it serves no practical purpose and only congests computer systems that are already nearly saturated.

Fuel on board, pilot's name, number aboard, color, and the like are entered in the same manner as a VFR flight plan.

FLIGHT PLAN FILING AND PROCESSING

IFR flight plans are filed like VFR flight plans. Basic filing procedures that are found in chapter 14 should be followed.

The *Airmen's Information Manual* states: "Pilots should file IFR flight plans at least 30 minutes prior to estimated time of departure to preclude possible delay in receiving a departure clearance from ATC"; however, with AFSS computer systems, IFR flight plans are transmitted virtually instantaneously. Because of a change in AFSS equipment, IFR flight plans can only be accepted up to a maximum of 22 hours in advance.

Similar to a VFR flight plan, when a pilot files with an FSS, the specialist is required to ask if the pilot has received any adverse condition reports for the route.

Upon receipt of an IFR flight plan, the FSS will usually transmit most of the information to the appropriate departure ARTCC no more than 2 hours prior to departure time. The computer checks the message for format and errors. Computer-error and reject messages arise for several reasons, but the largest percentage are "kicked back" due to incorrect routes.

When an error or reject is received, the FSS attempts to correct the problem. This can be time-consuming, especially during periods of heavy traffic. When the flight plan is correct, the computer generates an acknowledgment message (Fig. 15-2).

Thirty minutes prior to departure time, the computer sends a departure strip containing the flight plan information to the center departure sector and approach control or tower as appropriate. The information will not be readily available before this time. Pilots should advise the controller of the filed departure time when calling for clearance more than 30 minutes before the filed proposed time. This will allow the controller to request the flight plan. Pilots calling for clearance between 1 and 2 hours before a filed departure time most often will not receive a clearance.

It is not unheard of for a pilot to change aircraft at the last minute and become very indignant when the controller cannot find the flight plan that was filed under a different identification. Centers retain flight plans from one to three hours after a proposed departure time. The exact time depends on individual centers and computer storage capacity. Because of these factors, it's to the pilot's advantage to file within several hours of the proposed departure time and revise the departure time when it differs by more than 1 hour.

Pilots filing with one FSS for a later departure from an airport served by another FSS have called to change information such as the proposed departure time only to find there is no flight plan. Pilots cannot expect ATC to have a flight plan prior to 2 hours before the proposed departure time. And some pilots routinely call the FSS to see if the center has a flight plan. This is a redundant procedure. It's usually easiest for all concerned, including the pilot, to file as close to 30 minutes prior to departure time as possible.

DUATS transmits the IFR proposal to the departure center one hour prior to proposed time; therefore, if a pilot wants to change any part of the flight plan, the change must be made through a DUATS terminal up to one hour prior to departure. Less than one hour prior to departure, any amendments will have to be made through an FSS.

To facilitate flight plan processing, pilots should file a separate IFR flight plan for each leg of a flight. Even if merely a low approach is intended, a separate flight plan to the next destination will result in faster, more efficient service.

There is no such thing as a round-robin IFR flight plan (IFR with landings en route); however, this does not prevent a pilot from filing a route that goes from the departure airport to an en route fix and back to the departure airport. This is often a valid procedure because it provides ATC with all essential flight plan information that is necessary to issue a clearance. The computer has been programmed to reject a route that is filed "HAF direct HAF"; the error message will come back ZERO DISTANCE. Referring to Fig. 15-5, a pilot who wishes to depart, shoot the approach, and land (all at Half Moon Bay) should file HAF HADLY HAF. This is a perfectly acceptable route.

Many metropolitan areas have established abbreviated flight plan procedures through letters of agreement between ATC facilities whereby controllers enter local flight plans. Pilots can request a *tower en route* clearance from ground or clearance delivery without filing an IFR flight plan; however, pilots should not expect this service in all areas. A check with local pilots or the FSS can determine if abbreviated procedures are available; otherwise, the pilot must plan to file a complete flight plan.

In effect, ATC has properly authorized a deviation from the FAR that requires the pilot to file a complete flight plan; however, this procedure does not relieve the pilot from complying with FARs that pertain to alternate airports and fuel requirements. How does a pilot comply with these regulations? By filing an IFR flight plan, if the pilot wants to be "on record." Most pilots are also aware that they can obtain a clearance directly from the control facility in-flight. The controller, although not required to do so, has authorized the pilot to deviate from the FAR requirement to file a flight plan. You must realize that when the controller gets busy, he or she might not have the time or inclination to accommodate pilots. The best procedure is to file a flight plan.

COMPOSITE FLIGHT PLANS

Composite flight plans (VFR-IFR or IFR-VFR) are accepted by the FSS as stated in the *Airmen's Information Manual*; however, the instructions in the AIM are oversimplified, incomplete, and do not reflect today's ATC system. From a practical point of view and for DUATS users, composite flight plans are actually two separate flight plans.

A composite VFR-IFR flight plan is fairly simple. The pilot files and specifies a point to pick up IFR; however, don't be surprised when the FSS wants two flight plans because each is handled separately. The VFR portion is suspended, and the IFR portion is transmitted to the appropriate center; the VFR portion must be closed by the pilot, preferably prior to picking up IFR. The exact procedure should be agreed upon between the pilot and FSS specialist at the time of filing. After departure, the VFR portion is opened. This allows transmission of the VFR flight notification message to the appropriate FSS.

The pilot files a composite IFR-VFR flight plan in the usual manner and specifies the point where IFR is to be terminated. Again, don't be surprised when the FSS wants two flight plans. The FSS will transmit the IFR portion to the center and the VFR proposal to the FSS responsible for the location where VFR will begin. The pilot will have to activate the VFR portion after canceling IFR. Here again, a procedure should be agreed upon with the FSS specialist when the composite flight plan is filed.

OBTAINING AND CANCELING IFR CLEARANCES

Clearances are normally obtained from the departure ground or clearance delivery controller. At uncontrolled airports, receiving a clearance can be as simple as a request through a remote frequency for center or FSS, or perhaps receiving a clearance will require a clearance-void time that is issued to the pilot over the phone. Because the controller must have the flight plan information, pilots must file prior to departure. Preflight and runup should be completed before requesting clearance. The pilot should provide the FSS or ATC specialist with aircraft ident, departure and destination, and how soon the pilot can depart after the clearance is received. This will allow ATC to provide a clearance with an appropriate "window" of time for the pilot to depart.

Clearance-void times tie up airspace and increase controller workload, so controllers are reluctant to issue times beyond about 15 minutes after the time that the clearance is issued. Pilots must be ready for an immediate departure. If the pilot subsequently decides not to go for any reason, he or she must advise the FSS or ATC immediately upon making the decision; otherwise, ATC must assume that the pilot has departed and is experiencing radio trouble. The airspace would have to be sterilized (cleared of all known traffic), which would result in extensive delays for other traffic.

IFR flight plans are automatically canceled by an operating tower upon arrival; however, the pilot must close an IFR flight plan upon arrival at other locations. The failure to close can needlessly tie up airspace and cause ATC delays for hours. Often the control facility will specify how the pilot is to close the IFR flight plan. In any case, the flight plan can always be closed with an FSS.

True story: A pilot flying a twin Cessna arrived after the airport's tower had closed. The FSS, center, and sheriff's department spend a good part of the night searching for this aircraft, which was neatly tucked away in its hangar.

USING IFR FLIGHT PLAN SERVICE

A topical subject these days is whether a pilot should routinely file IFR, even when weather conditions are VFR. An IFR flight plan has the advantage that ATC helps the pilot precisely fly from point A to point B. It relieves the pilot of knowing what type of airspace he or she is traversing and who to contact. (Class A, B, C, D and E airspace have different clearance and communications requirements.)

Finally at the destination, ATC usually sequences the arrival with other IFR and VFR traffic. Certain aviators consider radar traffic advisories an additional safety factor, but the price for an ATC clearance might be departure and arrival delays, out-of-the-way routings and approaches, less than advantageous cruising altitudes, and increased pilot workload.

A pilot should also consider his or her experience and aircraft capability. A newly rated instrument pilot without experience should carefully consider the advisability of filing IFR in congested airspace. Single-pilot operations in high-density traffic areas, even when the weather is VFR, are extremely demanding. The pilot must control and navigate the aircraft and communicate with ATC at the same time. Anyone without IFR experience in busy terminal areas should seek training from a competent instructor before venturing single-pilot IFR in these areas.

A Cessna 150 or Tomahawk pilot who demands his or her right to the ILS at a major, congested airport makes friends with no one. Pilots should consider their obligations as well as their rights when filing IFR. There is little advantage with struggling to 11,000 or 12,000 feet in a Cessna 172 or Piper Cherokee just to go IFR.

If you decide to go IFR, the first task is to file the flight plan as outlined in previous sections of this chapter. I planned a proficiency IFR training flight out of Livermore, CA. I had reserved an airplane and scheduled an instructor. My plan was to go to Stockton, then Sacramento, and return, and I filed three abbreviated flight plans with flight service. (A letter of agreement authorizes this procedure within this area.) The weather was clear and alternate airports were not required. Flight plan from blocks one through 10 were provided for each leg. During each request for clearance, I advised ground control that we had prefiled. Each clearance was readily available, with no delay. Had this been a longer training flight, with delays in excess of 1 hour, I would have filed the successive leg at the departure airport prior to that departure.

On this occasion, the LA Basin had a deeper than usual marine cloud layer. Tops were around 7,000 feet. Under such conditions, pilots who prefer "IFR-to-on-top" might have to file because they will have to enter the center's airspace. (Abbreviated flight plans are normally restricted to the airspace of adjoining approach control facilities.)

I filed IFR to the Gorman VOR. This procedure afforded me the IFR clearance through the Basin and out of the stratus clouds. If I had wanted VFR flight plan protection beyond Gorman, I would have filed a VFR flight plan from Gorman to my destination. After canceling IFR, I would have activated the VFR portion with the Bakersfield FSS. I determined from the *Airport/Facility Directory* that Bakersfield had flight plan responsibility for Gorman. This was a combined IFR-VFR flight plan.

As promised from chapter 14, let's go back to the Lovelock-to-Van Nuys odyssey. California coastal sections are often covered with stratus. My route had few direct airways, and available airways had minimum altitudes far above a reasonable cruise altitude for the Cessna 150. In any case, there was no need for IFR because the desert weather was clear.

I filed my usual VFR flight plan but changed the destination to the Palmdale VOR. In remarks, I noted "pickup IFR O/PMD-VNY." I filed a separate IFR flight plan from PMD to VNY. Over Palmdale, I canceled VFR with Riverside Radio, the tie-in for Palmdale, and contacted Los Angeles Center for my IFR clearance to Van Nuys. This is an example of a VFR-IFR flight plan combination.

Despite certain equipment restrictions, general aviation has complete access to the ATC system. There is no question that system complexity is growing. Only by knowing the system and working within it can a pilot realize its maximum benefit.

A
Abbreviations

MANY CONTRACTIONS ARE USED ON AVIATION WEATHER REPORTS AND forecasts to save space on telecommunication circuits, in computer equipment, and on charts. The contractions will normally be used for any derivative of the root word. If confusion would result, variations might be shown by adding the following letters to the contraction of the root word.

able	BL	ive	V
al	L	iest, est	ST
ally, erly, ly	LY	iness, ness	NS
ary, ery, ory	RY	ing	G
ance, ence	NC	ity	TY
der	DR	ment	MT
ed, ied	D	ous	US
ening	NG	s, es, ies	S
er, ier, or	R	tion, ation	N
ern	RN	ward	WD
ically	CLY		

The following contractions are normally used on aviation weather reports, PIREPs, forecasts, charts, and Notices to Airmen.

A

A	absolute (temperature)
A	Arctic air mass
A	hail
AADC	approach and departure control
ABV	above
A/C	approach control
AC	altocumulus
ACC	altocumulus castellanus
ACCUM	accumulate
ACFT	aircraft
ACLTG	accelerating
ACR	air carrier
ACSL	standing lenticular altocumulus
ACTV	active
ACTVT	activate
ACYC	anticyclonic
ADJ	adjacent
ADRNDCK	Adirondack
ADV	advise
ADVCT	advection
ADVY	advisory
ADZ	advise
A/FD (AFD)	Airport/Facility Directory
AFDK	after dark
AIM	Airman's Information Manual
ALF	aloft
ALGHNY	Allegheny
ALNOT	alert notice
ALQDS	all quadrants
ALS	approach light system
ALSTG	altimeter setting
ALSEC	all sectors
ALT	altitude
ALTA	Alberta
ALTM	altimeter
ALTN	alternate
ALUTN	Aleutian
ALWF	actual wind factor
AMDT	amendment

AMPLTD	amplitude
AMS	air mass
ANLYS	analysis
APCH	approach
APL	airport lights
APLCN	Appalachian
ARFF	airport rescue and fire fighting
ARPT	airport
ARR	arrive (arrival)
ARSR	air route surveillance radar
AS	altostratus
ASDE	airport surface detection equipment
ASPH	asphalt
ASR	airport surveillance radar
ATCSCC	Air Traffic Control System Command Center
ATCT	airport traffic control tower
ATLC	Atlantic
AUTH	authority
AURBO	Aurora Borealis
AVBL	available
AWIPS	Advance Weather Interactive Processing System
AWY	airway
AWW	alert weather watch
AZM	azimuth

B

B	beginning of precipitation (time in minutes)
BC	back course
BC	British Columbia
BC	patches
BCH	beach
BCN	beacon
BCKG	backing
BD	blowing dust
BDA	Bermuda
BERM	snowbank/s containing earth/gravel
BFDK	before dark
BINOVC	breaks in overcast
BKN	broken
BL	between layers
BL	blowing
BLD	build
BLDUP	buildup

BLKHLS	Black Hills
BLKT	blanket
BLO	below
BLZD	blizzard
BN	blowing sand
BND	bound
BNDRY	boundary
BOVC	base of overcast
BR	mist
BRAF	braking action fair
BRAG	braking action good
BRAN	braking action nil
BRAP	braking action poor
BRF	brief
BRG	bearing
BRK	break
BRKHIC	breaks in higher overcast
BRKSHR	Berkshire
BRM	barometer
BS	blowing snow
BTWN	between
BY	blowing spray
BYD	beyond
C	
C	continental air mass
CA	clear above
CAAS	class A airspace
CAN	Canada
CARIB	Caribbean
CASCDS	Cascades
CAT	category
CAVOK	cloud and visibility OK
CAVU	clear or scattered clouds and visibility greater than 10 miles
CAVUOK	no significant clouds or weather
CBAS	class B airspace
CBMAM	cumulonimbus mamma
CBSA	class B surface area
CC	cirrocumulus
CCAS	class C airspace
CCLKWS	counterclockwise
CCSA	class C surface area
CCSL	standing lenticular cirrocumulus
CD	clearance delivery

CDFNT	cold front
CDAS	class D airspace
CDSA	class D surface airspace
CEAS	class E airspace
CESA	class E surface airspace
CFA	controlled firing area
CFP	cold front passage
CGAS	class G airspace
CHC	chance
CHG	change
CHSPK	Chesapeake
CI	cirrus
CIG	ceiling
CLD	cloud
CLKWS	clockwise
CLNC	clearance
CLR	clear
CLRS	clear and smooth
CLSD	closed
CMB	climb
CMSND	commissioned
CNCL	cancel
CNDN	Canadian
CNTRLN	centerline
CNVTV	convective
CONC	concrete
COND	condition
CONFDC	confidence
CONT	continuous
CONTDV	Continental Divide
CONTRAILS	condensation trails
COP	changeover point
CRS	course
CS	cirrostratus
CST	coast
CTAF	common traffic advisory frequency
CTC	contact
CTGY	category
CTL	control
CTSKLS	Catskills
CU	cumulus
CUFRA	cumulus fractus
CWA	center weather advisory

CYC	cyclonic
CYCLGN	cyclogenesis

D

D	dust
DABRK	daybreak
DALGT	daylight
DCMSND	decommissioned
DCT	direct
DEGS	degrees
DEP	depart/departure
DF	direction finder
DFUS	diffuse
DH	decision height
DKTS	Dakotas
DLAD	delayed
DLT	delete
DLY	daily
DMSH	diminish
DMSTN	demonstration
DNS	dense
DNSLP	downslope
DNSTRM	downstream
DP	deep
DPCR	departure procedure
DPNG	deepening
DPTH	depth
DR	low drifting
DRCT	direct
DRFT	drift
DRZL	drizzle
DS	dust storm
DSIPT	dissipate
DSPLCD	displaced
DSCT	distance
DTRT	deteriorate
DU	dust
DURGC	during climb
DURGD	during descent
DVV	downward vertical velocity
DWNDFTS	downdrafts
DWPNT	dew point
DZ	drizzle

E

E	ending of precipitation (time in minutes)
E	equatorial air mass
E	estimated ceiling
EBND	eastbound
ELEV	elevation
ELNGT	elongate
EMBDD	embedded
ENERN	east-northeastern
ENEWD	east-northeastward
ENTR	entire
ENRT	en route
EOF	expected operations forecast
ESERN	east-southeastern
ESEWD	east-southeastward
EXCP	except
EXTRAP	extrapolate
EXTRM	extreme

F

F	fog
FA	area forecast
FA	final approach
FAC	facility
FAF	final approach fix
FC	funnel cloud/tornado/waterspout
FD	wind and temperature aloft forecast
FILG	filling
FL	flight level
FLG	falling
FLRY	flurry
FM	fan marker
FNT	front
FNTGNS	frontogenesis
FNTLYS	frontolysis
FORNN	forenoon
FPM	feet per minute
FRH	fly runway heading
FRMG	forming
FROPA	frontal passage
FROSFC	frontal surface
FRST	frost
FRWF	forecast wind factor

FRZ	freeze
FRZN	frozen
FT	terminal forecast
FZ	freezing

G

G	gust reaching
GC	ground control
GCA	ground control approach
GF	ground fog
GFDEP	ground fog estimated (feet) deep
GICG	glaze icing
GLFALSK	Gulf of Alaska
GLFCAL	Gulf of California
GLFMEX	Gulf of Mexico
GLFSTLAWR	Gulf of St. Lawrence
GNDFG	ground fog
GOES	Geostationary Operational Environmental Satellite
GOESNEXT	Next Generation GOES
GP	glide path
GPS	Global Positioning System
GR	hail
GRAD	gradient
GRBNKS	Grand Banks
GRDL	gradual
GRTLKS	Great Lakes
GRVL	gravel
GS	glide slope
GS	small hail
GSTS	gusts
GSTY	gusty
GV	ground visibility

H

H	haze
HAA	height above airport
HAT	height above touchdown
HAZ	hazard
HCVIS	high clouds visible
HDG	heading
HDEP	haze layer estimated (feet) deep
HDFRZ	hard freeze
HDSVLY	Hudson Valley
HEL	helicopter
HELI	heliport

HF	high frequency
HI	high
HIFOR	high level forecast
HIRL	high intensity runway lights
HIWAS	Hazardous In-flight Weather Advisory Service
HLSTO	hailstones
HLTP	hilltop
HLYR	haze layer aloft
HOL	holiday
HP	holding pattern
HR	hour
HURCN	hurricane
HUREP	hurricane report
HX	high index
HZ	haze

I

IAF	initial approach fix
IAP	instrument approach procedure
IBND	inbound
IC	ice crystals
ICAO	International Civil Aviation Organization
ICG	icing
ICGIC	icing in clouds
ICGICIP	icing in clouds and precipitation
ICGIP	icing in precipitation
ID	identification
IDENT	identification
IF	ice fog
IF	intermediate fix
ILS	instrument landing system
IM	inner marker
IMDT	immediate
INDEFLY	indefinitely
INFO	information
INLD	inland
INOP	inoperative
INREQ	information request
INST	instrument
INSTBY	instability
INT	intersection
INTL	international
INTMT	intermittent
INTR	interior

INTRMTRGN	inter-mountain region
INTS	intense
INTSFY	intensify
INTST	intensity
INVRN	inversion
IOVC	in overcast
IP	ice pellets
IPV	improve
IR	ice on runway
ISA	International Standard Atmosphere
ISOLD	isolated
ITCZ	intertropical convergence zone

J

JTSTR	jet stream

K

K	cold air mass
K	smoke
KDEP	smoke layer estimated (feet) deep
KFRST	killing frost
KLYR	smoke layer aloft
KOCTY	smoke over city
KT	knots

L

L	drizzle
LAA	local airport advisory
LABRDR	Labrador
lat-long	latitude/longitude
LC	local control
LCL	lifted condensation level
LCL	local
LCTD	located
LDA	localizer type directional aid
LDIN	lead in lighting system
LFC	level of free convection
LFT	lift
LGRNG	long range
LGT	light
LIRL	low intensity runway lights
LK	lake
LLWAS	low level wind shear alert system
LLWS	low level wind shear
LMM	compass locator at the middle marker

LN	line
LNDG	landing
LOC	localizer
LOM	compass locator at the outer marker
LRN	loran
LSR	loose snow on runway
LT	left turn after takeoff
LTG	lightning
LTGCC	lightning cloud-to-cloud
LTGCG	lightning cloud-to-ground
LTGCW	lightning cloud-to-water
LTGIC	lightning in clouds
LTLCG	little change
LTNG	lightning
LVL	level
LWR	lower
LX	low index
LYR	layer or layered or layers
M	
M	maritime air mass
M	measured ceiling
M	minus
M	missing
MA	map analysis
MAG	magnetic
MAINT	maintenance
MALS	medium intensity approach lighting system
MALSF	medium intensity approach lighting system with sequence flashers
MALSR	medium intensity approach lighting system with runway alignment indicator lights
MAN	Manitoba
MAP	missed approach point
MB	millibar
MCA	minimum crossing altitude
MDA	minimum descent altitude
MDT	moderate
MEA	minimum en route altitude
MED	medium
MEGG	merging
METAR	weather observation
MEX	Mexico
MHKVL	Mohawk Valley
MI	shallow

MIDN	midnight
MIN	minute
MIRL	medium intensity runway lights
MIS	meteorological impact statement
MLTLVL	melting level
MM	middle marker
MNLD	mainland
MNM	minimum
MOA	military operations area
MOCA	minimum obstruction clearance altitude
MOGR	moderate or greater
MONTR	monitor
MOV	move
MRA	minimum reception altitude
MRGL	marginal
MRNG	morning
MRTM	maritime
MS	minus
MSA	minimum safe altitude
MSTLY	mostly
MSTR	moisture
MSAW	minimum safe altitude warning
MSL	mean sea level
MTN	mountain
MTR	military training route
MU	designate a friction value representing runway surface conditions
MULT	multiple
MUNI	municipal
MVFR	marginal VFR
MXD	mixed

N

NA	not authorized
NASA	National Aeronautics and Space Administration
NAS	National Airspace System
NAV	navigation
NAVAID	navigational aid
NB	New Brunswick
NBND	northbound
NCWX	no change in weather
NDB	nondirectional radio beacon
NELY	northeasterly
NERN	northeastern
NEW ENG	New England

NFLD	Newfoundland
NGT	night
NL	no layers
NMBR	number
NMR	nautical mile radius
NMRS	numerous
NNERN	north-northeastern
NNEWD	north-northeastward
NNWRN	north-northwestern
NNWWD	north-northwestward
NOPT	no procedure turn
NOSPL	no specials
NPRS	nonpersistent
NRW	narrow
NS	nimbostratus
NS	Nova Scotia
NS	nonsignificant radar echoes
NSW	no significant weather
NTAP	Notices to Airmen Publication
NVA	negative vorticity advection
NWLY	northwesterly
NWRN	northwestern

O

OAOI	on and off instruments
OAT	outside air temperature
OBS	observation
OBSC	obscure
OBSTN	obstruction
OCFNT	occluded front
OCLD	occlude
OCLN	occlusion
OCNL	occasional
OFP	occluded frontal passage
OFSHR	offshore
OM	outer marker
OMTNS	over mountains
ONSHR	on shore
ONT	Ontario
OPER	operate
OPN	operation
ORGPHC	orographic
ORIG	original
OTS	out of service

OTAS	on top and smooth
OTLK	outlook
OTP	on top
OTRW	otherwise
OVC	overcast
OVHD	overhead
OVR	over
OVRNG	overrunning

P

P	plus
P	polar air mass
P-	light precipitation of unknown type
P time	proposed departure time
PAC	Pacific
PAEW	personnel and equipment working
PAJA	parachute jumping
PAPI	precision approach path indicator
PAR	precision approach radar
PARL	parallel
PAT	pattern
PBL	probable
PCPN	precipitation
PCL	pilot controlled lighting
PDMT	predominant
PDMT	predominate
PDW	priority delayed weather
PE	ice pellets
PEN	peninsula
PERM	permanent
PGTSND	Puget Sound
PIBAL	pilot balloon observation
PLA	practice low approach
PLW	plow (snow)
PN	prior notice required
PNHDL	panhandle
PO	dust/sand whirls
PPR	prior permission required
PPINA	radar weather report not available
PPINE	radar weather report no echoes observed
PPINO	radar weather report equipment inoperative
PPIOK	radar weather report equipment operation resumed
PPIOM	radar weather report equipment out for maintenance
PRBLTY	probability

PRES	pressure
PRESFR	pressure falling rapidly
PRESRR	pressure rising rapidly
PREV	previous
PRIRA	primary radar
PRJMP	pressure jump
PROC	procedure
PROP	propeller
PROG	prognostic
PRSNT	present
PS	plus
PSG	passage
PSG	passing
PSGR	passenger
PSR	packed snow on runway
PT	procedure turn
PTCHY	patchy
PTLY	partly
PTN	portion
PVA	positive vorticity advection
PVT	private
PY	spray

Q

Q	squall
QSTNRY	quasistationary
QUAD	quadrant
QUE	Quebec

R

R	rain
RA	rain
RABAL	radiosonde balloon wind data
RADAT	radiosonde observation data
RAFRZ	radiosonde observation freezing levels
RAICG	radiosonde observation icing
RAIL	runway alignment indicator lights
RAOB	radiosonde observation
RAREP	radar weather report
RAWIN	upper winds observation
RCAG	remote communication facility
RCC	Rescue Coordination Center
RCD	radar cloud detection report
RCKY	Rockies (mountains)

RCL	runway centerline
RCLS	runway centerline light system
RCO	remote communication outlet
RCV	receive
RDG	ridge
REF	reference
REIL	runway end identifier lights
RELCTD	relocated
RESTR	restrict
RGD	ragged
RH	relative humidity
RHINO	radar echo height information not available
RHINO	radar range height indicator not operating on scan
RIOGD	Rio Grande
RMDR	remainder
RMRK	remark
RNFL	rainfall
ROBEPS	radar operating below prescribed standard
RPD	rapid
RPLC	replace
RPRT	report
RQRD	required
RS	record special observation
RSG	rising
RSVN	reservation
RT	right turn after takeoff
RTE	route
RTR	remote transmitter/receiver
RTS	return to service
RUF	rough
RVR	runway visual range
RVRM	runway visual range (midpoint)
RVRNO	RVR not available
RVRR	runway visual range (rollout)
RVRT	runway visual range (touchdown)
RVV	runway visibility value
RVVNO	RVV not available
RW	rain shower
RWY	runway
RY	runway
S	
S	snow
SA	sand

SA	surface aviation observation
SASK	Saskatchewan
SBND	southbound
SBSD	subside
SC	stratocumulus
SCSL	standing lenticular stratocumulus
SCT	scattered
SD	radar report (RAREP)
SDF	simplified directional facility
SECRA	secondary radar
SELS	severe local storms
SELY	southeasterly
SERN	southeastern
SFL	sequence flashing lights
SG	snow grains
SGD	Solar-Geophysical Data
SGFNT	significant
SH	showers
SHFT	shift
SHLW	shallow
SHRTLY	shortly
SHRTWV	shortwave
SHWR	shower
SI	straight-in approach
SIERNEV	Sierra Nevada
SID	standard instrument departure
SIMUL	simplified short approach light system with sequenced flashers
SIR	snow and ice on runway
SKC	sky clear
SKED	scheduled
SLD	solid
SLGT	slight
SLR	slush on runway
SLT	sleet
SMK	smoke
SMTH	smooth
SN	snow
SND	sanded
SNBNK	snowbank
SNFLK	snowflake
SNGL	single
SNOINC	snow depth increase in past hour
SNRS	sunrise

SNST	sunset
SNW	snow
SNWFL	snowfall
SP	special observation
SP	snow pellets
SP	station pressure
SPD	speed
SPECI	special observation
SPKL	sprinkle
SPLNS	South Plains
SPRD	spread
SQ	squall
SQAL	squall
SQLN	squall line
SR	sunrise
SS	sunset
SS	sandstorm
SSALF	simplified short approach lighting system with sequenced flashers
SSALR	simplified short approach lighting system with runway alignment indicator lights
SSALS	simplified short approach lighting system
SSERN	south-southeastern
SSEWD	south-southeastward
SSNO	no STARs, no SIDs
SSWRN	south-southwestern
SSWWD	south-southwestward
ST	stratus
STAGN	stagnation
STAR	standard terminal arrival
STFR	stratus fractus
STFRM	stratiform
STG	strong
STM	storm
STNRY	stationary
SVC	service
SVR	severe
SVRL	several
SW	snow showers
SWLG	swelling
SWLY	southwesterly
SWRN	southwestern
SX	stability index
SXN	section

SYNOP	synoptic
SYNS	synopsis
SYS	system
T	
T	thunderstorm
T	trace
T	tropical air mass
TACAN	tactical air navigational aid
TAF	terminal aerodrome forecast
TCU	towering cumulus
TDWR	Terminal Doppler Weather Radar
TDZ	touchdown zone
TDZL	touchdown zone lights
TEMP	temperature
TFC	traffic
TFR	temporary flight restriction
TGL	touch and go landing
THD	thunderhead
THDR	thunder
THK	thick
THN	thin
THR	threshold
TKOF	takeoff
TMPA	Traffic Management Program Alert
TMPRY	temporary
TOP	cloud top
TOVC	top of overcast
TPG	topping
TRIB	tributary
TRML	terminal
TRNG	training
TROF	trough
TROP	tropopause
TRPCD	tropical continental air mass
TRPCL	tropical
TRPLYR	trapping layer
TRRN	terrain
TRSN	transition
TS	thunderstorm
TSHWR	thundershower
TSNT	transient
TSQLS	thundersqualls
TSTM	thunderstorm

TURBC	turbulence
TURBT	turbulent
TWD	toward
TWR	tower
TWRG	towering
TWY	taxiway
U	
U	intensity unknown
UA	pilot report
UDDF	up and downdrafts
UNAVBL	unavailable
UNLGTD	unlighted
UNMKD	unmarked
UNMON	unmonitored
UNOFFL	unofficial
UNRELBL	unreliable
UNSBL	unseasonable
UNSTBL	unstable
UNSTDY	unsteady
UNSTL	unsettle
UNUSBL	unusable
UPDFTS	updrafts
UP	precipitation (automated Observation)
UPR	upper
UPSLP	upslope
UPSTRM	upstream
USP	urgent special observation
UTC	Coordinated Universal Time
UUA	urgent pilot report
UVV	upward vertical velocity
UWNDS	upper winds
V	
V	variable
VA	volcanic ash
VASI	visual approach slope indicator
VC	vicinity
VCNTY	vicinity
VDP	visual descent point
VIA	by way of
VICE	instead of
VLCTY	velocity
VLNT	violent
VLY	valley

VMC	visual meteorological conditions
VOL	volume
VOT	VOR test facility
VR	veer
VRBL	variable
VRISL	Vancouver Island, BC
VRT MOTN	vertical motion
VSBY	visibility
VSBYDR	visibility decreasing rapidly
VSBYIR	visibility increasing rapidly
W	
W	indefinite ceiling
W	warm air mass
WA	AIRMET
WDLY	widely
WDSPRD	widespread
WEA	weather
WFP	warm front passage
WI	within
WINT	winter
WK	weak
WKDAYS	Monday through Friday
WKEND	Saturday and Sunday
WMO	World Meteorological Organization
WND	wind
WNWRN	west-northwestern
WNWWD	west-northwestward
WPLTO	Western Plateau
WR	wet runway
WRM	warm
WRMFNT	warm front
WRNG	warning
WS	SIGMET
WSHFT	wind shift
WSR	wet snow on runway
WST	Convective SIGMET
WSTCH	Wasatch Range
WSWRN	west-southwestern
WSWWD	west-southwestward
WTR	water
WTSPT	waterspout
WV	wave
WW	Severe Weather Watch

WX	weather
WX NIL	no significant weather
X	
X	sky obscured
-X	sky partially obscured
XCP	except
XPC	expect
Z	
Z	Coordinated Universal Time (UTC)
ZL	freezing drizzle
ZR	freezing rain

IDENTIFIERS

This section contains major report (SA) and forecast (FT and FD) locations for the contiguous states, Alaska, and Hawaii. Many locations provide part-time or AWOS observations, therefore, observations and forecasts are not always available. Selected military locations are included; discrepancies will occur.

Depending on the retrieval system, some locations might not be available. Part-time and AWOS reporting locations might only be available locally. Reporting locations are continually being added and deleted.

Locations are identified by city and state (WESTFIELD, MA). When more than one airport is listed under the same city, the airport name precedes the city name (Jeffco, DENVER, CO). Military bases are identified by name, city, and state, except where the base name is the same as the associated city (NEW ORLEANS NAS, LA).

Number/Letter

03Y	HALLOCK, MN
0E4	PAYSON, AZ
0V1	CUSTER, SD
0Z0	DEERING, AK
13A	NOGALES, AZ
19D	MORA, MN
1O5	MONTAGUE, CA
1V1	RIFLE, CO
27U	SALMON, ID
2A3	BESSEMER, AL
2V9	GUNNISON, CO
3B1	GREENVILLE, ME
3B2	MARSHFIELD, MA
3BF	MERRILL PASS WEST, AK
3DU	DRUMMOND, MT
3G8	BATAVIA, NY
3HT	HARLOWTON, MT

3KM	James Jabara, WICHITA, KS
3OI	LAMONI, IA
3S2	AURORA, OR
3SE	SPENCER, IA
3TH	THOMPSON FALLS, MT
3V8	VENICE, LA
43F	LITCHFIELD, MN
39J	EVERGREEN, AL
4BK	BROOKINGS, OR
4BL	BLANDING, UT
4BQ	BROADUS, MT
4CR	CORONA, MN
4DG	DUGLAS, WY
4HV	HANKSVILLE, UT
4LJ	LAMAR, CO
4LW	LAKEVIEW, OR
4MY	MORIARTY, NM
4OM	OMAK, WA
4SL	CUBA, NM
4SV	STREVELL, ID
4V5	DURANGO, CO
59M	EASTPORT, MI
5BI	BIG RIVER LAKES, AK
5CE	CAPE ST ELIAS, AK
5EA	HEALY RIVER, AK
5GN	TAHNETA PASS LODGE, AK
5HR	HAYES RIVER, AK
5I3	PIKEVILLE, KY
5J0	JOHN DAY, OR
5PX	PAXSON, AK
5SO	SNOW SHOE LAKE, AK
5SZ	SLANA, AK
5WD	SEWARD, AK
5WT	WHITTIER, AK
60S	VANCOUVER, WA
61E	EAST CAMERON, LA
6T5	FREEPORT, TX
6V8	MONTROSE, CO
75S	BURLINTON MOUNT VERNON, WA
76S	TROY, MT
77M	MALTA, ID
7A9	PLAINS, GA
7R1	VENICE, LA

7R3	AMELIA, LA
7R4	INTRACOASTAL CITY, LA
7R5	CAMERON, LA
7TB	TOBYHANNA, PA
7T6	GULF OF MEXICO, LA
8Y8	CRANE LAKE, MN
9B2	NEWPORT, VT
9F2	FOURCHON, LA
9Z0	KLAWOCK, AK

A

AAP	Andrau Airpark, HOUSTON, TX
ABE	ALLENTOWN, PA
ABI	ABILENE, TX
ABQ	ALBUQUERQUE, NM
ABR	ABERDEEN, SD
ABY	ALBANY, GA
ACK	NANTUCKET, MA
ACT	WACO, TX
ACV	ARCATA, CA
ACY	ATLANTIC CITY, NJ
ADK	ADAK ISLAND, AK
ADM	ARDMORE, OK
ADQ	KODIAK, AK
ADS	Addison, DALLAS, TX
ADU	AUDUBON, IA
ADW	Andrews AFB, CAMP SPRINGS, MD
AEL	ALBERT LEA, MN
AEX	England AFB, ALEXANDRIA, LA
AFF	USAF Academy, COLORADO SPRINGS, CO
AFJ	WASHINGTON, PA
AFW	Alliance, FT WORTH, TX
AGC	Allegheny Co., PITTSBURGH, PA
AGN	ANGOON, AK
AGR	AVON PARK, FL
AGS	AUGUSTA, GA
AHN	ATHENS, GA
AHT	AMCHITKA, AK
AIA	ALLIANCE, NE
AID	ANTIGO, WI
AIN	WAINWRIGHT, AK
AIT	AITKIN, MN
AIZ	KAISER/LAKE OZARK, MO
AKN	KING SALMON, AK

AKO	AKRON, CO
ALB	ALBANY, NY
ALI	ALICE, TX
ALM	ALAMOGORDO, NM
ALN	St. Louis Regional, ALTON/ST. LOUIS, IL
ALO	WATERLOO, IA
ALS	ALAMOSA, CO
ALW	WALLA WALLA, WA
AMA	AMARILLO, TX
AMG	ALMA, GA
ANB	ANNISTON, AL
ANC	ANCHORAGE, AK
AND	ANDERSON, SC
ANE	Janes Field, MINNEAPOLIS, MN
ANI	ANIAK, AK
ANN	ANNETTE, AK
ANW	AINSWORTH, NE
AOO	ALTOONA, PA
APA	Centennial, DENVER, CO
APC	NAPA, CA
APF	NAPLES, FL
APG	Aberdeen Pvg Gnd, ABERDEEN, MD
APN	ALPENA, MI
AQH	QUINHAGAK, AK
AQQ	APALACHICOLA, FL
ARA	NEW IBERIA, LA
ARB	ANN ARBOR, MI
ARG	WALNUT RIDGE, AK
ARR	Aurora, CHICAGO, IL
ART	WATERTOWN, NY
ARV	MINOCQUA WOODRUFF, WI
ASE	ASPEN, CO
ASG	SPRINGDATE, AR
ASH	NASHUA, NH
AST	ASTORIA, OR
ATL	Hartsfield, ATLANTA, GA
ATW	APPLETON, WI
ATY	WATERTOWN, SD
AUG	AUGUSTA, ME
AUM	AUSTIN, MN
AUO	AUBURN, AL
AUS	AUSTIN, TX
AUW	WAUSAU, WI

AVL	ASHEVILLE, NC
AVP	WILKES-BARRE/SCRANTON, PA
AVX	Catalina, AVALON, CA
AWH	MOUNTAIN CITY, NV
AWY	ERIE, PA
AXN	ALEXANDRIA, MN
AXO	GRAND ISLE, LA
AYE	FT DEVENS AYER, MA
AYS	WAYCROSS, GA
AZO	KALAMAZOO, MI

B

B23	BATTLE MOUNTAIN, NV
BAB	Beal AFB, MARYSVILLE, CA
BAD	Barksdale AFB, BOSSIER CITY, LA
BAF	WESTFIELD, MA
BAK	COLUMBUS, IN
BBW	BROKEN BOW, NE
BCE	BRYCE CANYON, UT
BDE	BAUDETTE, MN
BDL	Bradley, WINDSOR LOCKS, CT
BDR	BRIDGEPORT, CT
BED	BEDFORD, MA
BET	BETHEL, AK
BEH	BENTON HARBOR, MI
BFD	BRADFORD, PA
BFF	SCOTTSBLUFF, NE
BFI	Boeing Field, SEATTLE, WA
BFL	BAKERSFIELD, CA
BFM	Brookley, MOBILE, AL
BGM	BINGHAMTON, NY
BGR	BANGOR, ME
BHB	BAR HARBOR, ME
BHM	BIRMINGHAM, AL
BID	BLOCK ISLAND, RI
BIE	BEATRICE, NE
BIG	DELTA JUNCTION/FT. GREELY, AK
BIH	BISHOP, CA
BIL	BILLINGS, MT
BIS	BISMARCK, ND
BIX	Keesler AFB, BILOXI, MS
BJC	Jeffco, DENVER, CO
BJI	BEMIDJI, MN
BKE	BAKER, OR

BKH	KEKAHA, KAUAI, HI
BKL	Lakefront, CLEVELAND, OH
BKW	Raleigh, BECKLEY, WV
BKX	BROOKINGS, SD
BLF	BLUEFIELD, WV
BLH	BLYTHE, CA
BLI	BELLINGHAM, WA
BLM	BELMAR FARMINGDALE, NJ
BLU	BLUE CANYON, CA
BLV	Scott AFB, BELLEVILLE, IL
BMG	BLOOMINGTON, IN
BMI	BLOOMINGTON/NORMAL, IL
BML	BERLIN, NH
BNA	NASHVILLE, TN
BNO	BURNS, OR
BNY	BURNEY, CA
BOI	BOISE, ID
BOS	BOSTON, MA
BOW	BARTOW, FL
BPI	BIG PINEY, WY
BPT	BEAUMONT/PORT ARTHUR, TX
BQK	BRUNSWICK, GA
BRD	BRAINERD, MN
BRL	BURLINGTON, IA
BRO	BROWNSVILLE, TX
BRW	BARROW, AK
BSM	Bergstrom AFB, AUSTIN, TX
BTI	BARTER ISLAND, AK
BTL	BATTLE CREEK, MI
BTM	BUTTE, MT
BTP	BUTLER, PA
BTR	BATON ROUGE, LA
BTT	BETTLES, AK
BTV	BURLINGTON, VT
BUF	BUFFALO, NY
BUO	BEAUMONT, CA
BUR	BURBANK, CA
BVE	BOOTHSVILLE, LA
BVI	BEAVER FALLS, PA
BVO	BARTLESVILLE, OK
BVY	BEVERLY, MA
BWD	BROWNWOOD, TX
BWG	BOWLING GREEN, KY

BWI	BALTIMORE, MD
BYH	Eaker AFB, BLYTHEVILLE, AR
BYI	BURLELY, ID
BZN	BOZEMAN, MT

C

C87	EAST CAMERON, LA
CAD	CADILLAC, MI
CAE	COLUMBIA, SC
CAG	CRAIG, CO
CAK	NORTH CANTON, OH
CAO	CLAYTON, NM
CAR	CARIBOU, ME
CBG	CAMBRIDGE, NM
CBM	Columbus AFB, COLUMBUS, MS
CCR	CONCORD, CA
CCY	CHARLES CITY, IA
CDB	COLD BAY, AK
CDC	CEDAR CITY, UT
CDH	CAMDEN, AR
CDR	CHADRON, NE
CDS	CHILDRESS, TX
CDV	CORDOVA, AK
CDW	Essex, CALDWELL, NJ
CEC	CRESCENT CITY, CA
CEF	SPRINGFIELD CHICOPEE, MA
CEW	CRESTVIEW, FL
CEZ	CORTEZ, CO
CGF	Cuyahoga, CLEVELAND, OH
CGI	CAPE GIRARDEAU, MO
CGX	Meigs, CHICAGO, IL
CGZ	CASA GRANDE, AZ
CHA	CHATTANOOGA, TN
CHH	CHATHAM, MA
CHO	CHARLOTTESVILLE, VA
CHS	CHARLESTON, SC
CIC	CHICO, CA
CID	CEDAR RAPIDS, IA
CIU	Chippewa, SAULT STE MARIE, MI
CKB	CLARKSBURG, WV
CKL	CENTREVILLE, AL
CKN	CROOKSTON, MN
CKV	CLARKSVILLE, TN
CLE	Hopkins, CLEVELAND, OH

CLL	COLLEGE STATION, TX
CLM	PORT ANGELES, WA
CLT	CHARLOTTE, NC
CMA	CAMARILLO, CA
CMH	COLUMBUS, OH
CMI	CHAMPAIGN/URBANA, IL
CMX	HANCOCK, MI
CNK	CONCORDIA, KS
CNM	CARLSBAD, NM
CNO	CHINO, CA
CNU	CHANUTE, KS
CNY	Canyonlands, MOAB, UT
COD	CODY, WY
COE	COEUR D ALENE, ID
COF	Patrick AFB, COCOA BEACH, FL
CON	CONCORD, NH
COQ	CLOQUET, NM
COS	COLORADO SPRINGS, CO
COT	COTULLA, TX
COU	COLUMBIA, MO
CPR	CASPER, WY
CPS	Cahokia, ST LOUIS, IL
CQV	COLVILLE, WA
CRE	NORTH MYRTLE BEACH, SC
CRG	Craig, JACKSONVILLE, FL
CRP	CORPUS CHRISTI, TX
CRQ	Palomar, CARLSBAD, CA
CRW	CHARLESTON, WV
CSG	COLUMBUS, GA
CSH	CAPE SARICHEF, AK
CSM	CLINTON, OK
CSV	CROSSVILLE, TN
CTB	CUT BANK, MT
CUL	CARMI, IL
CVG	COVINGTON/CINCINNATI, OH
CVM	ALTON, IL
CVN	CLOVIS, NM
CVO	CORVALLIS, OR
CVS	Cannon AFB, CLOVIS, NM
CWA	MOSINEE, WI
CWF	LAKE CHARLES, LA
CWI	CLINTON, IA
CXO	CONROE, TX

CXY	Capital, HARRISBURG, PA
CYS	CHEYENNE, WY
CYT	YAKATAGA, AK
CZD	COZAD, NE
CZK	CASCADE LOCKS, OR
CZZ	CAMPO, CA

D

D49	COLUMBUS, ND
DAA	FT BELVOIR, WA
DAB	DAYTONA BEACH, FL
DAG	DAGGETT, CA
DAL	Love Field, DALLAS, TX
DAN	DANVILLE, VA
DAY	DAYTON, OH
DBQ	DUBUQUE, IA
DCA	National, WASHINGTON, DC
DDC	DODGE CITY, KS
DEC	DECATUR, IL
DEN	Stapleton, DENVER, CO
DET	Detroit City, DETROIT, MI
DFW	DALLAS-FORT WORTH, TX
DHN	DOTHAN, AL
DHT	DALHART, TX
DIK	DICKINSON, ND
DLF	Laughlin AFB, DEL RIO, TX
DLG	DILLINGHAM, AK
DLH	DULUTH, MN
DLN	DILLON, MT
DLS	THE DALLES, OR
DMA	Davis-Monthan AFB, TUSCON, AZ
DMN	DEMING, NM
DNV	DANVILLE, IL
DOV	Dover AFB, DOVER, DE
DPA	Dupage, CHICAGO, IL
DPG	DUGWAY PROVING GROUND, UT
DRA	Desert Rock, MERCURY, NV
DRI	DE RIDDER, LA
DRO	DURANGO, CO
DRT	DEL RIO, TX
DSM	DES MOINES, IA
DTL	DETROIT LAKES, MN
DTN	Downtown, SHREVEPORT, LA
DTW	Wayne, DETROIT, MI

DUG	DOUGLAS BISBEE, AZ
DUJ	DU BOIS, PA
DUT	UNALASKA, AK
DUY	KONGIGANAK, AK
DVL	DEVILS LAKE, ND
DVT	Deer Valley, PHOENIX, AZ
DWH	David Wayne Hooks, HOUSTON, TX
DXR	DANBURY, CT
DYR	DYERBURG, TN
DYS	Dyess AFB, ABILENE, TX
E	
E23	TAOS, NM
E33	CHAMA, NM
E74	SAFFORD, AZ
EAR	KEARNEY, NE
EAT	WENATCHEE, WA
EAU	EAU CLAIRE, WI
EBS	WEBSTER CITY, IA
ECG	ELIZABETH CITY, NC
EDW	Edwards AFB, EDWARDS, CA
EED	NEEDLES, CA
EEN	KEENE, NH
EFD	Ellington AFB, HOUSTON, TX
EGE	EAGLE, CO
EKM	ELKHART, IN
EKN	ELKINS, WV
EKO	ELKO, NV
EKX	ELIZABETHTOWN, KY
ELD	EL DORADO, AR
ELM	ELMIRA, NY
ELN	ELLENSBURG, WA
ELO	ELY, MN
ELP	EL PASO, TX
ELV	ELFIN COVE, AK
ELY	ELY, NV
EMP	EMPORIA, KS
EMT	EL MONTE, CA
ENA	KENAI, AK
END	Vance AFB, ENID, OK
ENM	EMMONAK, AK
ENN	NEHANA, AK
ENV	WENDOVER, UT
EPH	EPHRATA, WA

ERI	ERIE, PA
ERV	KERRVILLE, TX
ESC	ESCANABA, MI
ESF	ALEXANDRIA, LA
EST	ESTHERVILLE, IA
ETH	WHEATON, MN
ETM	EAST SIMPSON, AK
EUG	EUGENE, OR
EVM	EVELETH, MN
EVV	EVANSVILLE, IN
EVW	EVANSTON, WY
EWB	NEW BEDFORD, MA
EWK	NEWTON, KS
EWN	NEW BERN, NC
EWR	NEWARK, NJ
EWU	NEWTOK, AK
EYW	KEY WEST, FL

F

F39	SHERMAN DENISON, TX
F60	Gregor, WACO, TX
FAF	FT EUSTIS, VA
FAI	FAIRBANKS, AK
FAR	FARGO, ND
FAT	FRESNO, CA
FAY	FAYETTEVILLE, NC
FBG	Simmons AAF, FORT BRAGG, NC
FBL	FARIBAULT, MN
FCA	KALISPELL, MT
FCL	FORT COLLINS, CO
FCM	Flying Cloud, MINNEAPOLIS, MN
FDY	FINDLAY, OH
FFM	FERGUS FALLS, MN
FFO	Wright-Patterson AFB, DAYTON, OH
FFZ	Falcon Field, MESA, AZ
FHR	FRIDAY HARBOR, WA
FHU	FT. HUACHUCA, AZ
FKL	FRANKLIN, PA
FLG	FLAGSTAFF, AZ
FLL	Hollywood, FORT LAUDERDALE, FL
FLO	FLORENCE, SC
FLV	FT. LEAVENWORTH, KS
FME	FT MEADE, MD
FMH	FALMOUTH, MA

FMN	FARMINGTON, NM
FMY	FORT MYERS, FL
FNB	FALLS CITY, NE
FNL	FORT COLLINS/LOVELAND, CO
FNR	FUNTER BAY, AK
FNT	FLINT, MI
FOD	FORT DODGE, IA
FOE	Forbes, TOPEKA, KS
FOK	Suffolk, WESTHAMPTON BEACH, NY
FOQ	FREEPORT, TX
FPR	FT PIERCE, FL
FRG	Republic, FARMINGDALE, NY
FRM	FAIRMONT, MN
FSD	SIOUX FALLS, SD
FSE	FOSSTON, MN
FSI	FT SILL, OK
FSM	FORT SMITH, AR
FTK	FT. KNOX, KY
FTW	FORT WORTH, TX
FTY	Fulton Co, ATLANTA, GA
FUL	FULLERTON, CA
FVE	FRENCHVILLE, ME
FVM	FIVE MILE, AK
FWA	FORT WAYNE, IN
FWH	Carswell AFB, FORT WORTH, TX
FWL	FAREWELL, AK
FWS	Spinks, FT WORTH, TX
FXE	Executive, FORT LAUDERDALE, FL
FYU	FORT YUKON, AK
FYV	FAYETTEVALE, AR
G	
GAD	GADSDEN, AL
GAG	GAGE, OK
GAI	GAITHERSBURG, MD
GAL	GALENA, AK
GAM	GAMBELL, AK
GBD	GREAT BEND, KS
GBG	GALESBURG, IL
GBH	GALBRAITH LAKE, AK
GBN	GILA BEND, AZ
GCC	GILLETTE, WY
GCK	GARDEN CITY, KS
GCN	GRAND CANYON, AZ

GDV	GLENDIVE, MT
GED	GEORGETOWN, DE
GEG	International, SPOKANE, WA
GFA	Malmstrom AFB, GREAT FALLS, MT
GFK	GRAND FORKS, ND
GFL	GLENS FALLS, NY
GGG	LONGVIEW, TX
GGW	GLASGOW, MT
GJT	GRAND JUNCTION, CO
GKN	GULKANA, AK
GLD	GOODLAND, KS
GLH	GREENVILLE, MS
GLS	GALVESTON, TX
GMU	GREENVILLE, SC
GNT	GRANTS, NM
GNV	GAINSVILLE, FL
GON	GROTON/NEW LONDON, CT
GPT	Biloxi, GULFPORT, MS
GPZ	Itasca Co., GRAND RAPIDS, MN
GRB	GREEN BAY, WI
GRF	Gray AAF, TACOMA, WA
GRI	GRAND ISLAND, NE
GRK	FT. HOOD, TX
GRR	Kent Co., GRAND RAPIDS, MI
GSB	Seymour Johnson AFB, GOLDSBORO, NC
GSO	GREENSBORO, NC
GSP	GREER, SC
GST	GUSTAVUS, AK
GTB	CUT BANK, MT
GTF	GREAT FALLS, MT
GTR	COLUMBUS/W POINT/STARKVILL, MS
GUC	GUNNISON, CO
GUP	GALLUP, NM
GUS	Grissom AFB, PERU, IN
GVT	GREENSVILLE, TX
GVW	Richards-Gebaur AFB, KANSAS CITY, MO
GWO	GREENWOOD, MS
GXY	GREELEY, CO
GYR	Goodyear, PHOENIX, AZ
GYY	GARY, IN

H

HAT	HATTERAS, NC
HBR	HOBART, OK

HCD	HUTCHINSON, MN
HDN	HAYDEN, CO
HDO	HONDO, TX
HEF	MANASSAS,VA
HEZ	NATCHEZ, MS
HFD	HARTFORD, CT
HGR	HAGERSTOWN, MD
HHF	CANADIAN, TX
HHR	HAWTHORNE, CA
HIB	HIBBING, MN
HIF	Hill AFB, OGDEN, UT
HIO	Hillsboro, PORTLAND, OR
HKY	HICKORY, NC
HLC	HILL CITY, KS
HLG	Ohio Co., WHEELING, WV
HLN	HELENA, MT
HMM	HAMILTON, MT
HMN	Halloman AFB, ALAMOGORDO, NM
HMS	HANFORD, WA
HNL	HONOLULU, HI
HNM	HANA, HI
HOB	HOBBS, NM
HOM	HOMER, AK
HON	HURON, SD
HOP	FT. CAMPBELL, KY
HOT	HOT SPRINGS, AR
HOU	Hobby, HOUSTON, TX
HPB	HOOPER BAY, AK
HPN	Weschester, WHITE PLAINS, NY
HQM	HOQUIAM, WA
HRL	HARLINGEN, TX
HRO	HARRISON, AR
HRT	MARY ESTHER, FL
HSI	HASTINGS, NE
HSP	HOT SPRINGS, VA
HSS	HOT SPRINGS, NC
HST	Homestead AFB, HOMESTEAD, FL
HSV	HUNTSVILLE, AL
HTH	HAWTHORNE, NV
HTL	HOUGHTON LAKE, MI
HTS	HUNTINGTON, WV
HUF	TERRE HAUTE, IN
HUL	HOULTON, ME

HUM	HOUMA, LA
HUT	HUTCHINSON, KS
HVN	Tweed, NEW HAVEN, CT
HVR	HAVRE, MT
HWD	HAYWARD, CA
HWV	SHIRLEY, NY
HYA	HYANNIS, MA
HYS	HAYS, KS

I

IAB	McConnel AFB, WICHITA, KS
IAD	Dulles, WASHINGTON, DC
IAG	NIAGARA FALLS, NY
IAH	Intercontinental, HOUSTON, TX
IAN	KIANA, AK
ICT	WICHITA, KS
IDA	IDAHO FALLS, ID
IGM	KINGMAN, AZ
IIK	KIPNUT, AK
IKK	KANKAKEE, IL
ILE	KILLEEN, TX
ILG	WILMINGTON, DE
ILI	ILIAMNA, AK
ILL	WILLMAR, MN
ILM	WILMINGTON, NC
ILN	WILMINGTON, OH
IML	IMPERIAL, NE
IMT	IRON MOUNTAIN/KINGSFORD, MI
IND	INDIANAPOLIS, IN
INK	WINK, TX
INL	INTERNATIONAL FALLS, MN
INT	WINSTON SALEM, NC
INW	WINSLOW, AZ
IPL	IMPERIAL, CA
IPT	WILLIAMSPORT, PA
IRK	KIRKSVILLE, MO
ISM	Kissimmee, ORLANDO, FL
ISN	WILLISTON, ND
ISO	KINSTON, NC
ISP	MacArthur, ISLIP, NY
ISW	WISCONSIN RAPIDS, WI
ITH	ITHACA, NY
ITO	HILO, HI
IWA	Williams, CHANDLER, AZ

IWD	IRONWOOD, MI
IWS	Lakeside, HOUSTON, TX
IXD	OLATHE, KS
IYK	INYOKERN, CA

J

JAC	Jackson Hole, JACKSON, WY
JAN	JACKSON, MS
JAX	JACKSONVILLE, FL
JBR	JONESBORO, AR
JEF	JEFFERSON CITY, MO
JFK	JFK, NEW YORK, NY
JHM	LAHAINA, HI
JHW	JAMESTOWN, NY
JKL	JAKSON, KY
JLN	JOPLIN, MO
JMS	JAMESTOWN, ND
JNU	JUNEAU, AK
JNW	NEWPORT, OR
JOH	JOHNSTONE POINT, AK
JST	JOHNSTOWN, PA
JVL	JANESVILLE, WI
JXN	JACKSON, MI

K

KAE	KAKE, AK
KLS	KELSO, WA
KOA	KAILUA/KONA, HI
KPC	PORT CLARENCE, AK
KSM	ST MARY'S, AK
KTN	KETCHIKAN, AK

L

LAF	Purdue, LAFAYETTE, IN
LAL	LAKELAND, FL
LAM	LOS ALAMOS, NM
LAN	LANSING, MI
LAR	LARAMIE, WY
LAS	LAS VEGAS, NV
LAW	LAWTON, OK
LAX	International, LOS ANGELES, CA
LBB	LUBBOCK, TX
LBE	LATROBE, PA
LBF	NORTH PLATTE, NE
LBL	LIBERAL, KS

LBX	ANGLETON/LAKE JACKSON, TX
LCH	LAKE CHARLES, LA
LCI	LACONIA, NH
LCK	Rickenbacker, COLUMBUS, OH
LEB	LEBANON, NH
LEW	AUBURN/LEWISTON, ME
LEX	LEXINGTON, KY
LFI	Langley AFB, HAMPTON, VA
LFK	LUFKIN, TX
LFT	LAFAYETTE, LA
LGA	La Guardia, NEW YORK, NY
LGB	LONG BEACH, CA
LGD	LA GRANDE, OR
LGU	LOGAN, UT
LHW	FT. STEWART, GA
LHX	LA JUNTA, CO
LIC	LIMON, CO
LIH	LIHUE, HI
LIT	LITTLE ROCK, AR
LIZ	Loring AFB, LIMESTONE, ME
LKV	LAKEVIEW, OR
LMS	LOUISVILLE, MS
LMT	KLAMATH FALLS, OR
LND	LANDER, WY
LNK	LINCOLN, NE
LNN	WILLOUGHBY, OH
LNP	WISE, VA
LNR	LONE ROCK, WI
LNS	LANCASTER, PA
LNY	LANAI CITY, HI
LOL	LOVELOCK, NV
LOU	Bowman Field, LOUISVILLE, KY
LOZ	LONDON, KY
LPC	LOMPOC, CA
LRD	LAREDO, TX
LRF	Little Rock AFB, JACKSONVILLE, AR
LRU	LAS CRUCES, NM
LSE	LA CROSSE, WI
LSF	FORT BENNING, GA
LSV	Nellis AFB, LAS VEGAS, NV
LTS	ALTUS, OK
LUF	Luke AFB, GLENDALE, AZ
LUK	Lunken, CINCINNATI, OH

LVK	LIVERMORE, CA
LVM	LIVINGSTON, MT
LVS	LAS VEGAS, NV
LWB	LEWISBURG, WV
LWM	LAWRENCE, MA
LWS	LEWISTON, ID
LWT	LEWISTON, MT
LXL	LITTLE FALLS, MN
LXV	LEADVILLE, CO
LYH	LYNCHBURG, VA

M

MAF	MIDLAND, TX
MAZ	MAYAGUEZ, PR
MBL	MANISTEE, MI
MBS	SAGINAW, MI
MCB	MC COMB, MS
MCC	McClellan AFB, SACRAMENTO, CA
MCE	MERCED, CA
MCF	MacDill AFB, TAMPA, FL
MCG	MC GRATH, AK
MCI	International, KANSAS CITY, MO
MCK	MC COOK, NE
MCN	MACON, GA
MCO	International, ORLANDO, FL
MCW	MASON CITY, IA
MDH	CARBONDALE/MURPHYSBORO, IL
MDO	MIDDLETON ISLAND, AK
MDT	Harrisburg, MIDDLETOWN, PA
MDW	Midway, CHICAGO, IL
MEI	MERIDIAN, MS
MEM	International, MEMPHIS, TN
MER	Castle AFB, MERCED, CA
MEV	MINDEN, NV
MFD	MANSFIELD, OH
MFE	MC ALLEN, TX
MFI	MARSHFIELD, WI
MFR	MEDFORD, OR
MGE	MARIETTA, GA
MGM	MONTGOMERY, AL
MGR	MOULTRIE, GA
MGW	MORGANTOWN, WV
MHE	MITCHELL, SD
MHK	MANHATTAN, KS

MHM	MINCHUMINA, AK
MHN	MULLEN, NE
MHT	MANCHESTER, NH
MHV	MOJAVE, CA
MIA	International, MIAMI, FL
MIB	Minot AFB, MINOT, ND
MIC	Crystal, MINNEAPOLIS, MN
MIE	MUNCIE, IN
MIV	MILLVILLE, NJ
MJQ	JACKSON, MN
MKC	Downtown, KANSAS CITY, MO
MKE	Mitchell, MILWAUKEE, WI
MKG	MUSKEGON, MI
MKK	Molokai, KAUNAKAKAI, HI
MKL	JACKSON, TN
MKT	MANKATO, MN
MKY	MARCO ISLAND, FL
MLB	MELBORNE, FL
MLC	MC ALESTER, OK
MLD	MALAD CITY, ID
MLF	MILFORD, UT
MLI	MOLINE, IL
MLS	MILES CITY, MT
MLT	MILLINOCKET, ME
MLU	MONROE, LA
MLY	MANLEY HOT SPRINGS, AK
MMH	MAMMOTH LAKES, CA
MML	MARSHALL, MN
MMO	MARSEILLES, IL
MMT	Mc Entire, COLUMBIA, SC
MMU	MORRISTOWN, NJ
MMV	MC MINNVILLE, OR
MNM	MENOMINEE, MI
MOB	MOBILE, AL
MOD	MODESTO, CA
MOT	MINOT, ND
MOX	MORRIS, MN
MPV	BARRE/MONTPELIER, VT
MQI	MANTEO, NC
MQM	MONDIA, MT
MQT	MARQUETTE, MI
MQY	SMYRNA, TN
MRB	MARTINSBURG, WV

MRF	MARFA, TX
MRI	Merrill Field, ANCHORAGE, AK
MRY	MONTEREY, CA
MSL	MUSCLE SHOALS, AL
MSN	MADISON, WI
MSO	MISSOULA, MT
MSP	International, MINNEAPOLIS, MN
MSS	MASSENA, NY
MSV	MONTICELLO, NY
MSY	Moisant, NEW ORLEANS, LA
MTC	MOUNT CLEMENS, MI
MTH	MARATHON, FL
MTJ	MONTROSE, CO
MTN	Martin State, BALTIMORE, MD
MTW	MANITOWOC, WI
MUE	Waimeakohala, KAMUELA, HI
MUO	MOUNTAIN HOME, ID
MVE	MONTEVIDEO, MN
MVN	MOUNT VERNON, IL
MVY	Vineyard Haven, MARTHAS VINEYARD, MA
MWA	MARION, IL
MWC	Timmerman, MILWAUKEE, WI
MWH	MOSES LAKE, WA
MWL	MINERAL WELLS, TX
MWN	MOUNT WASHINGTON, NH
MWS	MOUNT WILSON, CA
MXF	Maxwell AFB, MONTGOMERY, AL
MXY	MC CARTHY, AK
MYF	Montgomery, SAN DIEGO, CA
MYL	MC CALL, ID
MYR	Myrtle Beach AFB, SC
MYU	MEKORYUK, AK
MYV	MARYSVILLE, CA
MZZ	MARION, IN
N	
NBC	Beaufort MCAS, SC
NBE	Hensley NAS, DALLAS, TX
NBG	New Orleans NAS, LA
NBU	Glenview NAS, IL
NCA	New River MCAS, JACKSONVILLE, NC
NEL	LAKEHURST, NJ
NEW	Lakefront, NEW ORLEANS, LA
NFG	Camp Pendleton, OCEANSIDE, CA

NFL	Fallon NAS, NV
NGP	Corpus Christi NAS, TX
NGU	Norfolk NAS, VA
NGZ	Alameda NAS, CA
NHK	Patuxent River NAS, MD
NHZ	Brunswick NAS, ME
NID	China Lake NAS, CA
NIP	Jacksonville NAS, FL
NIR	Chase Field NAS, BEEVILLE, TX
NKT	Cherry Point MCAS, NC
NKX	Miramar NAS, SAN DIEGO, CA
NLC	Lemoore NAS, CA
NMM	MERIDIAN, MS
NPA	Pensacola NAS, FL
NQA	Memphis NAS, MILLINGTON, TX
NQI	Kingsville NAS, TX
NRB	Mayport NAS, FL
NSE	MILTON, FL
NSI	SAN NICOLAS ISLAND, CA
NTD	Point Nugu NAS, CA
NTU	Oceana NAS, VIRGINIA BEACH, VA
NUC	SAN CLEMENTE ISLAND, CA
NUQ	Moffett NAS, CA
NUW	Whidbey Is NAS, OAK HARBOR, WA
NXX	Willow Grove NAS, PA
NYG	Quantico MCAF, VA
NZC	Cecil NAS, FL
NZJ	El Toro MCAS, SANTA ANA, CA
NZW	South Weymouth NAS, MA
NZY	North Island NAS, SAN DIEGO, CA

O

OAJ	JACKSONVILLE, NC
OAK	OAKLAND, CA
OAR	Ft. Ord, MONTEREY, CA
OCF	OCALA, FL
OCH	NACOGDOCHES, TX
OCV	BERING SEA, AK
ODX	ORD, NE
OFF	Offutt AFB, OMAHA, NE
OFK	NORFOLK, NE
OGD	OGDEN, UT
OGG	KAHULUI, HI
OGS	OGDENSBURG, NY

OJC	OLATHE, KS
OKC	Will Rogers, OKLAHOMA CITY, OK
OKK	KOKOMO, IN
OLE	OLEAN, NY
OLF	WOLF POINT, MT
OLI	OLIKTOK POINT, AK
OLM	OLYMPIA, WA
OLU	COLUMBUS, NE
OMA	OMAHA, NE
OME	NOME, AK
ONA	WINONA, MN
ONL	O NEILL, NE
ONM	SOCORRO, NM
ONO	ONTARIO, OR
ONP	NEWPORT, OR
ONT	ONTARIO, CA
OPF	Opa Locka, MIAMI, FL
OQU	NORTH KINGSTOWN, RI
ORD	O Hare, CHICAGO, IL
ORF	NORFOLK, VA
ORG	ORANGE, TX
ORH	WORCESTER, MA
ORL	Executive, ORLANDO, FL
ORT	NORTHWAY, AK
OSC	Wurtsmith AFB, OSCODA, MI
OSH	OSHKOSH, WI
OSU	Ohio State, COLUMBUS, OH
OTG	WORTHINGTON, MN
OTH	NORTH BEND, OR
OTM	OTTUMWA, IA
OTZ	KOTZEBUE, AK
OUN	NORMAN, OK
OVN	WSO, OMAHA, NE
OWA	OWATONNA, MN
OWB	OWENSBORO, KY
OWD	NORWOOD, MA
OWY	OWYHEE, NV
OXC	OXFORD, CT
OXR	OXNARD, CA
OZR	Cairns AAF, FT RUCKER, AL
P	
PAE	Paine Field, EVERETT, WA
PAH	PADUCAH, KY

PAM	Tindall AFB, PANAMA CITY, FL
PAO	PALO ALTO, CA
PAQ	PALMER, AK
PBF	PINE BLUFF, AR
PBG	Plattsburgh AFB, NY
PBH	PHILLIPS, WI
PBI	WEST PALM BEACH, FL
PDK	Peachtree, ATLANTA, GA
PDT	PENDLETON, OR
PDX	PORTLAND, OR
PFC	PACIFIC CITY, OR
PFN	PANAMA CITY, FL
PGA	PAGE, AZ
PHF	NEWPORT NEWS, VA
PHL	Int'l, PHILADELPHIA, PA
PHO	POINT HOPE, AK
PHX	Sky Harbor, PHOENIX, AZ
PIA	PEORIA, IL
PIB	LAUREL/HATTIESBURG, MS
PIE	ST. PETERSBURG/CLEARWATER, FL
PIH	POCATELLO, ID
PIR	PIERRE, SD
PIT	International, PITTSBURGH, PA
PIZ	POINT LAY, AK
PKB	PARKERSBURG, WV
PKD	PARK RAPIDS, MN
PLN	PELLSTON, MI
PMD	PALMDALE, CA
PMP	POMPANO BEACH, FL
PNC	PONCA CITY, OK
PNE	Northeast, PHILADELPHIA, PA
PNM	PRINCETON, MN
PNS	PENSACOLA, FL
POB	Pope AFB, FAYETTEVILLE, NC
POC	Brackett, LA VERNE, CA
POE	Polk AAF, LA
POU	POUGHKEEPSIE, NY
PPC	PROSPECT CREEK, AK
PPF	PARSONS, KS
PQI	PRESQUE ISLE, ME
PQN	PIPESTONE, MN
PRB	PASO ROBLES, CA
PRC	PRESCOTT, AZ

PRX	PARIS, TX
PSB	PHILIPSBURG, PA
PSC	PASCO, WA
PSF	PITTSFIELD, MA
PSG	PETERSBURG, AK
PSK	DUBLIN, VA
PSM	Pease AFB, PORTSMOUTH, NH
PSP	PALM SPRINGS, CA
PSX	PALACIOS, TX
PTH	PORT HEIDEN, AK
PTI	PUNTILLA LAKE, AK
PTK	PONTIAC, MI
PTN	PATTERSON, LA
PTU	PLATINUM, AK
PTV	PORTERVILLE, CA
PUB	PUEBLO, CO
PUC	PRICE, UT
PUO	PRUDHOE BAY, AK
PUW	PULLMAN/MOSCOW, WA
PVC	PROVINCETOWN, MA
PVD	PROVIDENCE, RI
PVG	PORTSMOUTH, VA
PVU	PROVO, UT
PVW	PLAINVIEW, TX
PWA	Wiley Post, OKLAHOMA CITY, OK
PWM	PORTLAND, ME
PWT	BREMERTON, WA
PYM	PLYMOUTH, MA
R	
RAD	WARROAD, MN
RAL	RIVERSIDE, CA
RAP	RAPID CITY, SD
RBD	Redbird, DALLAS, TX
RBG	ROSEBURG, OR
RBL	RED BLUFF, CA
RBY	RUBY, AK
RCA	Ellsworth AFB, RAPID CITY, SD
RDD	REDDING, CA
RDG	READING, PA
RDM	REDMOND, OR
RDR	Grand Forks AFB, ND
RDU	RALEIGH/DURHAM, NC
REE	Reese AFB, LUBBOCK, TX

RFD	ROCKFORD, IL
RGK	RED WING, MN
RHI	RHINELANDER, WI
RHV	Reid-Hillview, SAN JOSE, CA
RIC	RICHMOND, VA
RIE	RICE LAKE, WI
RIL	RIFLE, CO
RIV	March AFB, RIVERSIDE, CA
RIW	RIVERTON, WY
RKD	ROCKLAND, ME
RKP	ROCKPORT, TX
RKS	ROCK SPRINGS, WY
RME	Griffiss AFB, ROME, NY
RMG	ROME, GA
RND	Randolph AFB, UNIVERSAL CITY, TX
RNO	RENO, NV
RNT	RENTON, WA
ROA	ROANOKE, VA
ROC	ROCHESTER, NY
ROG	ROGERS, AR
ROW	ROSWELL, NM
ROX	ROSEAU, MN
RPE	SABRINE PASS, TX
RSL	RUSSELL, KS
RST	ROCHESTER, MN
RSW	SW Regional, FORT MYERS, FL
RUI	RUIDOSO, NM
RUT	RUTLAND, VT
RWF	REDWOOD FALLS, MN
RWI	ROCKY MOUNT, NC
RWL	RAWLINS, WY
RYV	WATERTOWN, WI

S

SAC	SACRAMENTO, CA
SAF	SANTA FE, NM
SAN	Lindbergh, SAN DIEGO, CA
SAT	SAN ANTONIO, TX
SAV	SAVANNAH, GA
SAW	K.I. Sawyer AFB, GWINN, MI
SBA	SANTA BARBARA, CA
SBN	SOUTH BEND, IN
SBP	SAN LUIS OBISPO, CA
SBS	STEAMBOAT SPRINGS, CO

SBY	SALISBURY, MD
SCC	DEADHORSE, AK
SCH	SCHENECTADY, NY
SCK	STOCKTON, CA
SDF	Standiford, LOUISVILLE, KY
SDL	SCOTTSDALE, AZ
SDM	Brown Field, SAN DIEGO, CA
SDP	SAND POINT, AK
SDY	SIDNEY, MT
SEA	Seattle-Tacoma, SEATTLE, WA
SEE	Gillespie Field, SAN DIEGO, CA
SEP	STEPHENVILLE, TX
SEZ	SEDONA, AZ
SFB	SANFORD, FL
SFF	Felts Field, SPOKANE, WA
SFM	SANFORD, ME
SFO	SAN FRANCISCO, CA
SFZ	PAWTUCKET, RI
SGF	SPRINGFIELD, MO
SGJ	SAINT AUGUSTINE, FL
SGT	STUTTGART, AR
SGU	ST. GEORGE, UT
SGY	SKAGWAY, AK
SHD	STAUNTON/WAYNESBORO/HARRISONBURG, VA
SHH	SHISHMAREF, AK
SHN	SHELTON, WA
SHR	SHERIDAN, WY
SHV	Regional, SHREVEPORT, LA
SIT	SITKA, AK
SIY	Siskiyou, MONTAGUE, CA
SJC	SAN JOSE, CA
SJN	ST. JOHNS, AZ
SJT	SAN ANGELO, TX
SKA	Fairchild AFB, SPOKANE, WA
SKF	Kelly AFB, SAN ANTONIO, TX
SKW	SKWENTNA, AK
SKX	TAOS, NM
SLC	SALT LAKE CITY, UT
SLE	SALEM, OR
SLG	SILOAM SPRINGS, AR
SLI	LOS ALAMITOS, CA
SLK	SARANAC LAKE, NY
SLN	SALINA, KS

SME	SOMERSET, KY
SMF	Metropolitan, SACRAMENTO, CA
SMO	SANTA MONICA, CA
SMP	STAMPEDE PASS, WA
SMX	SANTA MARIA, CA
SNA	SANTA ANA, CA
SNP	ST. PAUL ISLAND, AK
SNS	SALINAS, CA
SNY	SIDNEY, NE
SOP	SOUTHERN PINES, NC
SOW	SHOW LOW, AZ
SPA	SPARTANBURG, SC
SPF	SPEARFISH, SD
SPG	Whitted, ST. PETERSBURG, FL
SPI	SPRINGFIELD, IL
SPS	WICHITA FALLS, TX
SQI	STERLING ROCK FALLS, IL
SQL	SAN CARLOS, CA
SRQ	SARASOTA/BRADENTON, FL
SRR	RUIDOSO, NM
SSC	Shaw AFB, SUMTER, SC
SSF	Stinson, SAN ANTONIO, TX
SSI	BRUNSWICK, GA
SSU	WHITE SULPHUR SPRINGS, WV
STC	ST. CLOUD, MN
STJ	ST. JOSEPH, MO
STL	Lambert, ST. LOUIS, MO
STP	ST. PAUL, MN
STS	SANTA ROSA, CA
SUE	STURGEON BAY, WI
SUN	Sun Valley Friedman, HAILEY, ID
SUS	Spirit of St. Louis, ST. LOUIS, MO
SUU	Travis AFB, FAIRFIELD, CA
SUX	SIOUX CITY, IA
SVA	SAVOONGA, AK
SVC	SILVER CITY, NM
SVE	SUSANVILLE, CA
SVN	Hunter AAF, SAVANNAH, GA
SWF	NEWBURGH, NY
SWO	STILLWATER, OK
SXQ	SOLDOTNA, AK
SYR	SYRACUSE, NY
SZL	Whiteman AFB, KNOB NOSTER, MO
S06	MULLAN, ID

T

TAD	TRINIDAD, CO
TAL	TANANA, AK
TBN	Forney AAF, FT. LEONARD WOOD, MO
TCC	TUCUMCARI, NM
TCL	TUSCALOOSA, AL
TCM	McChord AFB, TACOMA, WA
TCS	TRUTH OR CONSEQUENCES, NM
TDO	TOLEDO, WA
TEB	TETERBORO, NJ
TEX	TELLURIDE, CO
TIK	Tinker AFB, OKLAHOMA CITY, OK
TIW	Tacoma Narrows, TACOMA, WA
TIX	TITUSVILLE, FL
TKA	TALKEETNA, AK
TLH	TALLAHASSEE, FL
TMB	Tamiami, MIAMI, FL
TOA	TORRANCE, CA
TOG	TOGIAK VILLAGE, AK
TOI	TROY, AL
TOL	TOLEDO, OH
TOP	TOPEKA, KS
TPA	TAMPA, FL
TPH	TONOPAH, NV
TPL	TEMPLE, TX
TRI	BRISTOL/JOHNSON/KINGSPORT, TN
TRK	TRUCKEE, CA
TRM	THERMAL, CA
TSP	TEHACHAPI, CA
TTD	Troutdale, PORTLAND, OR
TTN	TRENTON, NJ
TUL	TULSA, OK
TUP	TUPELO, MS
TUS	TUCSON, AZ
TVC	TRAVERSE CITY, MI
TVF	THIEF RIVER FALLS, MN
TVL	SOUTH LAKE TAHOE, CA
TWF	TWIN FALLS, ID
TXK	TEXARKANA, AR
TYR	TYLER, TX
TYS	KNOXVILLE, TN

U

UCA	UTICA, NY

UES	WAUKESHA, WI
UGN	Waukegan, CHICAGO, IL
UIL	QUILLAYETE, WA
UIN	QUINCY, IL
UKI	UKIAH, CA
ULM	NEW ULM, MN
UMM	SUMMIT, AK
UMT	UMIAT, AK
UNK	UNALAKLEET, AK
UNV	STATE COLLEGE, PA
UOX	OXFORD, MS

V

VAD	Moody AFB, VALDOSTA, GA
VBG	Vandenberg AFB, LOMPOC, CA
VCT	VICTORIA, TX
VDZ	VALDEZ, AK
VEL	VERNAL, UT
VER	BOONVILLE, MO
VIH	ROLLA/VICHY, MO
VIS	VISALIA, CA
VLD	VALDOSTA, GA
VNY	VAN NUYS, CA
VOK	CAMP DOUGLAS, WI
VPS	Eglin AFB, VALPARAISO, FL
VRB	VERO BEACH, FL
VRX	GULF OF MEXICO, LA
VSF	SPRINGFIELD, VT
VTN	VALENTINE, NE
VUW	EUGENE ISLE, LA
VWS	WSO, VALDEZ, AK

W

WAL	WALLOPS ISLAND, VA
WCR	CHANDALAR LAKE, AK
WDG	ENID, OK
WEY	WEST YELLOWSTONE, MT
WHP	Whiteman, LOS ANGELES, CA
WJF	LANCASTER, CA
WLK	SELAWIK, AK
WMC	WINNEMUCCA, NV
WRB	Robins AFB, WARNER ROBINS, GA
WRG	WRANGELL, AK
WRI	McGuire AFB, WRIGHTSTOWN, NJ

WRL	WORLAND, WY
WVL	WATERVILLE, ME
WWD	WILDWOOD, NJ
WYS	WEST YELLOWSTONE, MT

Y

YAK	YAKUTAT, AK
YIP	Willow Run, DETROIT, MI
YKM	YAKIMA, WA
YKN	YANKTON, SD
YNG	YOUNGSTOWN, OH
YUM	YUMA, AZ

Z

ZZV	ZANESVILLE, OH

AIRCRAFT DESIGNATORS

This listing contains FAA civil, general aviation, aircraft type designators. Type designators consist of two to four character alphanumeric groups, required for proper flight plan processing. Listings are alphabetical by aircraft manufacturer. Because some models have been built by more than one company, types are cross-referenced when necessary.

Most types are simple and self-explanatory. A Cessna 150's type designator is C150. Others might not be as obvious. Certain manufacturers produce a basic aircraft with several different engine options. A Piper Cherokee or Warrior might be equipped with one of several different engines. The manufacturer might designate the aircraft as PA-28-151 and PA-28-181. The FAA designator, however, for both Cherokee and Warrior, with either engine, is PA28. A similar situation exists for the Mooney Mark 20 series. Mooney Mark 20, 201, 231, and 252 airplanes are designated M020.

Manufacturer's designators often contain references to turbocharging, retractable landing gear, and T-tails; FAA designators most often do not contain these references. A Piper Cherokee Six, Lance, and Saratoga, regardless of landing gear, turbocharging, or T-tail configuration, would be designated PA32. Likewise, Beech might designate the airplane as a Bonanza V35 or Bonanza A36TC; FAA designators are BE35 and BE36 respectively.

FAA aircraft type designators are continuously revised. FAA Handbook 7340.1 *Contractions* contains the latest information.

Consult a flight service station for aircraft type designators not contained within this listing, or to resolve questions about correct codes.

Aero Commander, see Rockwell

Aeronca

AR58	Aeronca Champion
AR11	Chief/Super Chief
AR15	Sedan

Aerospatiale

HR30	Alouette II
HR60	Alouette III
HR36	Dauphin
HR35	Ecureuil/Astar
HR34	Gazelle
S880	Rallye
S892	Rallye Commodore
S894	Rallye Minerva
TB10	Tabago
TB9	Tampico
TB20	Trinidad
TB21	Trinidad TC
HR55	Twin Star

Alon

F02	Ercoupe

Beech

BE99	Airliner
BE55	Baron 55
BE58	Baron 58
BE02	Beech 1900
BE40	Beech 400
BE9F	Beech F90
BE35	Bonanza 35 (V-Tail)
BE36	Bonanza 36
BE33	Debonair/Bonanza 33
BE76	Duchess 76
BE60	Duke 60
BE10	King Air 100
BE90	King Air 90
BE45	Mentor (T34)
BE65	Queen Air 65/70
BE80	Queen Air 80
BE88	Queen Air 88
BE24	Sierra 24
BE77	Skipper 77
BE19	Sport/Musketeer 19
BE17	Staggerwing 17
BE23	Sundowner/Musketeer 23
BE8S	Super 18
BE20	Super King Air 200
BE30	Super King Air 300

BE3L	Super King Air 300LW
BE95	Travelair 95
BE18	Twin Beech 18
BE50	Twin Bonanza

Bell

BH14	Biglifter
BH04	Iroquois 204
BH05	Iroquois 205
BH12	Iroquois/Twin
BH22	Model 222
BH06	Jet/Long/Sea Ranger
BH47	Sioux/Trooper 47
BHST	Super Transport 214ST

Bellanca

CH8	Challenger
CH5	Champion
CH10	Citabria
CH9	Citabria 7ECA
BL14	Cruisemaster
BL30	Decathlon
CH40	Lancer 402
BL28	Scout
CH7	Traveler 7EC
BL31	Turbo-Viking
BL26	Viking

Boeing

B75	Stearman
B105	Model 105
HV07	Seaknight

Brantly

HB43	Model 305
HB42	Model B-2A/B-2B

British Aerospace

BA31	Jetstream

Britten-Norman

BN2	Islander
BN3	Trislander

Canadair

CL60	Challenger (AV)
CL61	Challenger (GE)

Cessna

C188	Agwagon 188
C305	Bird Dog
C208	Caravan I
C177	Cardinal
C120	Cessna 120
C14	Cessna 140
C150	Cessna 150
C152	Cessna 152
C170	Cessna 170
C172	Cessna 172/Skyhawk/Cutlass
C175	Cessna 175/Skylark
C180	Cessna 180/Skywagon
C182	Cessna 182/Skylane
C185	Cessna 185/Skywagon
C190	Cessna 190
C195	Cessna 195
C205	Cessna 205
C206	Cessna 206/Super Skywagon
C207	Cessna 207/Stationair
C208	Caravan
C210	Cessna 210/Centurion
C303	Cessna 303/Crusader
C310	Cessna 310
C320	Cessna 320/Skynight
C335	Cessna 335
C336	Cessna 336/Skymaster
C337	Cessna 337/Super Skymaster
C340	Cessna 340
C401	Cessna 401
C402	Cessna 402
C404	Cessna 404/Titan
C411	Cessna 411
C414	Cessna 414/Chancellor
C421	Cessna 421/Golden Eagle
C425	Cessna 425/Conquest
C441	Cessna 441/Conquest
C500	Cessna 500/Citation I
C501	Cessna 501/Citation I/SP
C550	Cessna 550/Citation II/ S/2
C551	Cessna 551/Citation II/SP
C560	Cessna 560/Citation V
C650	Cessna 650/Citation III

Champion, see Aeronca/Bellanca

Commander, see Rockwell

Dessault-Breguet

DA50	Falcon
DA10	Falcon 10
DA20	Falcon 20/Fan Jet

DeHavilland

DH2	Beaver
DH5	Buffalo
DH4	Caribou
DH1	Chipmunk
DH7	DASH 7
DH8	DASH 8
DH10	Dove
DH11	Heron
DH3	Otter
DH2T	Turbo Beaver
DH6	Twin Otter

Ercoupe, see Alon

Experimental

HXA	IAS less than 100 knots
HXB	IAS 100 to 200 knots
HXC	IAS more than 200 knots

Glassair, see Stoddard Hamilton

Grumman American

G164	Ag-Cat
G28	Cheetah
GA7	Cougar AA7
AA5	Traveler
AA1	Yankee/Trainer

Gulfstream Aerospace

GA84	Commander 840/900/980/1000
G159	Gulfstream I
G2	Gulfstream II
G3	Gulfstream III
G4	Gulfstream IV

Hawker-Siddeley

HS25	Model HS/DH/BH125

Helio

HE25	Courier
HE29	Super Courier

Homebuilt, see Experimental

Israel Aircraft

WW25	Astra 1125
WW23	Westwind 1123
WW24	Westwind 1124

Lake

LA4	Skimmer/Buccaneer

Lancair, see Neico Aviation

Learjet

LR23	Learjet 23
LR24	Learjet 24
LR25	Learjet 25
LR28	Learjet 28
LR29	Learjet 29
LR31	Learjet 31
LR35	Learjet 35
LR36	Learjet 36
LR54	Learjet 54
LR55	Learjet 55
LR60	Learjet 60

Lear Jet, see Learjet

Lockheed

L329	Jetstar
L18	Lodestar

Luscombe, see Silvaire

Martin

M202	Model 202
M404	Model 404

Maule

ML4	Rocket
ML5	Clunar/Star Rocket
ML6	Super Rocket

McDonnell-Douglas

DC3	Skytrain

Mitsubishi

MU3	Diamond
MU2	Marquise/Solitaire
HU30	Model 269/300
HU50	Pawnee 369/500

Mooney

MO10   Cadet
MO20   Mark 20
MO21   Mark 21
MO22   Mark 22

Navion

NA1   Rangemaster
NA16   Twin Navion
(see also Rockwell International; Navion)

Neico Aviation

LC20   Lancair II & III
LC30   Lancair IV

Piaggio

P136   Royal Gull
P166   Super Gull
P808   Vespa Jet

Piper

PA60   Aerostar
PA23   Apache
PA29   Archer
PAZT   Aztec
PA28   Cherokee
PARO   Cherokee Arrow
PA32   Cherokee Six/Lance/Saratoga
PA41   Cheyenne 400
PAYE   Cheyenne I/II
PA42   Cheyenne III/IV
PA16   Clipper
PA24   Comanche
PA5   Cruiser
PA11   Cub Special
PA2   Cub Trainer
PA14   Family Cruiser
PA46   Malibu
PA31   Navajo
PA20   Pacer
PA44   Seminole
PASE   Seneca
PA12   Super Cruiser
PA18   Super Cub
PA38   Tomahawk
PA22   Tri-Pacer/Colt

PA30	Twin Comanche
PA15	Vagabond Trainer
PA17	Vagabond
PA28	Warrior

Riley

RY65	Rocket/Model 65
RY21	Eagle 21
RY40	Turbo-Executive

Robinson

RH22	Robinson R22

Rockwell International

AC85	Aero Commander 685
AC72	Air Cruiser
AC12	Commander 112
AC2A	Commander 112A
AC2T	Commander 112TC
AC14	Commander 114
AC20	Commander 200
AC50	Commander 500
AC52	Commander 520
AC56	Commander 560
AC10	Darter 100/150
AC60	Grand Commander 680
AC21	Jet Commander
AC69	Jet Prop
LARK	Lark
N145	Navion
N265	Subreliner
AC68	Super Commander
AC6T	Turbo Commander

Short

SHD6	Model 360
SHD3	Shorts 330
SH7	Skyvan

Sikorsky

SK52	Model S52
SK58	Choctaw/Seahorse
SK59	Model S59
SK62	Model S62
SK76	Model S76
SK64	Skycrane

Silvaire
SL8   Observer/Luscombe

Stearman, see Boeing

Stinson
ST77   Reliant
ST75   Voyager 108

Stoddard Hamilton
GL20   Glassair II
GL30   Glassair III

Swearingen
SW2   Merlin IIA/B
SW3   Merlin IIIA/B/C
SW4   Merlin IV/Metro III

Swift, see Vought

Taylorcraft
TC19   Sportsman
TC20   Topper
TC15   Tourist

Ted Smith
TS60   Aerostar

Temco, see Vought

Varga
VG21   Kachina 2150A

Vertol, see Boeing

Vought
GC1   Swift

Westwind, see Israel Aircraft

B
Forecast and report locations

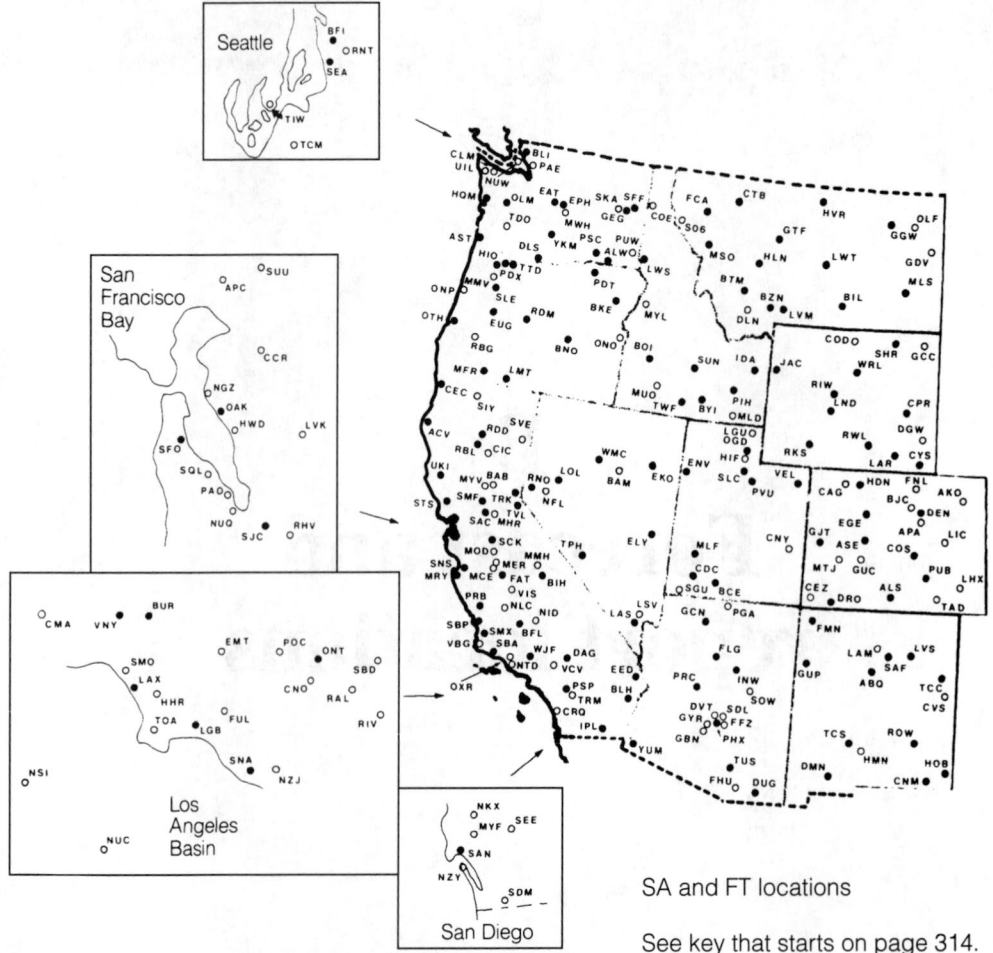

SA and FT locations

See key that starts on page 314.

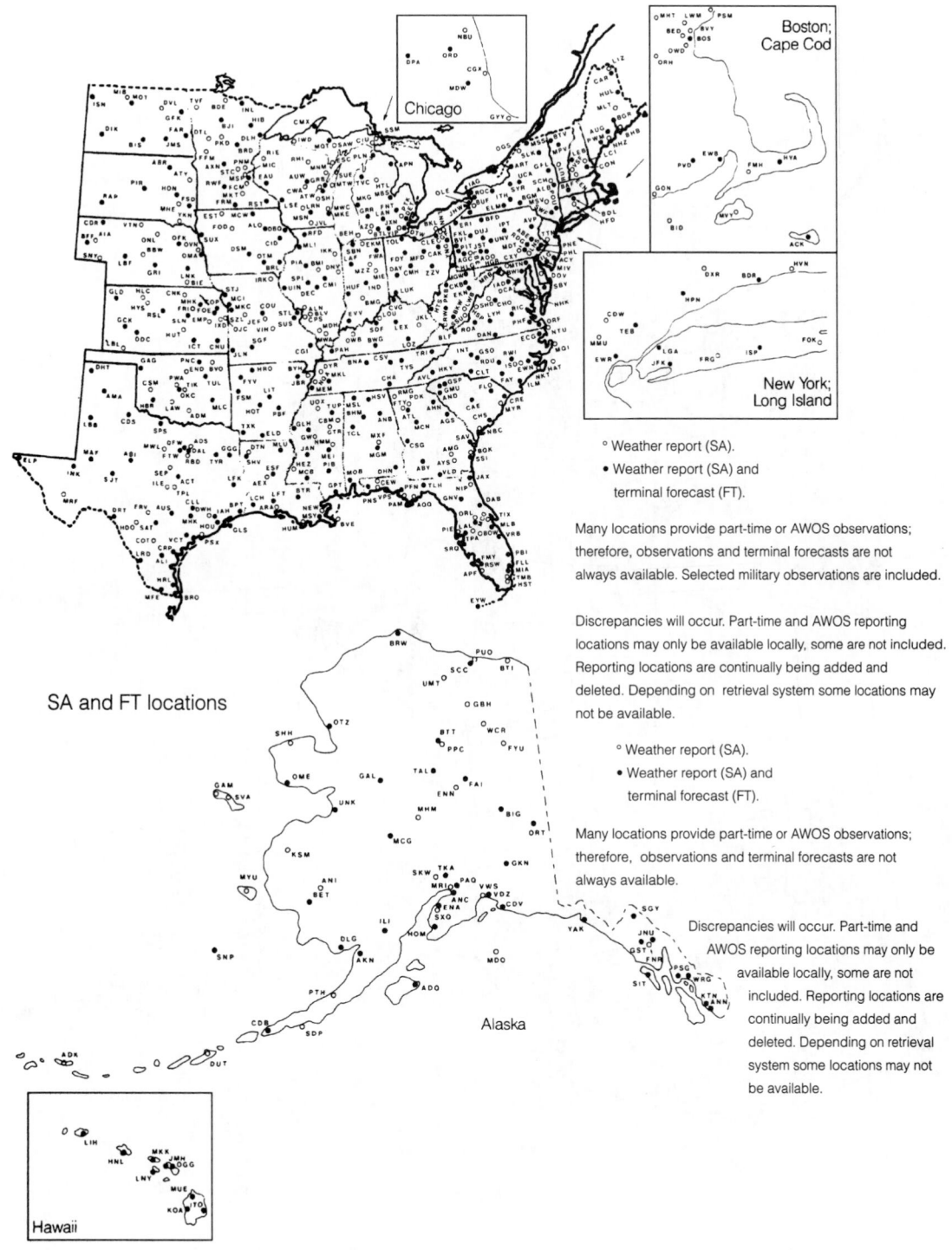

Forecast and report locations

Chicago

Boston; Cape Cod

New York; Long Island

° Weather report (SA).
• Weather report (SA) and terminal forecast (FT).

Many locations provide part-time or AWOS observations; therefore, observations and terminal forecasts are not always available. Selected military observations are included.

Discrepancies will occur. Part-time and AWOS reporting locations may only be available locally, some are not included. Reporting locations are continually being added and deleted. Depending on retrieval system some locations may not be available.

SA and FT locations

° Weather report (SA).
• Weather report (SA) and terminal forecast (FT).

Many locations provide part-time or AWOS observations; therefore, observations and terminal forecasts are not always available.

Discrepancies will occur. Part-time and AWOS reporting locations may only be available locally, some are not included. Reporting locations are continually being added and deleted. Depending on retrieval system some locations may not be available.

Alaska

Hawaii

TWEB ROUTES

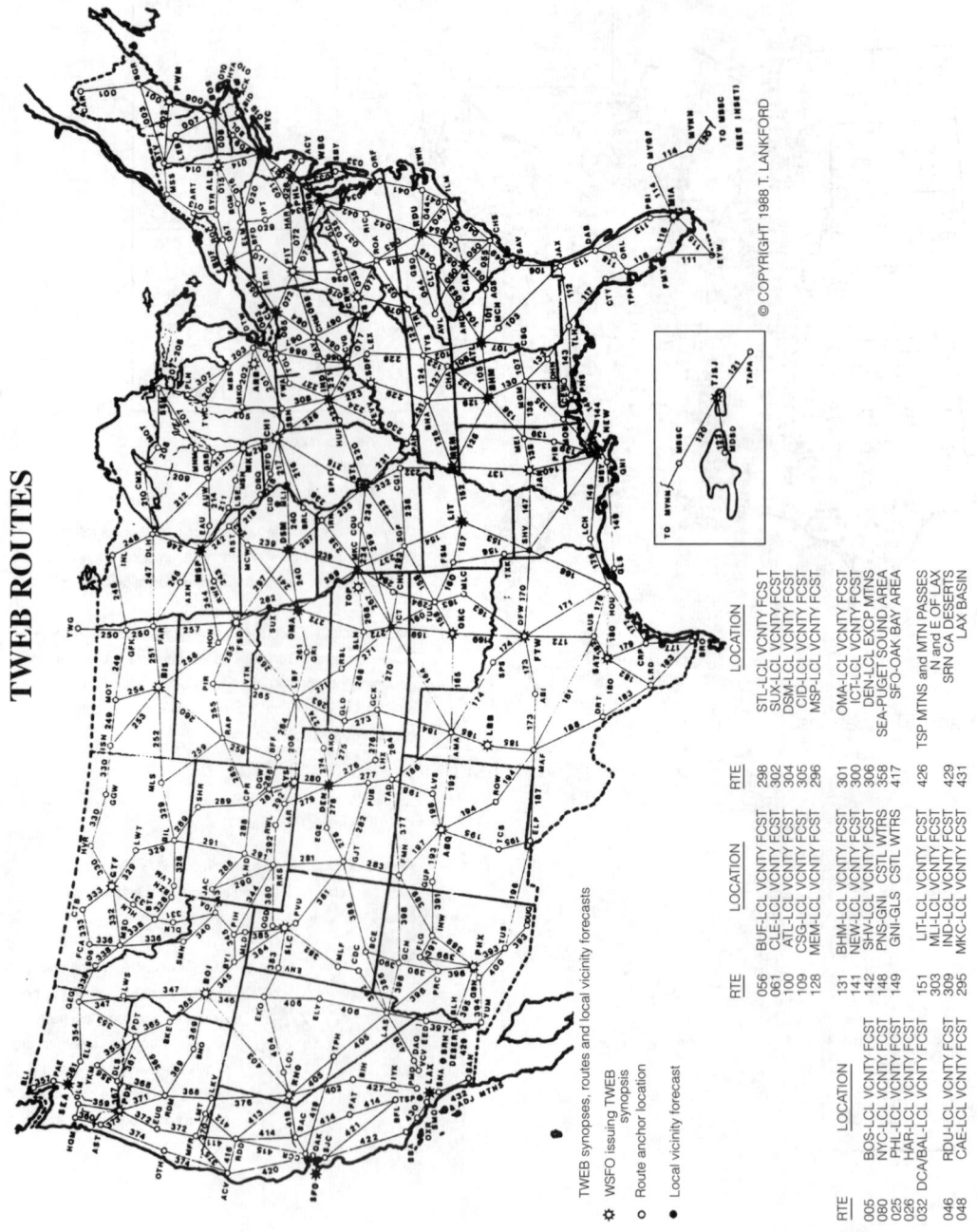

© COPYRIGHT 1988 T. LANKFORD

TWEB synopses, routes and local vicinity forecasts

☀ WSFO issuing TWEB synopsis

○ Route anchor location

● Local vicinity forecast

RTE	LOCATION
005	BOS-LCL VCNTY FCST
080	NYC-LCL VCNTY FCST
025	PHL-LCL VCNTY FCST
026	HAR-LCL VCNTY FCST
032	DCA/BAL-LCL VCNTY FCST
046	RDU-LCL VCNTY FCST
048	CAE-LCL VCNTY FCST

RTE	LOCATION
056	BUF-LCL VCNTY FCST
061	CLE-LCL VCNTY FCST
100	ATL-LCL VCNTY FCST
109	CSG-LCL VCNTY FCST
128	MEM-LCL VCNTY FCST
131	BHM-LCL VCNTY FCST
141	NEW-LCL VCNTY FCST
142	SHV-LCL VCNTY FCST
148	PNS-GNI CSTL WTRS
149	GNI-GLS CSTL WTRS
151	LIT-LCL VCNTY FCST
303	MLI-LCL VCNTY FCST
309	IND-LCL VCNTY FCST
295	MKC-LCL VCNTY FCST

RTE	LOCATION
298	STL-LCL VCNTY FCS T
302	SUX-LCL VCNTY FCST
304	DSM-LCL VCNTY FCST
305	CID-LCL VCNTY FCST
296	MSP-LCL VCNTY FCST
301	OMA-LCL VCNTY FCST
300	ICT-LCL VCNTY FCST
306	DEN-LCL EXCP MTNS
358	SEA-PUGET SOUND AREA
417	SFO-OAK BAY AREA
426	TSP MTNS and MTN PASSES
429	N and E OF LAX
431	SRN CA DESERTS
	LAX BASIN

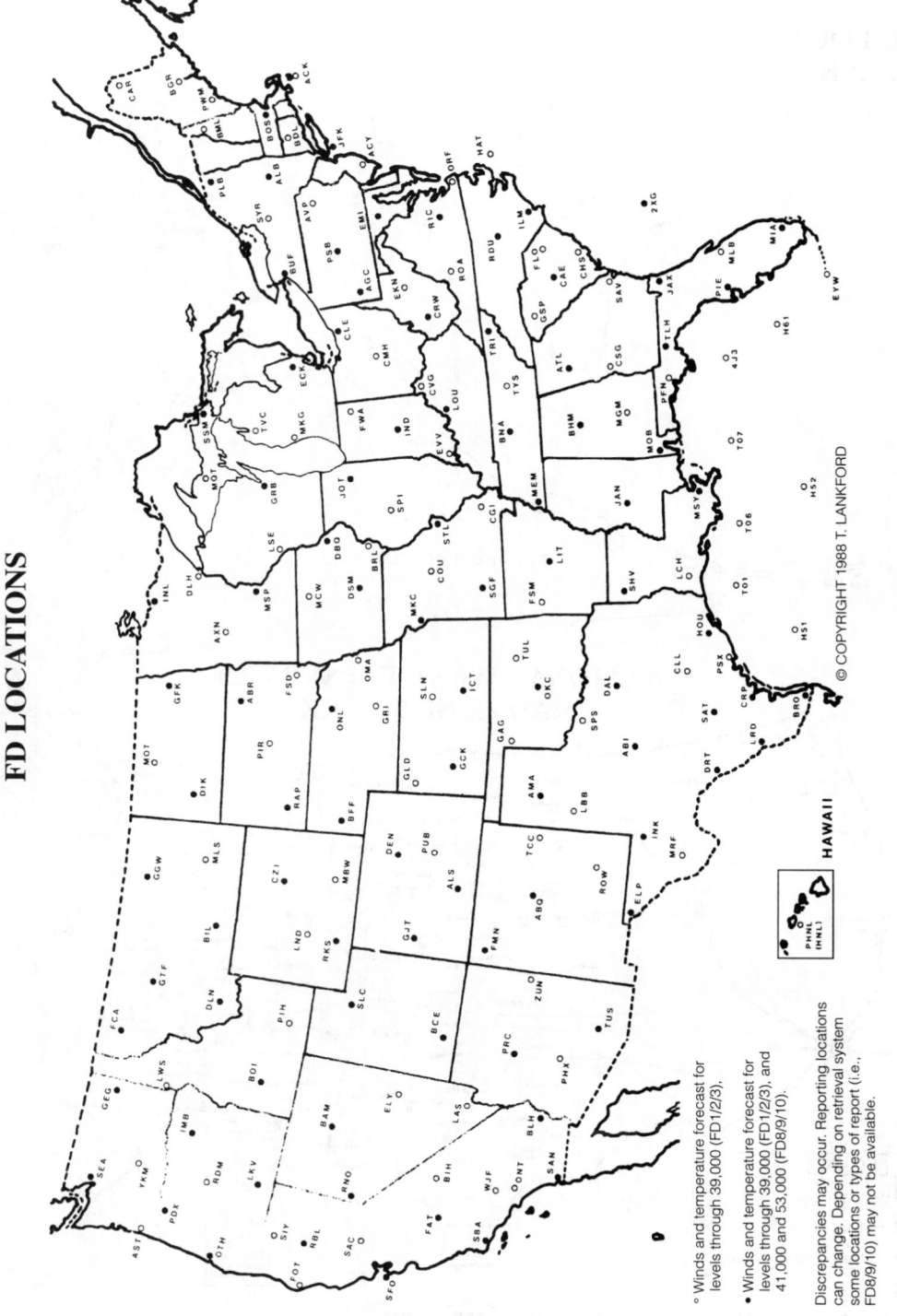

FD LOCATIONS

HAWAII

PHNL
(HNL)

° Winds and temperature forecast for
levels through 39,000 (FD1/2/3).

• Winds and temperature forecast for
levels through 39,000 (FD1/2/3), and
41,000 and 53,000 (FD8/9/10).

Discrepancies may occur. Reporting locations
can change. Depending on retrieval system
some locations or types of report (i.e.,
FD8/9/10) may not be available.

© COPYRIGHT 1988 T. LANKFORD

FD LOCATIONS
ALASKA

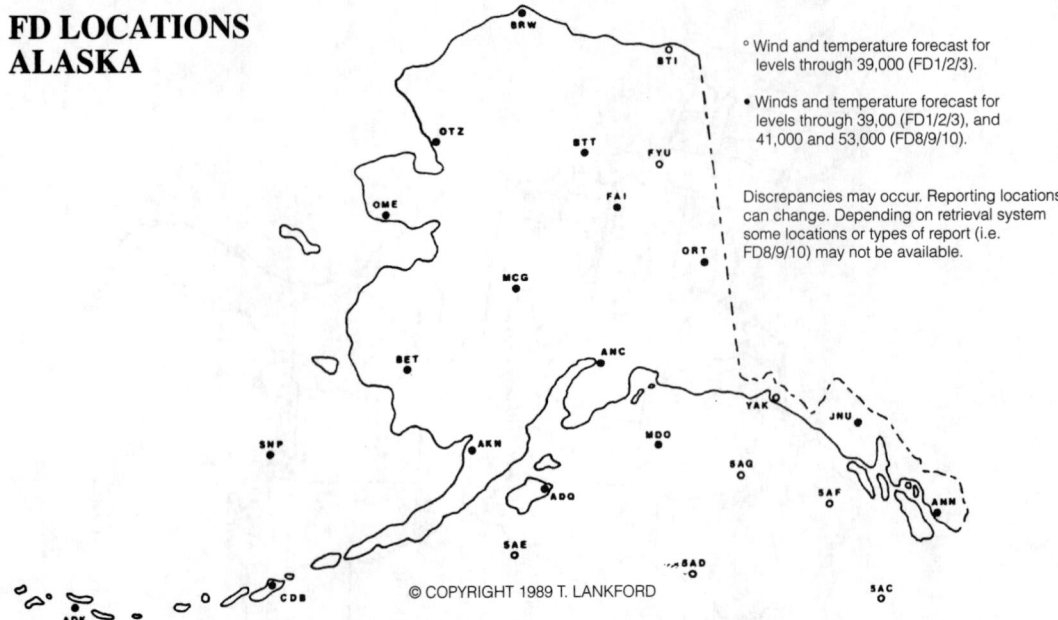

° Wind and temperature forecast for
 levels through 39,000 (FD1/2/3).

• Winds and temperature forecast for
 levels through 39,00 (FD1/2/3), and
 41,000 and 53,000 (FD8/9/10).

Discrepancies may occur. Reporting locations
can change. Depending on retrieval system
some locations or types of report (i.e.
FD8/9/10) may not be available.

© COPYRIGHT 1989 T. LANKFORD

NATIONAL WEATHER SERVICE
RADAR NETWORK

LEGEND:

WSR 57	☼
WSR 74S	⊠

LOCAL WARNING RADARS
  WSR 74C
  OTHER
	■
	●

FAA artc radar (remoted)	✕

C
Plotting identifiers

The following location identifiers appear on the in-flight advisory plotting chart.

A

ABI	ABILENE, TX
ABQ	ALBUQUERQUE, NM
ABR	ABERDEEN, SD
ABY	ALBANY, GA
ACK	NANTUCKET, MA
ACT	WACO, TX
ACY	ATLANTIC CITY, NJ
ADM	ARDMORE, OK
AEX	ALEXANDRIA, LA
AGS	AUGUSTA, GA
AHN	ATHENS, GA
AKO	AKRON, CO
ALB	ALBANY, NY
ALS	ALAMOSA, CO
AMA	AMARILLO, TX
AMG	ALMA, GA

AND   ANDERSON, SC
APN   ALPENA, MI
ARG   WALNUT RIDGE, AR
ATL   ATLANTA, GA
AUS   AUSTIN, TX
AVP   SCRANTON, PA
AXN   ALEXANDRIA, MN

B
BCE   BRYCE CANYON, UT
BDF   BRADFORD, IL
BDR   BRIDGEPORT, CT
BFF   SCOTTSBLUFF, NE
BFL   BAKERSFIELD, CA
BGM   BINGHAMTON, NY
BGR   BANGOR, ME
BHM   BIRMINGHAM, AL
BIL   BILLINGS, MT
BIS   BISMARCK, ND
BJI   BEMIDJI, MN
BKW   BECKLEY, WV
BLI   BELLINGHAM, WA
BML   BERLIN, NH
BNA   NASHVILLE, TN
BNO   BURNS, OR
BOI   BOISE, ID
BRL   BURLINGTON, IA
BRO   BROWNSVILLE, TX
BTR   BATON ROUGE, LA
BTY   BEATTY, NV
BUF   BUFFALO, NY
BUM   BUTLER, MO
BVL   BONNEVILLE, UT
BWG   BOWLING GREEN, KY
BWI   BALTIMORE, MD
BZN   BOZEMAN, MT

C
CAR   CARIBOU, ME
CDS   CHILDRESS, TX
CEW   CRESTVIEW, FL
CHA   CHATTANOOGA, TN
CHE   HAYDEN, CO
CHO   CHARLOTTESVILLE, VA
CHS   CHARLESTON, SC

CID	CEDAR RAPIDS, IA
CLE	CLEVELAND, OH
CLT	CHARLOTTE, NC
CMH	COLUMBUS, OH
COD	CODY, WY
CON	CONCORD, NH
COU	COLUMBIA, MO
CPR	CASPER, WY
CRP	CORPUS CHRISTI, TX
CTY	CROSS CITY, FL
CUU	CHIHUAHUA, MEX
CVG	COVINGTON, KY

D

DAG	DAGGETT, CA
DBQ	DUBUQUE, IA
DEC	DECATUR, IL
DEN	DENVER, CO
DFW	FORT WORTH, TX
DHT	DALHART, TX
DIK	DICKINSON, ND
DLH	DULUTH, MN
DMN	DEMING, NM
DPR	DUPREE, SD
DRT	DEL RIO, TX
DSM	DES MOINES, IA
DTA	DELTA, UT
DTW	DETROIT, MI
DVC	DOVE CREEK, CO
DYR	DYERSBURG, TN

E

EAT	WENATCHEE, VA
EAU	EAU CLAIRE, WI
EED	NEEDLES, CA
EEY	WINCHESTER, VA
EGC	ELIZABETH CITY, NC
EKN	ELKINS, WV
EKO	ELKO, NV
ELD	EL DORADO, AR
ELP	EL PASO, TX
ELY	ELY, NV
ERI	ERIE, PA
EUG	EUGENE, OR
EVV	EVANSVILLE, IN

EWR	NEWARK, NJ
EYW	KEY WEST, FL

F

FAM	FARMINGTON, MO
FAR	FARGO, ND
FAT	FRESNO, CA
FCA	KALISPELL, MT
FDY	FINDLAY, OH
FLO	FLORENCE, SC
FMN	FARMINGTON, NM
FMY	FORT MYERS, FL
FOT	FORTUNA, CA

G

GCC	GILLETTE, WY
GCK	GARDEN CITY, KS
GEG	SPOKANE, WA
GFK	GRAND FORKS, ND
GGG	LONGVIEW, TX
GGW	GLASGOW, MT
GJT	GRAND JUNCTION, CO
GLD	GOODLAND, KS
GRB	GREEN BAY, WI
GRI	GRAND ISLAND, NE
GRW	GREENWOOD, MS
GSO	GREENSBORO, NC
GTF	GREAT FALLS, MT
GUC	GUNNISON, CO
GUP	GALLUP, NM

H

HAR	HARRISBURG, PA
HAT	HATTERAS, NC
HNN	HENDERSON, WV
HQM	HOQUIAM, WA
HRO	HARRISON, AR
HVE	HANKSVILLE, UT
HVR	HAVRE, MT
HYS	HAYS, KS

I

IAH	HOUSTON, TX
ICT	WICHITA, KS
IGB	COLUMBUS, MS
ILC	WILSON CREEK, NV

ILM	WILMINGTON, NC
IND	INDIANOPOLIS, IN
INL	INTERNATIONAL FALLS, MN
INW	WINSLOW, AZ
IRK	KIRKSVILLE, MO
ISN	WILLISTON, ND
IWD	IRONWOOD, MI

J

JAC	JACKSON, WY
JAN	JACKSON, MS
JAX	JACKSONVILLE, FL
JCT	JUNCTION, TX
JST	JOHNSTOWN, PA

L

LAA	LARMAR, CO
LAF	LAFAYETTE, IN
LAR	LARAMIE, WY
LAS	LAS VEGAS, NV
LAX	LOS ANGELES, CA
LBB	LUBBOCK, TX
LBF	NORTH PLATTE, NE
LBL	LIBERAL, KS
LCH	LAKE CHARLES, LA
LFK	LUFKIN, TX
LIT	LITTLE ROCK, AR
LKT	SALMON, ID
LKV	LAKEVIEW, OR
LOU	LOUISVILLE, KY
LOZ	LONDON, KY
LSE	LA CROSSE, WI
LVS	LAS VEGAS, NM
LWS	LEWISTON, ID
LWT	LEWISTOWN, MT
LYH	LYNCHBURG, VA

M

MAF	MIDLAND, TX
MBS	SAGINAW, MI
MĊB	MCCOMB, MS
MCN	MACON, GA
MCW	MASON CITY, IA
MEI	MERIDIAN, MS
MEM	MEMPHIS, IN

MFR	MEDFORD, OR
MGM	MONTGOMERY, AL
MIA	MIAMI, FL
MKC	KANSAS CITY, MO
MKE	MILWAUKEE, WI
MKG	MUSKEGON, MI
MLC	MC ALESTER, OK
MLP	MULLAN PASS, ID
MLS	MILES CITY, MT
MLT	MILLINOCKET, ME
MLU	MONROE, LA
MOB	MOBILE, AL
MOD	MODESTO, CA
MOT	MINOT, IN
MOV	MONCLOVA, MEX
MPV	MONTPELIER, VT
MQT	MARQUETTE, MI
MRF	MARFA, TX
MSL	MUSCLE SHOALS, AL
MSN	MADISON, WI
MSO	MISSOULA, MT
MSP	MINNEAPOLIS, MN
MSY	NEW ORLEANS, LA
MTU	MYTON, UT
MYL	MC CALL, ID

O

OKC	OKLAHOMA CITY, OK
OMA	OMAHA, NE
ONL	O NEILL, NE
ONP	NEWPORT, OR
ORD	CHICAGO, IL
ORF	NORFOLK, VA
ORL	ORLANDO, FL
OSW	OSWEGO, KS

P

PBI	WEST PALM BEACH, FL
PDT	PENDLETON, OR
PDX	PORTLAND, OR
PEQ	PECOS CITY, TX
PGS	PEACH SPRINGS, AZ
PHX	PHOENIX, AZ
PIT	ST. PETERSBURG, FL
PIH	POCATELLO, ID

PIR	PIERRE, SD
PIT	PITTSBURGH, PA
PKB	PARKERSBURG, WV
PPE	PUNTA PENASCO, MEX
PSP	PALM SPRINGS, CA
PSX	PALACIOS, TX
PUB	PUEBLO, CO
PVD	PROVIDENCE, RI
PWE	PAWNEE CITY, NE
PWM	PORTLAND, ME

R

RAP	RAPID CITY, SD
RBL	RED BLUFF, CA
RDM	REDMOND, OR
REO	ROME, OR
RHI	RHINELANDER, WI
RIC	RICHMOND, VA
RIW	RIVERTON, WY
RKS	ROCK SPRINGS, WY
RNO	RENO, NV
ROW	ROSWELL, NM
RWF	REDWOOD FALLS, MN
RWI	ROCKY MOUNT, NC

S

SAC	SACRAMENTO, CA
SAN	SAN DIEGO, CA
SAT	SAN ANTONIO, TX
SAV	SAVANNAH, GA
SBA	SANTA BARBARA, CA
SBN	SOUTH BEND, IN
SBY	SALISBURY, MD
SEA	SEATTLE, WA
SFO	SAN FRANCISCO, CA
SGF	SPRINGFIELD, MO
SHR	SHERIDAN, WY
SJN	ST. JOHNS, AZ
SJT	SAN ANGELO, TX
SLC	SALT LAKE CITY, UT
SLK	SARANAC LAKE, NY
SLN	SALINA, KS
SLT	SLATE RUN, PA
SNS	SALINAS, CA
SPS	WICHITA FALLS, TX

SSM	SAULT STE MARIE, MI
SSO	SAN SIMON, AZ
STL	ST. LOUIS, MO
SUX	SIOUX CITY, IA
SYR	SYRACUSE, NY

T

TBC	TUBA CITY, AZ
TBE	TOBE, CO
TCC	TUCUMCARI, NM
TCS	TRUTH OR CONSEQUENCES, NM
TLH	TALLAHASSEE, FL
TOU	TATOOSH, WA
TPH	TONOPAH, NV
TRI	BRISTOL, TN
TUL	TULSA, OK
TUS	TUCSON, AZ
TVC	TRAVERSE CITY, MI
TWF	TWIN FALLS, ID
TXK	TEXARKANA, AR
TYS	KNOXVILLE, TN

U

UKI	UKIAH, CA

V

VBI	SIOUX NARROWS, ONT
VIH	ROLLA/VICHY, MO
VRB	VERO BEACH, FL
VTN	VALENTINE, NE

W

WMC	WINNEMUCCA, NV

Y

YDC	PRINCETON, BC
YDR	BROADVIEW, SASK
YKM	YAKIMA, WA
YOW	OTTAWA, ONT
YQB	QUEBEC, QUE
YQL	LETHBRIDGE, ALTA
YQT	THUNDER BAY, ONT
YSC	SHERBROOKE, QUE
YSJ	ST. JOHN, NB
YUM	YUMA, AZ
YVV	WIARTON, ONT

YWG   WINNIPEG, MAN
YXH   MEDICINE HAT, ALTA
YYN   SWIFT CURRENT, SASK
YYZ   TORONTO, ONT

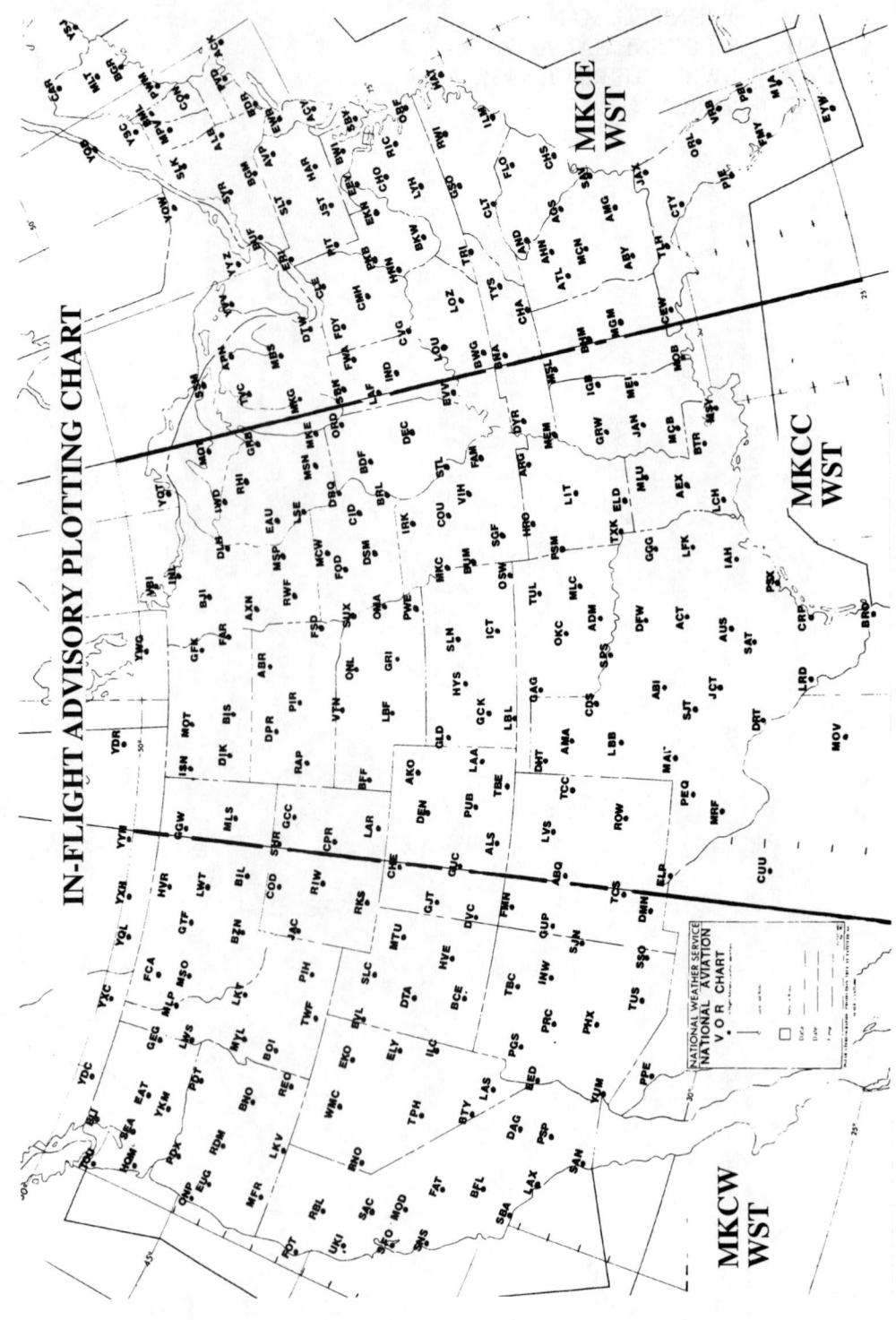

IN-FLIGHT ADVISORY PLOTTING CHART

D
Area designators

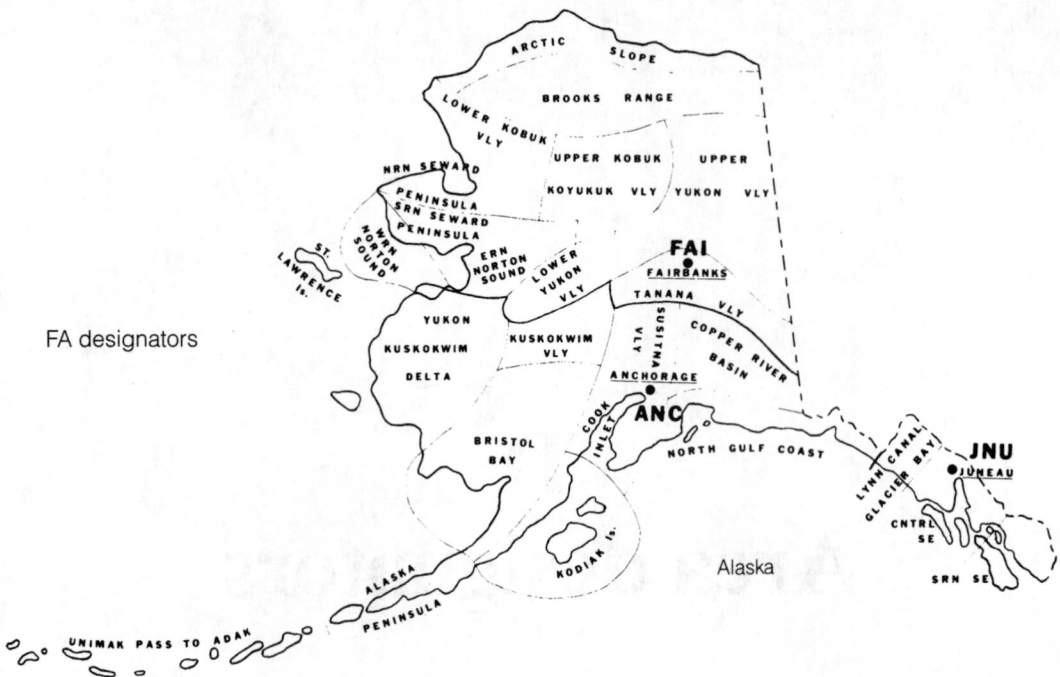

FA designators

Alaska

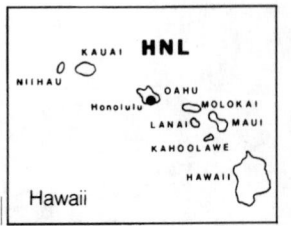

Hawaii

© COPYRIGHT 1988 Terry T. Lankford

COMMON GEOGRAPHICAL AREA DESIGNATORS

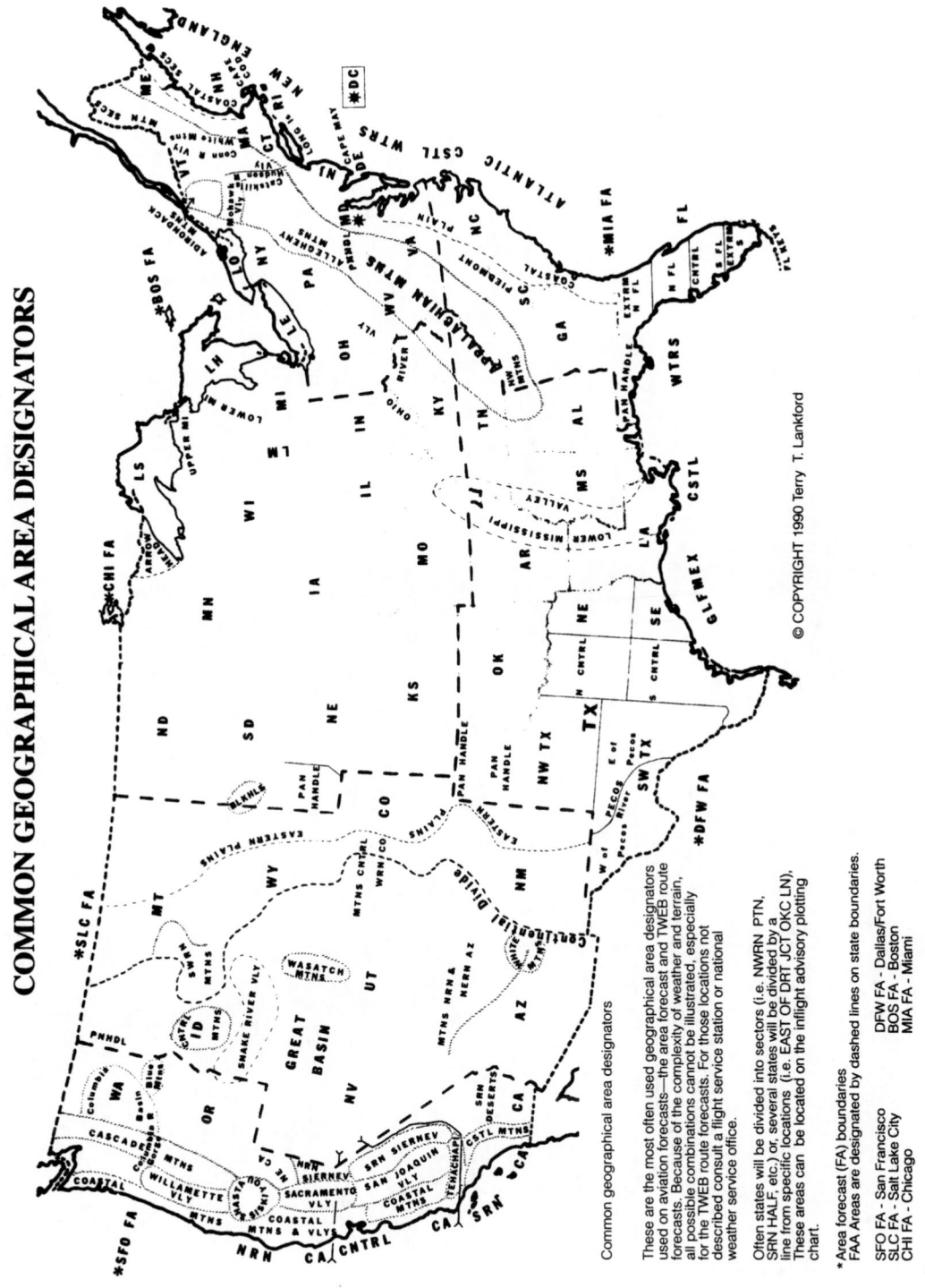

Common geographical area designators

These are the most often used geographical area designators used on aviation forecasts—the area forecast and TWEB route forecasts. Because of the complexity of weather and terrain, all possible combinations cannot be illustrated, especially for the TWEB route forecasts. For those locations not described consult a flight service station or national weather service office.

Often states will be divided into sectors (i.e. NWRN PTN, SRN HALF, etc.) or, several states will be divided by a line from specific locations (i.e. EAST OF DRT JCT OKC LN). These areas can be located on the inflight advisory plotting chart.

* Area forecast (FA) boundaries
FAA Areas are designated by dashed lines on state boundaries.

SFO FA - San Francisco   DFW FA - Dallas/Fort Worth
SLC FA - Salt Lake City   BOS FA - Boston
CHI FA - Chicago   MIA FA - Miami

© COPYRIGHT 1990 Terry T. Lankford

371

E
Weather products

SURFACE OBSERVATIONS (SA)

ACY SA 2050 250 -SCT 5H 108/87/75/2009/985/ 722 1001
JFK SA 2050 -X 5H 102/78/73/1918/983/H4/ 730
LGA SA 2250 40 SCT E70 BKN 130 OVC 4TRW+ 119/78/70/2720G23/988/
WSHFT 27 TB13 W MOVG SE FQT LTGICCCCG W-NW PRESRR RB30
ABE SA 2253 E45 BKN 120 BKN 6F 112/75/73/E1705/987

NWS observers, like the military, have a talent for seeing a multitude of cloud layers. This SFO report contains five:

SFO RS 1854 11 SCT 23 SCT M30 BKN 85 BKN 100 OVC 8
122/46/44/1310/989/RE43 WSHFT 30

LAWRS OBSERVATIONS

MDT SA 2145 E50 OVC 3H 82/76/2405/986/PRESRR
RDG SA 2245 8 SCT E25 OVC 7 76/73/1610/988/TE05
STS SA 2048 CLR 30 110/54/1405/990
COE SA 2350 12 SCT E30 OVC 20 38/35/1412/994
APF SA 2345 20 SCT E20 BKN 7 76/M/0000/010
VIS SA 1645 W1 X 1/4F 39/0000/004
APC SA 0045 CLR 10 3605/998

AUTOMATED OBSERVATIONS

This AWOS observation from Sanford, ME (SFM), reports clear above 12,000. Obstructions to visibility are not available with this AWOS unit. The remarks indicate that precipitation has occurred.

FOD SA 2349 AWOS M12 OVC 10 24/19/0604/009
SFM SA 0013 AWOS 120 CLR 5 27/24/0607/006/ P001

This AWOS observation from Bremerton, WA, indicates that the wind direction is missing (M), speed 7 knots.

PWT SA 2349 AWOS M6 OVC 10 44/42/M07/998

At Block Island, RI, the peak wind reported is 18 knots, and the pressure tendency and change during the past 3 hours is 020.

BID SP 0029 AMOS 39/-27/0513/998 PK WIND 18 020

WEATHER ADVISORIES

SIGMETs

This is an example of a Fairbanks, AK, SIGMET for volcanic ash released by Mt. Redoubt in December 1989.

FAIA UWS 151625
PAZA SIGMET ALPHA 1 VALID 151630/152030 PAFA-
VOLCANIC ASH BLO FL350 FCST TO MOV INTO AREA BOUNDED BY 160 SW
FAI TO TAL TO 80 N EAA TO 200 S EAA TO 40 S BIG TO 90 S FAI TO 160SW
FIA.

Convective SIGMETs

MKCC WST 022155
CONVECTIVE SIGMET 46C
VALID UNTIL 2355Z
AR MS LA AL
FROM 40 S MEM-30ESE IGB-MOB-50ME MUL-40S MEM
AREA TSTSM MOVG FROM 2410. TOPS ABV 450.

CONVECTIVE SIGMET 47C
VALID UNTIL 2355Z
KS OK TX
FROM 30NNW OSW-30SSW SPS-60WNW LBB
DVLPG LINE SVR TSTMS 40 MI WIDE MOVG FROM 2620. TOPS ABV 450.
HAIL TO 3 IN...WIND GUSTS TO 65 KT PSBL.

CONVECTIVE SIGMET 49C
VALID UNTIL 2355Z
NM
30NNW ROW
DVLPG ISOLD SVR TSTM D25 MOVG FROM 2720. TOPS ABV 450
TORNADOES...HAIL T0 3 IN...WIND GUSTS TO 70KT PSBL.

OUTLOOK VALID UNTIL 0355Z
FROM MCW-COU-FSM-DFW-SJT-LVS-VTN-MCW
REF WW 374. REF WW 375. VERY UNSTBL AMS HAS MOVED NWD INTO
CNTRL PLAINS AREA. SFC LI IN -10 RANGE IN SOME AREAS. MSTR
CONVG ANAL SHOWS SEVERAL AREAS OF CONCERN...FIRST IN CDFNT
MOVG INTO NRN NE WHERE MDT-HIGH VALUES OBSVD ALSO INTO NRN IA.
MDT MSTR CONVG XTNDS SWD INTO WRN KS. OTHR AREAS OF MDT MSTR
CONVG OVER ERN OK AND W CRNTRL TX WHERE TSTMS ARE ALSO VERY
STG ATTM. TSTMS OVER LA/MS ARE XPCD TO DMSH NEXT 1-2 HOURS AS
OUTFLOWS HAVE STBLIZED MUCH OF THE AMS THERE.

Center weather advisories

ZAN1 UCWA 03 201630-202030
ADK-CNTL ALEUTIANS
ALTHOUGH NO OVERNIGHT PILOT REPORTS ARE AVAILABLE FROM
KANAGA VOLCANO AREA...UNCONFIRMED REPORTS OF SMALL
EARTHQUAKES IN THE REGION SUGGEST CONTINUED ERUPTIVE ACTIVITY.
ANY ASH EJECTED BY THIS VOLCANO CAN BE EXPECTED TO REACH FL100
AND BE MIXED WITH THE CLOUDS OF AN APPROACHING COLD FRONT.
FORECAST WINDS IN THE CENTRAL ALEUTIANS FROM FL050-FL180 ARE
WRLY AT 30-40 KTS.
KANAGA VOLCANO... ELEVATION 4416 FT... IS LOCATED AT
51.56N 177.10W AND HAS A HISTORY OF MINOR ERUPTIVE ACTIVITY.
THE NEXT STATEMENT CONCERNING KANAGA VOLCANO WILL BE ISSUED
AT 2030 UTC.

Alert weather watch

MKC AWW 021930
WW 374 SVERE TSTM OK KS 022000Z - 030200Z
AXIS..70 STATUTE MILES EAST AND WEST OF LINE..
20 SW ADM/ARDMORE OK/ - 30W CNU/CHANUTE KS/
HAIL SURFACE AND ALOFT..3 INCHES. WIND GUSTS..65 KNOTS.
MAX TOPS TO 550. MEAN WIND VECTOR 230/25.

MKC AWW 022003
WW 375 TORNADO TX NM 022020Z - 030300Z
AXIS..80 STATUTE MILES NORTH AND SOUTH OF LINE..
20WSW ROW/ROSWELL NM/ - 60SSE CDS/CHILDRESS TX/
HAIL SURFACE AND ALOFT..3 INCHES. WIND GUST..70 KNOTS.
TAX TOPS TO 550. MEAN WIND VECTOR 230/30.

Severe weather watch bulletin

MKC WW 021935
BULLETIN - IMMEDIATE BROADCAST REQUESTED
SEVERE THUNDERSTORM WATCH NUMBER 374
NATIONAL WEATHER SERVICE KANSAS CITY MO
2:35 PM CDT FRI JUN 02 1989

   A...THE NATIONAL SEVERE STORMS FORECAST CENTER HAS ISSUED A
SEVERE THUNDERSTORM WATCH FOR
MUCH OF CENTRAL AND EASTERN OKLAHOMA
PARTS OF SOUTH-CENTRAL AND SOUTHEAST KANSAS

FROM 3:00 PM CDT UNTIL 9:00 PM CDT THIS FRIDAY AFTERNOON AND
EVENING
LARGE HAIL...DANGEROUS LIGHTNING AND DAMAGING THUNDERSTORM
WINDS ARE POSSIBLE IN THESE AREAS.

THE SEVERE THUNDERSTORM WATCH AREA IS ALONG 70 STATUTE MILES
EAST AND WEST OF A LINE FROM 20 MILES SOUTHWEST OF ARDMORE
OKLAHOMA TO 30 MILES WEST OF CHANUTE KANSAS.

REMEMBER...A SEVERE THUNDERSTORM WATCH MEANS CONDITIONS ARE
FAVORABLE FOR SEVERE THUNDERSTORMS IN AND CLOSE TO THE WATCH
AREA. PERSONS IN THESE AREAS SHOULD BE ON THE LOOKOUT FOR
THREATENING WEATHER CONDITIONS AND LISTEN FOR LATER
STATEMENTS AND POSSIBLE WARNINGS.

   C...A FEW SVR TSTMS WITH HAIL SFC AND ALOFT TO 3 IN. EXTRM TURBC
AND SFC WND GUST TO 65 KN. A FEW CBS WITH MAX TOPS TO 550. MEAN
WIND VECTOR 230/25.

   D...TSTMS CONG TO DVLP OVR N CNTRL OK AND EXTRM S CNTRL KS IN
VERY MOIST AND UNSTABL AMS. SFC BASED LIFTED INDEX IS MINUS 10 OR
LESS. TCU/SML CBS ALSO BEGINNING TO DVLP OVR S CNTRL OK. TSTMS
EXPCD TO INTSFY FURTHER DURG AFTN HEATING PD WITH STGR CELLS
LIKELYREACHING SVR LVLS INTO EVE HRS.

   E...OTR TSTMS...CONT VALID PTN OF WW 737.
...WEISS

Hurricane bulletin

WINT32 KMIA 131842
BULLETIN
HURRICANE GILBERT INTERMEDIATE ADVISORY
NATIONAL WEATHER SERVICE MIAMI FL
3 PM EDT TUE SEP 13 1988

...GILBERT NOW PACKING 160 MPH WINDS...
HURRICANE WARNINGS ARE IN EFFECT FOR THE CAYMAN
ISLANDS...NORTHEASTERN YUCATAN FROM FELIPE CARILLO PUERTO TO
PROGRESO INCLUDING COZUMEL AND CANCUN...THE ISLE OF YOUTH
AND WESTERN CUBA WEST OF THE CITY OF PINAR DEL RIO.

AT 3 PM EDT...1900Z...THE CENTER OF HURRICANE GILBERT WAS LOCATED
NEAR LATITUDE 19.3 NORTH...LONGITUDE 82.3 WEST...OR ABOUT 280
MILES...450 KM...EAST SOUTHEAST OF THE ISLAND OF COZUMEL MEXICO.
GILBERT IS MOVING TOWARD THE WEST NORTHWEST NEAR 15 MPH...24
KM/HR. THIS MOTION IS EXPECTED TO CONTINUE THIS AFTERNOON AND
TONIGHT

NOAA RECONNASSANCE NOW REPORTS WINDS IN GILBERT HAVE
REACHED 160 MPH WITH A CENTRAL PRESSURE OF 903 MILLIBARS...26.66
INCHES. GILBERT MAY STILL INCREASE A LITTLE MORE IN STRENGTH.
TROPICAL STORM FORCE WINDS EXTEND OUTWARD UP TO 250 MILES...400
KM...TO THE NORTH AND 200 MILES...320 KM...TO THE SOUTH OF THE
CENTER.

PEOPLE IN THE WARNED AREAS OF CUBA AND MEXICO SHOULD HAVE
NEARLY COMPLETED PREPARATIONS FOR THIS SEVERE HURRICANE.

TIDES OF UP TO 8 TO 12 FEET AND HIGH WAVES ARE LIKELY NEAR AND TO
THE RIGHT OF WHERE THE CENTER CROSSES THE COAST. LESSER VALUES
WILL OCCUR ELSEWHERE IN THE WARNED AREAS. THESE CONDITIONS
COULD UNDERMINE BUILDINGS ALONG THE BEACH.

SQUALLS ARE CURRENTLY PASSING THROUGH THE FLORIDA KEYS.
SMALLCRAFT SHOULD REMAIN IN PORT FROM KEY LARGO TO DRY
TORTUGAS. TIDES OF1 TO 2 FEET ABOVE NORMAL WILL CONTINUE OVER
THE FLORIDA KEYS.

RAINFALL TOTALS OF 5 TO 10 INCHES ACCOMPANY THE HURRICANE.

AIRMET bulletin

Alaska
ANCS WA 280745
AIRMET SIERRA FOR IFR AND MTN OBSCN VALID UNTIL 281400.

COPPER RIVER BASIN AC
OCNL CIGS BLO 10 VSBY BLO 3R-S-F. CONTG BYD 14Z.

.

AK PEN AI
OCNL CIGS BLO 10 VSBYS BLO 3R-F. DMSHG BY 14Z.

.

ANCT WA 280745
AIRMET TANGO FOR TURBC/STG SFC WINDS VALID UNTIL 281400.

.

NONE.

.

ANCZ WA 280745
AIRMET ZULU FOR ICING VALID UNTIL 281400.
COPPER RIVER BASIN AC
OCNL MDT RIME ICGIC 040-140. FRZLVL 035. CONTG BYD 14Z.

AREA FORECASTS

Alaska

In Alaska, in addition to geographical area designators, there is a two-letter code for each area to facilitate computer processing (COOK INLET AND SUSITNA VLY "AB"). (A listing is contained in appendix D.)

ANCH FA 280745
AK SRN HLF EXCP SE AK...

.

AIRMETS VALID UNTIL 281400
TSTMS IMPLY PSBL SVR OR GTR TURBC SVR ICG LLWS AND IFR CONDS.
NON MSL HGTS NOTED BY AGL OR CIG.

.

SYNOPSIS VALID UNTIL 290200
A LOW 400 S YAK WL DRFT NE 200 W ANN BY 02Z. A SCND LOW NR ANN
WL MOV N AND WKN. HIGH PRES RDG DVR WRN ALUTNS WL RMN.

.

COOK INLET AND SUSITNA VLY AB...VALID UNTIL 282000
...CLDS/WX...
ISOLD CIGS BLO 10 VSBYS BLO 3R-F. OTRW 15 SCT 30 BKN 80 DVC LYRD
150. OCNL VSBY 3-5R-F. OTLK VALID 282000-291400 VFR.
PASSES...LK CLARK..MERRILL..RAINY VFR OCNL MVFR CIG E ENDS. WINDY
MVFR CIG RW. CHICKALOON IFR CIG R S F. PORTAGE MVFR OCNL IFR CIG
F.
...TURBC...
NONE SGFNT.
...ICG AND FRZLVL...
ISOLD MDT RIME ICGICIP 035-140. FRZLVL 035.

.

COPPER RIVER BASIN AC...VALID UNTIL 282000
...CLDS/WX...

***AIRMET IFR/MTN OBSCN ***OCNL CIGS BLO 10 VSBY BLO 3R-S-F. CONTG
BYD 14Z...OTRW 15 SCT 25 BKN 50 DVC. LYRD 160. OCNL VSBY 3-5R-S-F.
OTLK VALID 282000-291400 MVFR CIG R S BCMG VFR AFT 00Z.
...TURBC...
NONE SGFNT.
...ICG AND FRZLVL...
***AIRMET ICG ***OCNL MDT RIME ICGIC 040-140. FRZLVL 035. CONTG
BYD 14Z...

.

CNTRL GLF CST AD...VALID UNTIL 282000
...CLDS/WX...
SCT CIG BLO 10 VSBY BLO 3R-F. MNLY W SIDE. OTRW 25 SCT LCLY BKN 40
BKN 60 OVC LYRD 180. SCT VSBY 3-5R-F.
OTLK VALID 282000-291400 VFR RW.
...TURBC...
NONE SGFNT.
...ICG AND FRZLVL...
ISOLD MDT RIME ICGIC 035-140. FRZLVL 035.

.

KODIAK IS AE...VALID UNTIL 282000
...CLDS/WX...
ISOLD CIGS BLO 10 VSBYS BLO 3R-F. OTRW 15 SCT 25 SCT OCNL BKN 60.
SCT VSBY 3-5R-F.
OTLK VALID 282000-291400 VFR.
...TURBC...
NONE   SGFNT.
...ICG AND FRZLVL...
ISOLD MDT RIME ICGICIP 025-120. FRZLVL 025.

.

KUSKOKWIM VLY AF...VALID UNTIL 282000
...CLDS/WX...
40 SCT 70 BKN 100 BKN 130. CI ABV. ISOLD TCU/CB RW- TOPS 270.
OTLK VALID 282000-291400 VFR.
...TURBC...
NONE SGFNT.
...ICG AND FRZLVL...
NONE SGFNT. FRZLVL 060.

.

YKN-KUSKOKWIM DELTA AG...VALID UNTIL 282000
...CLDS/WX...
OTRW 70 SCT 90 BKN LYRS TO 180. ISOLD CB/TRW TOPS 220.
OTLK VALID 282000-291400 VFR.
...TURBC...
NONE SGFNT.
...ICG AND FRZLVL...
NONE SGFNT. FRZLVL 050.

.

BRISTOL BAY AH...VALID UNTIL 282000

...CLDS/WX...
40 SCT 80 BKN LYRD 180. ISOLD RW-.
OTLK VALID 282000-291400 VFR.
...TURBC...
NONE SGFNT.
...ICG AND FRZLVL...
NONE SGFNT. FRZLVL 045.

.

AK PEN AI...VALID UNTIL 282000
...CLDS/WX...
***AIRMET IFR/MTN OBSCN *** OCNL CIGS BLO 10 VSBYS BLO 3R-F. DMSHG
BY 14Z...OTRW 20 BKN 30 OVC LYRD 160. CI ABV. OCNL VSBY 3-5R-F.
OTLK VALID 282000-291400 MVFR CIG RW.
...TURBC...
ISOLD MDT BLO 030 AGL.
...ICG AND FRZLVL...
SCT MDT RIME ICGICIP 020-120. FRZLVL 030.

.

UNIMAK PASS TO ADAK AJ...VALID UNTIL 282000
...CLDS/WX...
ISOLD CIGS BLO 10 VSBYS BLO 3R-F. OTRW 15 SCT 25 BKN 35 OVC 080. CI
ABV. SCT VSBY 3-5RW-.
OTLK VALID 282000-291400 MVFR CIG WND.
...TURBC...
ISOLD MDT BLO 030 AGL.
...ICG AND FRZLVL...
NONE SGFNT. FRZLVL 020 W SLPG 030 E.

Hawaii

SYNOPSIS AND VFR CLOUDS/WEATHER
HNLC FA 142140
SYNOPSIS AND VFR CLDS/WX
SYNOPSIS VALID UNTIL 151600

CLDS/WX VALID UNTIL 151000...OTLK VALID 151000-151600
SEE AIRMET SIERRA IFR CLDS AND MTN OBSCN.
TSTMS IMPLY SVR OR GTR TURBC SVR ICING LLWS AND IFR CONDS.
NON MSL HGTS DENOTED BY AGL OR CIG.

SYNOPSIS...SFC RDG FAR N OF ISLANDS NRLY STNRY.

.

OVR KAUAI AND ADJACENT CSTL WTRS.
20 SCT 45 BKN 60 120 OVC 150 OCNL 20 BKN RW-. OTLK...VFR.

.

WNDWD/MTN SXNS AND WNDWD CSTL WTRS OF OTR ISLANDS.
20 SCT 40 SCT-BKN 70 ISOL 15 BKN RW-. OTLK...VFR.

.

KONA AND KAU SLPS.
30 BKN 80 ISOL RW-. 08Z 30 SCT. OTLK...VFR.

.
ELSW.
25 SCT 45 SCT. OTLK...VFR.

Gulf of Mexico

FAGXOL KMIA 281031
FAGXOL KNHC 281040
281100Z-282300Z
OTLK 282300Z-291100Z
AMDTS NOT AVBL 02Z-11Z

...
GULF OF MEXICO W OF 85W AND N OF 27N..CSTL PLAINS AND CSTL WTRS
AGO-BRO..HGTS MSL UNLESS NOTED.
...
TSTMS IMPLY PSBL SVR OR GTR TURBC..SVR ICG..LLWS..STG SFC WNDS..HI
WVS..CIG BLO 10..AND VSBY BLO 3MI.
...
01 SYNS...
WEAK COLD FRONT EXTENDING FROM 29N85W TO 29N90W TO NEAR LCH
WILL DRIFT S AND GRADUALLY DISSIPATE.

...
02 FLT PRCTNS...NONE.

...
03 MARINE PRCTNS...NONE.

...
04 SGFNT CLDS AND WEA...
CSTL PLAINS BRO-HOU...
10-15 SCT/BKN 25-35 SCT VSBY 4-6F LCL BLO 1F. 281500Z 20-30 BKN/SCT
VSBY UNRSTD. 281900Z 30-40 SCT/BKN VSBY UNRSTD. OTLK...VFR.

.
CSTL PLAINS HOU-AGO CSTL WTRS RPE-AGO AND OFSHR WTRS E OF
90W...8-12 SCT/BKN 25-40 SCT/BKN 90-110 SCT/BKN VSBY UNRSTD LCL 4-
6F. 281500Z 20-30 BKN/SCT 90-120 BKN LYR TO ABV 120 VSBY UNRSTD.
CONDS LWR IN SCT TRW. OTLK...VFR.

.
CSTL WTRS BRO-RPE AND OFSHR WTRS W OF 90W...
15-25 SCT/BKN VSBY UNRSTD. 281600Z 25-35 SCT/BKN 80-120 SCT/BKN
VSBY UNRSTD. ISOLD TRW. OTLK...VFR.

...
05 ICG AND FRZLVL BLO 120...
NONE. FRZLVL ABV 120.

...
06 TURBC BLO 120...
NONE.
...

07 WNDS BLO 120...
CSTL PLAINS CSTL AND OFSHR WTRS N OF COLD FRONT...SFC-020 NE 10-15 KT EXCP OVER LAND BLO 010 NE 5-10 KT. 020-040 VRBL 10 KT. 040-080 W 10-15 KT. 080-120 W 15-25 KT.   OTLK...LTLCG.
CSTL PLAINS CSTL AND OFSHR WTRS S OF COLD FRONT...SFC-040 SE-E 10-15 KT. 040-080 VRBL 10 KT. 080-120 W 10-15 KT.   OTLK...LTLCG.
...
08 WWS...
CSTL WTRS...AQQ-PSX...1-3 FT.   OTLK...LTLCG.
CSTL WTRS...PSX-BRO...2-4 FT.   OTLK...LTLCG.
OFSHR WTRS W OF 92W...5 FT OR LESS.   OTLK...LTLCG.
OFSHR WTRS E OF 92W...LESSTHAN 5 FT.   OTLK..LTLCG.

Glossary

advection The process of moving an atmospheric property from one location to another. Advection usually refers to the horizontal movement of properties (temperature, moisture, vorticity, etc.).

air mass A large body of air, with homogeneous horizontal temperature and moisture characteristics.

anticyclonic A clockwise rotation in the Northern Hemisphere, associated with the circulation around an anticyclone (high-pressure area).

arc cloud An arc-shaped line of convective clouds often observed in satellite imagery moving away from a dissipating thunderstorm area.

atmospheric phenomena As reported on SAs, atmospheric phenomena are weather occurring at the station and any obstructions to vision. Obstructions to vision are only reported when the prevailing visibility is less than 7 miles.

Automated Weather Observing System AWOS is a computerized system that measures sky condition, visibility, precipitation, temperature, dew point, wind, and altimeter setting. It has a voice synthesizer to report minute-by-minute observations over radio frequencies, through telephone lines, or on local video displays.

baroclinic A state of the atmosphere where isotherms (lines of equal temperature) cross contours, temperature and pressure gradients are steep, and temperature advection takes place. A baroclinic atmosphere enhances the formation and strengthens the intensity of storms. It is characterized by an upper level wave that is one-quarter wave length behind the surface front.

barotropic Barotropic is an absence of or the opposite of baroclinic. Theoretically, an entirely barotropic atmosphere would yield constant pressure charts with no height or temperature gradients or vertical motion.

boundaries Zones in the lower atmosphere characterized by sharp gradients or discontinuities of temperature, pressure, or moisture and often accompanied by convergence in the wind field. Examples include surface fronts, dry lines, and outflow boundaries. In the latter case, the boundary is produced by a surge of rain-cooled air flowing outward near the surface from the originating area of convection. In an unstable air mass, thunderstorms tend to develop along these zones and especially at intersections of two or more boundaries.

BWER/WER/LEWP (Bounded Weak Echo Region/Weak Echo Region/Line Echo Wave Pattern) All of these weather-radar terms are indicators of strong thunderstorms and the development of severe weather.

chop Reported with turbulence, chop refers to a slight, rapid, and somewhat rhythmic bumpiness without appreciable changes in altitude or attitude.

clear air turbulence (CAT) Nonconvective wind-shear turbulence occurring at or above 15,000 feet, although it usually refers to turbulence above 25,000 feet.

closed cell stratocumulus Common over open water, this term is used to describe the satellite view of oceanic stratocumulus associated with an inversion. The clouds are associated with high pressure.

cloud band A nearly continuous cloud formation with a distinct long axis, a length-to-width ratio of at least 4-to-1, and a width greater than 1° of latitude (60 nm).

cloud element The smallest cloud form that can be resolved on satellite imagery from a given satellite system.

cloud line A narrow cloud band in which individual elements are connected and the line is less than 1° of latitude in width. Indicates strong winds, often 30 knots or more over open water.

cloud shield A broad cloud formation that is not more than four times as long as it is wide. Often it is formed by cirrus clouds associated with a ridge or the jet stream.

cloud streets A series of aligned cloud elements that are not connected. Several cloud streets usually line up parallel to each other and each street is not more than 10 miles wide.

cold-core low A low-pressure area that intensifies aloft. When this type of low contains closed contours at the 200 mb level, its movements tend to be slow and erratic.

cold pool Generally refers to an area at 500 mbs in which the air temperature is colder than adjacent areas. Other atmospheric conditions being equal, thermodynamic instability is greater beneath cold pools, which makes thunderstorm development more likely.

comma cloud system A cloud system that resembles the comma punctuation mark (,). The shape results from differential rotation of the cloud border and upward- and downward-moving air.

comma head The rounded portion of the comma cloud system. This region often produces most of the steady precipitation.

comma tail The portion of the comma cloud that lies to the right of, and often nearly parallel to, the axis of maximum winds.

confluence A region where streamlines converge. The speed of the horizontal flow will often increase where there is confluence. It is the upper-level equivalent of surface convergence.

Coordinated Universal Time Formerly Greenwich Mean Time, also known as Z or ZULU time, Coordinated Universal Time (UTC) is the international time standard. UTC is used in all aviation time references, except a control zone's active times, which are expressed in local time.

convergence Air flowing together near the surface is forced upward due to convergence. It is a vertical-motion producer that tends to destabilize the atmosphere near the surface.

cyclonic A counterclockwise rotation in the Northern Hemisphere, associated with the circulation around a cyclone or low-pressure area.

deformation zone An area within the atmospheric circulation where air parcels contract in one direction and elongate in the perpendicular direction. The narrow zone along the axis of elongation is called the deformation zone. Deformation is a primary factor in frontogenesis and frontolysis.

dew point front See dry line.

difluence The spreading apart of adjacent streamlines. The speed of horizontal flow often decreases with a difluent zone. It is the upper-air equivalent of surface divergence and activates or perpetuates thunderstorm development.

direct user access terminal A computer terminal where pilots can directly access meteorological and aeronautical information, plus file a flight plan without the assistance of an FSS. Typically abbreviated as DUAT or DUATS.

divergence Subsiding air diverges, or spreads, at the surface. Divergence is a downward-motion producer that tends to stabilize the atmosphere near the surface.

dry line An area within an air mass that has little temperature gradient, but significant differences in moisture. The boundary between the dry and moist air produces a lifting mechanism. Although not a true front, it has the potential to produce hazardous weather. It is also known as a dew point front.

dry slot A satellite meteorology term used to describe a cloud feature associated with an upper-level short wave trough. Generally speaking, the cloud system is shaped like a large comma. As the system develops, sinking air beneath the jet stream causes an intrusion of dry, relatively cloud-free air on the upwind side of the comma cloud. The air of the intrusion is the dry slot. It is commonly the location where lines of thunderstorms subsequently develop.

embedded thunderstorm A thunderstorm that occurs within nonconvective precipitation. A thunderstorm that is hidden in stratiform clouds.

enhanced infrared (IR) imagery A process by which infrared imagery is enhanced to provide increased contrast between features to simplify interpretation. This is done by assigning specific shades of gray to specific temperature ranges.

enhanced V A cloud top signature sometimes seen in enhanced infrared imagery in which the coldest cloud top temperatures form a V shape. Storms that show this cloud top feature are often associated with severe weather.

fine line/thin line At times weather radar picks up dust or debris that appears as a fine or thin line caused by a dry front or gust front. The line indicates the presence of low-level wind shear.

flight level Pressure altitude that is read off an altimeter set to standard pressure of 29.92. Altitude references used in the United States when flying above 17,999 feet.

Global Positioning System A network of Earth satellites that provides high-accuracy position, velocity, and time information to ground-based or airborne receivers.

GOES Geostationary Operational Environmental Satellites, normally located about 22,000 nm above the equator at 75°W and 135°W. The satellites provide visible and infrared imagery every 30 minutes.

gust front A low-level windshift line created by the downdrafts associated with thunderstorms and other convective clouds. Acting like a front, these features might produce strong gustiness, pressure rises, and low-level wind shear.

Hadley Cell A circulation theory describing how low pressure at the equator rises, moves toward the pole, sinks in high pressure at the pole, and then moves toward the equator.

hail shaft A shaft of hail detected on weather radar.

hook echo A bona fide hook echo indicates the existence of a mesolow associated with a large thunderstorm cell. Such mesolows are often associated with severe thunderstorms and tornadoes. Hook echoes are not seen on ATC radars.

impulse A weak, mid- to upper-level, and fast moving short wave feature that can kick off thunderstorms.

international standard atmosphere (ISA) A hypothetical vertical distribution of atmospheric properties (temperature, pressure, and density). At the surface, the ISA has a temperature of 15°C (59°F), pressure of 1013.2 mb (29.92 in.), and a lapse rate of approximately 2°C in the troposphere.

intertropical convergence zone (ITCZ) The dividing line between the southeast trade winds in the Southern Hemisphere and the northeast trade winds of the Northern Hemisphere.

level of free convection (LFC) The level at which a parcel of air lifted adiabatically until saturation would become warmer than its surrounding air and become unstable.

LEWP See BWER/WER/LEWP.

lifted index A measure of atmospheric instability that is computed on a thermodynamic chart by lifting a parcel of air near the surface to the 500 mb level and subtracting the temperature of the parcel from the temperature of the environment. A negative index means the lifted parcel is buoyant at 500 mb and will continue to rise, which is an unstable condition.

lifted condensation level (LCL) The level at which a parcel of air lifted adiabatically would cool and become saturated. The level at which clouds would form.

location identifier Consisting of three to five alphanumeric characters, location identifiers are contractions used to identify geographical locations, navigational aids, and intersections.

long wave See Rossby wave.

low-level wind shear See wind shear.

mean wind vector The mean wind vector is the direction and magnitude of the mean winds from 5,000 feet AGL to the tropopause. It can be used to estimate cell movement.

mesolow Also known as mesocyclone, a mesolow is a small area of low pressure within a severe thunderstorm. Tornadoes can develop within the vortex.

mesoscale Small-scale meteorological phenomena that can range in size from that of a single thunderstorm to an area the size of the state of Oklahoma.

mesoscale convective complex (MCC) A large homogeneous convective weather system on the order of 100,000 square miles. They tend to form during the morning hours.

microburst A small-scale and severe downburst of air less than 2.5 miles across. Reaching the ground, the burst continues as an expanding outflow producing severe wind shear.

moisture convergence An objective analysis field combining wind flow convergence and moisture advection. Under certain circumstances, this field is useful for forecasting areas of thunderstorm development.

negative tilt Refers to troughs with an axis in the horizontal plane tilting from northwest to southeast. These systems tend to cause more weather in California because they bring in warm, moist air.

outflow boundary An outflow boundary is a surface boundary left by the horizontal spreading of thunderstorm-cooled air. The boundary is often the lifting mechanism needed to generate new thunderstorms.

overrunning A condition in which air flow from one air mass is moving over another air mass of greater density. The term usually applies to warmer air flowing over cooler air as in a warm frontal situation. It implies a lifting mechanism that can trigger convection in unstable air.

positive tilt Refers to a trough with an axis in the horizontal plane tilting from northeast to southwest.

positive vorticity advection (PVA) Positive vorticity advection usually applies to the 500 mb level and refers to areas where the wind flow implies advection from higher values of absolute vorticity to lower ones. These areas are presumed to mark zones where upward vertical motion will be supported or enhanced. A vertical-motion producer.

Rossby waves Also known as major waves, planetary waves, or long waves, they are characterized by their long length and significant amplitude. They tend to be slow moving.

scud Shreds of small detached clouds moving rapidly below a solid deck of higher clouds.

severe thunderstorm A thunderstorm that produces winds of 50 knots or more, or hail ¾ in. in diameter or greater.

severe wind shear See wind shear.

shear axis An axis indicating maximum lateral change in wind direction, as in an elongated circulation. This lateral change or shear might be either cyclonic or anticyclonic.

short wave With a wave length that is shorter than a long wave, a short wave tends to move rapidly through the long wave circulation. A short wave can intensify or dampen weather systems.

standard atmosphere See International Standard Atmosphere.

stratosphere The atmospheric layer above the tropopause. It is characterized by a slight increase in temperature with height and the near absence of water vapor. Occasionally, severe thunderstorms will break through the tropopause into the stratosphere.

streamline A line that represents the wind flow pattern and is parallel to the instantaneous velocity. Streamlines indicate the trajectory of the flow.

sublimation The process by which ice changes directly to water vapor, or water vapor directly to ice. The sublimation of ice to vapor is a cooling process, and water vapor to ice is a warming process.

synoptic scale Large-scale weather patterns that are the size of the migratory high- and low-pressure systems of the lower troposphere with wave lengths on the order of 1,000 miles.

thermal high An area of high atmospheric pressure caused by the cooling of air by a cold surface. It will remain relatively stationary over the cold ground.

thermal low An area of low atmospheric pressure caused by intense surface heating. It is common to the continental subtropics in summer, remains stationary, and cyclonic circulation is generally weak and diffuse.

thin line See fine line/thin line.

transverse cirrus banding Irregularly spaced bandlike cirrus clouds that form nearly perpendicular to a jet stream axis. The band indicates turbulence associated with the jet stream.

tropopause The boundary between the troposphere and the stratosphere is called the tropopause. It consists of several discrete overlapping leaves, rather than a single continuous surface, and acts as a lid trapping almost all water vapor in the troposphere. It is marked by a decrease in wind speed and constant temperature at increased heights.

troposphere The lower layer of the atmosphere, extending from the surface to an average of 7 miles. Temperature normally decreases with height, and winds increase with height. It is the layer of the atmosphere where almost all weather occurs.

vort lobe An abbreviation for "vorticity lobe." It usually applies to the 500 mb level and identifies an area of relatively higher values of vorticity. It is synonymous with a short wave trough or upper-level impulse. Generally speaking, there is rising air ahead of the vort lobe and sinking air behind it.

vort max An abbreviation for vorticity maximum. It usually applies to the 500 mb level and refers to a point along a vorticity lobe where the absolute vorticity reaches a maximum value.

vortex In the most general use, any flow possessing vorticity. More often the term refers to a flow with closed streamlines.

vorticity Indicates a circulation or rotation within the atmosphere.

warm air advection A condition in the atmosphere characterized by air flowing from a relatively warmer area to a cooler area. It is often accompanied by upward vertical motion that in the presence of sufficient instability leads to thunderstorm development.

warm core low An area of low pressure that is warmer at its center than at its periphery. Thermal lows and tropical cyclones are examples.

wave A pattern of ridges and troughs in the horizontal flow as depicted on upper-level charts. At the surface, a wave is characterized by a break along a frontal boundary. A center of low pressure is frequently located at the apex of the wave.

WER See BWER/WER/LEWP.

wind field Winds plotted at a specified level in the atmosphere are referred to as a wind field (surface, 5,000 feet, 300 mb, etc.). Wind fields show areas of convergence, divergence, and advection, which provide meteorologists with a valuable forecast tool.

wind shear Any rapid change in wind direction or velocity. Low-level wind shear (LLWS) generally occurs within about 2,000 feet of the surface. LLWS is classified "severe" when a rapid change in wind direction or velocity causes an airspeed change greater than 15 knots or vertical speed change greater than 500 feet per minute.

zonal flow A wind flow that is generally in a west to east direction.

Index

About the author

Terry T. Lankford earned a private pilot certificate in 1967 through a U.S. Air Force aero club in England. He added commercial and flight instructor certificates in 1968 and holds a Gold Seal flight instructor certificate, which has airplane single-engine, multiengine, and instrument ratings. The commercial pilot certificate has airplane single-engine and multiengine land and instrument ratings. Lankford also has a ground instructor certificate with advanced and instrument ratings. He has more than 4,000 hours of flight time: 1,300 hours as a flight instructor. Lankford is also an accident prevention counselor for the FAA.

He became a flight service specialist in 1974 and works at the Oakland, California, Flight Service Station. Lankford is certified at all positions, including Flight Watch. He is a graduate of the pilot weather briefer refresher course and en route flight advisory course (flight watch) at the FAA Academy in Oklahoma City. Lankford has consulted with Jeppesen-Sanderson regarding its educational video tapes about aviation weather and pilot briefing services.

Lankford has owned two Cessna 150s. One was flown from Van Nuys, California, to upstate New York and a year later to Orlando, Florida. He has flown in Mexico, Canada, and Europe. Lankford was a speaker at the EAA's Oshkosh Fly-In in 1992 and 1993. He made presentations about pilot weather briefings to the National Weather Association's annual meeting in 1994 and the American Meteorological Society's meeting in 1995.

Other Bestsellers
of Related Interest

Weather Patterns and Phenomena: A Pilot's Guide
—Thomas P. Turner

A pilot must have the latest, most complete information about weather conditions before taking off. Learn how to assess aviation weather hazards correctly and confidently. Chapters cover weather theory in depth; specific hazards such as thunderstorms, turbulence, reduced visibility, and ice; and everything else pilots need to know to make the critical 'go/no go' decision.

0-07-065602-9 $16.95 Paper

The Pilot's Radio Communications Handbook, Fourth Edition
—Paul E. Illman

This book covers everything VFR pilots need to know to communicate with confidence from the cockpit. Readers will find the latest information on airspace reclassification, radio communication procedures for in-flight emergencies, requirements for Mode C transponders in Class B (TCA) airspace, and more.

0-07-031757-7 $29.95 Hard
0-07-031756-9 $18.95 Paper

Aviator's Guide to GPS, Second Edition
—Bill Clarke

The new Global Positioning System (GPS) satellites provide the most accurate means of geographical locating available, and pilots must know what this system can do and how to operate the avionics in order to fly safely in the air traffic system. Bill Clarke explains it clearly, concisely, and completely. Although aimed primarily at pilots, this book also reveals other uses for GPS by hikers, mountain climbers, search and rescue teams, and more.

0-07-011545-1 $39.95 Hard
0-07-011546-X $24.95 Paper

Flying in Adverse Conditions
—R. Randall Padfield

Even the most cautious pilot will occasionally find himself flying in adverse conditions. It may be a sudden weather change, a bird strike, overwater turbulence, or fatigue, this book prepares the pilot for these unexpected situations. Learn how to avoid these conditions when possible, and how to handle them if they arise. First person accounts and examples help to illustrate points.

0-07-048139-3 $29.95 Hard
0-07-048140-7 $19.95 Paper

The Art of Instrument Flying, Third Edition
—J.R. Williams

This comprehensive primer on instrument flying is improved with updated material on GPS, changes in U.S. airspace classifications, and other recent developments in the world of aviation. Author J.R. Williams delivers a clearly written, instructional guide to flying aircraft by instruments alone.

0-07-070598-4 $34.95 Hard
0-07-070599-2 $25.95 Paper

Van Sickle's Modern Airmanship, Seventh Edition
—John F. Welch, Ed.

This book has long been considered the definitive sourcebook on essential flight principles, techniques, and performance standards. It has been updated and expanded to reflect the latest advances in airplane and aerospace structural designs; engines, instruments, and avionics; aeromedicine; satellite-based navigation; helicopters and gyrocopters; and more.

0-07-069184-3 $44.95 Hard